Comprehensible
Viva & Practical
BIOCHEMISTRY

Comprehensible
Viva & Practical
BIOCHEMISTRY

[For Medical Science, General Science, Veterinary Science, Home Science, Life Science, Agriculture, and Homeopathy Courses]

Dr A C Deb PhD

Ex-Professor and Head, Department of Biochemistry
J L N Medical Collge, Bhagalpur, Bihar

New Central Book Agency (P) Ltd

LONDON

HYDERABAD ERNAKULAM BHUBANESWAR
NEW DELHI KOLKATA PUNE GUWAHATI

NCBA

REGD OFFICE
8/1 Chintamoni Das Lane, Kolkata 700 009, India
email: ncbapvtltd@eth.net

OVERSEAS
NCBA (UK) Ltd, 149 Park Avenue North, Northampton, NN3 2HY, UK
email: ncbauk@yahoo.com

EXPORT
NCBA Exports Pvt. Ltd, 212 Shahpur Jat, New Delhi 110 049
email: ncbaexports@ncbapvtltd.com

BRANCHES

2/27 Ansari Road, 1st Floor
Daryaganj, **New Delhi** 110 002
email: ncbadel@ncbapvtltd.com

Shop Nos. 3 & 4, Vinayak Towers
681/B Budhwar Peth, Appa Balwant Chowk
Pune 411 002
email: ncbapune@ncbapvtltd.com

Shop No. 15, Bharati Towers, Ground Floor
Forest Park
Bhubaneswar 751 001
email: ncbabub@ncbapvtltd.com

House No. 3-1-315, 1st Floor
Nimboliadda, Kachiguda, **Hyderabad** 500 027
email: ncbahydb@ncbapvtltd.com

GSS Shopping Complex, 1st Floor
Opposite College Post Office, Convent Road
Ernakulam 682 035
0484-4021054 / 53

Radhamayee Complex [Second Floor]
Opp. Axis Bank, Panbazar, Jaswanta Road
Guwahati 781 001
email: ncbaguwahati@ncbapvtltd.com

COMPREHENSIBLE VIVA & PRACTICAL BIOCHEMISTRY ● A C Dev

First Published: 1996
Reprinted: 1997
Second Edition: 2000
Third Edition: 2005
Reprinted: 2007, 2008, 2009, 2012, 2013, 2017, 2019

PUBLISHER AND TYPESETTER
New Central Book Agency (P) Ltd
8/1 Chintamoni Das Lane, Kolkata 700 009

PRINTER
New Central Book Agency (P) Ltd
Web-Offset Division, Dhulagarh, Sankrail, Howrah

TECHNICAL EDITOR
Dr Mita Sen, MBBS

PROJECT COORDINATORS
Prabir Ghosh and Prabhat Jas

ISBN: 978 81 7381 208 8

Price ₹ 350.00

ॐ

Narayanam Namaskritya Naranchaiba Narottamam
Debing Saraswatinchaiba Tatojayamodirayat

Dedicated to those who are keen to know and adopt Biochemistry during humansufferings

Third Edition

PREFACE

Both the author and the publisher are extremely pleased to present this third edition of `Comprehensible Viva and Practical Biochemistry' within such a short period of time. It is rather significant to note that six more essential chapters have been incorporated in this edition for providing updated knowledge. Recently modified normal laboratory values are introduced too in a separate chapter for all students to have the latest concepts in medical affairs. The author tried his best to make this book more advanced in the best interest of the students, so that, they do not face any unpleasant situation to attend the viva-voce examination before highly qualified and respectable examiners. He is all expectations about the best performance of his beloved students in all spheres of life by the grace of God.

Further permit me to add that the author cordially accepted different suggestions from all quarters, and it would be his pleasure to incorporate further suggestions from the respectable teachers as well as students for future development of this book.

The author is grateful to his wife Mrs Bela Deb, M.Sc., B.Ed. for her valuable and vivid suggestions for early completion of the task without much worries and anxieties. He profusely thanks his son Dr Pradipta Deb, MBBS (Cal) and daughter-in-law Mrs Debarna Deb, MBA for their valuable assistance throughout the preparation of this edition. Finally, he offers his best wishes to the publisher Mr Amitabha Sen for his active cooperation and the typesetter Mr Prabir Ghosh for the finest possible page design of the book.

Calcutta
16th October, 2005

DR. A. C. DEB

First Edition PREFACE

It is generally observed that the students face acute difficulties in accurately answering the questions raised in viva-voce examination in Biochemistry. This is true not only for questions related to theory but also related to practical problems as well. It has also been observed that most students are not confident about securing good marks in their practical examinations. Keeping these problems of the students in mind I have attempted to prepare simple, short and accurate answers to the questions generally asked in viva-voce examinations of Biochemistry. The practical portion has also been written in a systematic and lucid manner so as to enable the students to grasp the subject easily and quickly. I am confident that by making appropriate use of this title, students will be able to answer any question to the complete satisfaction of the honourable examiners and thus derive the maximum benefit from their honest endeavour.

The preparation of such a title requires help from various quarters. In particular I am indebted to my wife Mrs. Bela Deb, M. Sc. (Botany), B. Ed. for her inspirative and valuable suggestions for accomplishing the task of compiling this purposive manual. I would also like to thank my son Pradipta Deb (a medical student) for his efficient assistance throughout the preparation of this title. Finally I must also express the sincere cooperation of the publisher Mr. Arunabha Sen and the typesetter Mr. Prabir Ghosh for the nice get up of the book.

Of course, no title is perfect in its first print and I welcome suggestions of my esteemed colleagues and beloved students to make this title more utilitarian in its next and subsequent editions.

Calcutta
1st June, 1996

DR. A. C. DEB

TABLE OF CONTENTS

VIVA — Part I

PRACTICAL — Part II

PART-I
VIVA

CHAPTER 1

BIOPHYSICS

Q1. Define diffusion.

The process by which the particles move from a region of higher concentration to a lower concentration to spread uniformly.

Q2. What is concentration gradient?

It is the rate of change in solute concentration per unit distance.

The concentration gradient $= \frac{\Delta n}{\Delta x}$

where, Δn = the change in molar concentration of the solute.

Δx = the small distance.

Q3. What do you mean by diffusion coefficient?

It is the mass of solute diffusing across 1 cm^2 area in 1 second down a concentration gradient of unity.

Q4. What is diffusional flux?

It is a linear function of the difference in molar concentration of the solute between two regions. If a solute A diffuses through a thickness Δx of a membrane from an outer aqueous phase to an inner aqueous phase and has a diffusion coefficient D in the membrane,

$$J = D \frac{[A]_{out} - [A]_{in}}{\Delta x} = \frac{D}{\Delta x} \Delta n = P \Delta n$$

Q5. State the clinical importance of diffusion.

(a) The alveolar surface area per unit body weight is larger in children than in adult humans. So the resting uptake of O_2 per unit body weight by its diffusion across the alveolar surface is higher in children than in adults.

(b) Diffusion of respiratory gases across the erythrocytes membrane is much faster in smaller erythrocytes of mammals than in larger erythrocytes owing to higher surface : volume ratios in the smaller cells.

(c) The enzyme lysozyme is polar and water soluble. Since it is insoluble in the outer lipoprotein layer of gram-negative bacterial cell wall it cannot diffuse through that layer to penetrate into the cell wall for digesting the latter. Thus, gram negative bacterial cells are unaffected by lysozyme.

(d) The diffusion of O_2 and CO_2 between the capillary blood and the tissues is facilitated due to the thinness of capillary walls. The alveolar membrane is also thin which helps in the diffusion of respiratory gases across it. But this membrane is thickened in pneumonia resulting in the decreased diffusion of O_2 showing decline in the oxygenation of blood.

(e) Many small alveoli fuse to form fewer and larger alveoli due to chronic smoking. As a result, the total alveolar surface is reduced and the diffusion of gases between the alveolar air and the blood is decreased leading to asphyxiating effects.

(f) Diffusion is involved in the intestinal absorption of pentoses, some mineral ions and some water-soluble vitamins and the renal reabsorption of urea.

Q6. Define P^H

It is the negative logarithm of hydrogen ion concentration or it is the logarithm of the reciprocal of hydrogen ion concentration

$$P^H = -\text{Log}\,[H^+] = \text{Log}\frac{1}{[H^+]}$$

[The P^H scale ranges from 0 to 14 for dilute aqueous solutions like biological fluids. P^H 7 is considered the neutral P^H. Acidic solution is below 7 and alkaline solution is above 7 to 14. The greater the acidity the lower the P^H.]

Q7. Mention the P^H of adult's gastric juice, infant's gastric juice, saliva, pure water, milk, tears, blood, intestinal juice, pancreatic juice.

Adult's gastric juice	= 1.2 to 1.5	Milk	= 7.1
Infant's gastric juice	= 5.0	Tears	= 7.2
Saliva	= 6.8	Blood	= 7.4
Pure Water	= 7.0	Intestinal juice	= 7.8
		Pancreatic juice	= 8.0

Q8. Discuss the clinical importance of P^H.

(a) P^H maintains the natural three-dimensional forms of nucleic acid molecules.

(b) It influences the keq, spontaneity of metabolic oxidation—reductions and some nonenzymatic acid-base catalysis.

(c) The optimum P^H of juice of the body has important role for enzyme maximum activities.

(d) It influences the ionization of ionizable polar groups of amino acids, proteins, nucleic acids, phospholipids and mucopolysaccharides.

(e) Abrupt changes in P^H may destroy the ionic and hydrogen bonds and change the three-dimensional structure of a protein resulting in its inactivation and denaturation. Activity of an enzyme is declined due to changes in its three-dimensional conformation, ionic state of its substrate — binding sites and the ionized state of its substrate. Similarly, P^H affects the membrane structure by influencing the ionization of membrane proteins and lipids.

Q9. Define buffer.

It is a mixture of weak acids with their salts of strong bases and resists sudden change in hydrogen ion concentration even after the addition of hydrogen or hydroxyl ions

eg. $(CH_3COOH + CH_3COONa)$

Q10. What is Henderson-Hasselbalch equation?

$$P^H = PKa + \log \frac{[Salt]}{[Acid]}$$

Q11. What should be the ratio of the concentration of bicarbonate and carbonic acid to make the blood P^H 7.4?

20 : 1.

Q12. Give some examples of buffer pairs.

H_2CO_3	H. Hemoglobin	H. Protein	KH_2PO_4
$BHCO_3$	B. Haemoglobin	B. Protein	K_2HPO_4

Q13. State the significance of buffer in the body.

Acids produced in the body as a bye-product of metabolism tend to decrease the P^H of different fluids like intracellular fluid, interstitial fluid, blood etc. But the P^H remains steady due to the presence of buffer.

Q14. Define Osmosis and Osmotic pressure.

Osmosis is defined as the spontaneous flow of water from a dilute solution to a more concentrated solution when they are separated by a semi-permeable membrane.

[The most selective artificial semi-permeable memebrane is $Cu_2Fe(CN)_6$]

Osmotic pressure may be defined as the pressure which must be applied on the concentrated solution to prevent the osmotic inflow into the solution when they are separated by a semi-permeable membrane.

Q15. Show the difference between osmotic pressure and oncotic pressure.

The osmotic pressure will be greater if the molecular weight of a substance is low.

The oncotic pressure increases when the molecular weight of the substance is high. The substance human albumin having high molecular weight does not increase the osmotic pressure since it is impermeable to the membrane. Hence it increases the oncotic pressure.

Q16. Name the factors regulating osmotic pressure.

ADH, aldosterone, the third factor.

Q17. State Vant Hoff's laws of osmotic pressure.

(a) The osmotic pressure of a solution varies directly with the concentration of the solute in the solution.

(b) It also varies directly with the absolute temperature.

Q18. What do you mean by the word electroendosmosis?

The movement of a liquid through a capillary under the influence of an electric potential is termed electroendosmosis.

Q19. Give the equation for the calculation of osmotic pressure.

$$\pi V = nRT \quad \therefore \pi = \frac{nRT}{V}$$

$$\text{where, } n = \frac{G}{M} \quad \therefore \pi V = \frac{g}{M} RT$$

where, G = Grams of solute, V = the volume in litres, M = the molecular weight of the solute, R = the molar gas constant, T = absolute temperature.

Q20. What do you mean by Haemolysis and Crenation?

If RBCs are kept in hypotonic solution, the cell walls of RBC are ruptured due to the flow of water into it. This is said to be haemolysis.

But if the cells are kept in hypertonic solution, water comes out of the cells causing shrinking of the cells. This is said to be crenation.

Q21. What is the clinical importance of osmosis?

(a) The clinical application of osmotic force is the injection of hypertonic solution of magnesium sulphate to reduce the volume of the brain or lower the pressure of C.S.F.

(b) The purgative action of epsom salt ($MgSO_4$). The strong solution of this salt prevents the absorption of water in the intestine causing dilution of the intestinal contents.

(c) Hemolysis of RBC is caused by the osmotic phenomenon.

(d) Urine excretion is assisted by osmosis to keep the blood isotonic with the cells.

(e) The pain due to the application of salt on the cut of the skin is the result of osmotic phenomenon.

Q22. Define dialysis and state its clinical importance.

It is the process of separating crystalloids from colloids by diffusion through a membrane adopting osmotic force.

Clinical importance :

(a) Dialysis is essential to separate proteins from a mixture of salts and to separate out macromolecules of cell extract.

(b) It is also required for preventing enzymatic or metabolic reactions by removing small cofactor molecules from the cell extract.

(c) By dialysis, proteins do not appear in the ultrafiltrate and retained in the plasma for maintaining normal health.

(d) Patients with uremia and acute kidney failure, dialysis of blood is performed in artificial kidneys to eliminate non-electrolyte waste products and the excess of electrolytes.

Q23. What is surface tension?

The force with which the surface molecules of a liquid are held together is called surface tension.

Q24. Name the substances that increase and lower surface tension.

Bile salts, fatty acids, ammonia, potassium permanganate, strong mineral acids decrease surface tension. Alkalis, and most inorganic salts increase surface tension.

Q25. State the Gibbs-Thomson Principle in relation to surface tension.

(a) The substances that lower the surface tension become concentrated at the interface.

(b) The substances that increase surface tension tend to move away from the interface.

(c) Lipids and proteins that are effective in lowering surface tension are found concentrated in the cell wall.

Q26. Discuss the clinical importance of surface tension.

(a) Surface tension is involved in the process of digestion. Bile salts assist emulsification with lipids and thus reduce the surface tension of lipids facilitating the lipase activity on it.

(b) The inward force generated by alveolar surface tension increases diffusion of fluids into alveoli from alveolar capillaries. Lung surfactant prevents oedema by lowering surface tension.

(c) In postnatal life, type 2 alveolar epithelial cells in lungs secrete a surfactant called DPL (Di-palmityl lecithin) which prevents collapse of lung alveoli in expiration.

Q27. What is adsorption?

It is a surface phenomenon. This is the process of taking up substances from solution.

Q28. State the biomedical importance of adsorption.

(a) Surface adsorption helps to combine enzymes with substrates to give the reaction product.

(b) Adsorption is applied in the purification of enzymes.

(c) Adsorption on the cell membranes promotes many vital chemical reactions.

(d) Drugs and poisons which are adsorbed on cell surface exert their effects from that location.

Q29. Define hydrotropy.

It is the process by which water-insoluble substances (phospholipids, sterols, uric acid etc.) are made water-soluble by hydrotropic substances (cholic acid, benzoic acid, hippuric acid, soap of higher fatty acids etc.)

Q30. Enumerate the biological importance of hydrotropy.

(a) Substances dissolved by hydrotropy are diffusible through membranes.

(b) The hydrotropic substances have the capacity to decrease surface tension.

(c) Hydrotropic substances are useful in helping absorption and transport of insoluble substances.

Q31. Define viscosity.

The resistance experienced by one layer of a liquid in moving over another layer is called viscosity.

The unit of viscosity is the "Poise".

$$\eta \text{ (eta)} = \frac{\pi P r^4}{8VL} t$$

Q32. Mention the clinical importance of viscosity.

(a) The viscosity of blood offers resistance to the heart during circulation. The heart muscle functions best while working against a certain resistance.

(b) The viscosity of blood is increased with the increase in the number of blood cells. So in polycythemia viscosity is high but viscosity is low in anemia.

(c) Blood viscosity helps in streamlining blood flow.

Q33. What are the effects of temperature and colloids on viscosity?

Viscosity decreases with the rise in temperature and increases with the increase in colloids.

Q34. What are colloids?

The substances (Proteins, Polysaccharides) which do not diffuse through parchment or animal membrane although form homogeneous or heterogeneous solution are called colloids.

Q35. What is the size of each colloidal particle?

10-100 nm (submicrons) only visible under ultramicroscope.

Q36. What are the methods of determining the size of colloidal particle?

(a) By ultrafiltration, (b) By diffusion rate, (c) By ultracentrifuge, (d) By light scattering, (e) By X-ray analysis, (f) By osmotic pressure.

Q37. What do you mean by salting out?

The precipitation of emulsoids by adding large amounts of soluble salts is referred to as salting out.

Q38. What are suspensoids and emulsoids?

Suspensoids :

(a) The suspensoid particles carry a definite electric charge.

(b) They are easily precipitated if the charge is neutralised.

(c) Since they are not hydrated they are said to be hydrophobic or lyophobic colloids (water-fearing colloids).

(d) Once they are precipitated, they are not brought back into colloidal solution again.

Emulsoids :

(a) They are very stable and not easily precipitated by salts.

(b) All colloids of the living cell exist as emulsoids. They are hydrophilic (water-loving colloids).

(c) An emulsoid may be changed to suspensoid by dehydration and then is precipitated.

(d) They carry electric charges. Some carry positive and negative charges simultaneously, eg. Proteins.

Q39. What are Sol and Gel?

Sol :

(a) It is a dilute solution.

(b) It can be converted to Gel by change of temperature.

(c) Addition of $CaCl_2$ to an alkaline solution of caseinogen gives a sol.

Gel :

(a) Gels are quasi-solids.

(b) High pressure is required to squeeze water out of a gel.

(c) Formaldehyde can remove water from the gel and hence it is used in the histological process of fixing tissues.

(d) Tissues are not dehydrated due to the existence of gel.

Q40. Define imbibition.

Dried emulsoids take up (imbibe) water and swell considerably. This process is called imbibition. Heat is liberated during this process.

Q41. What do you mean by protective colloids?

The colloids which protect other substances from precipitation.

Example : Calcium phosphates of plasma are held in colloidal suspension by the protective action of proteins present in it.

Q42. Explain Brownian Movement.

(a) The continuous motion of the colloidal particles is known as Brownian movement.

(b) The movement is quite haphazard.

(c) The rate of Brownian movement depends on the size of the particles, the smaller particles have higher movement.

(d) It accounts for the permanency of colloidal solutions.

Q43. State Donnan membrane Equilibrium.

In case the membrane is freely permeable, equilibrium is reached easily. But if one of the ions is colloid, the result becomes different.

Let us consider the interaction between the salts NaR and water through a membrane. The equilibrium will be as follows :

Initial state		Equilibrium	
Na^+	H^+	Na^+	Na^+
R^-	OH^-	R^-	OH^-
		H^+	

Here the P^H is changed. The NaR side is acid and the alkali will be "excreted" across the membrane.

In the body, equilibrium may never be established owing to the removal of "excreted" ions in other reactions. This rather helps to accelerate "excretion".

Q44. Mention the clinical importance of colloids.

(a) Bile salts prevent precipitation of cholesterol and bilirubin salts. In absence of bile salts, cholesterol and calcium bilirubinate get precipitated to form gall stones.

(b) Protective colloids present in urine prevent formation of urinary stones.

Q45. What do you mean by facilitated transfer?

Certain substances cannot pass easily through membranes. But these substances are facilitated to be transported through the membranes by means of carriers, eg. the mitochondrial membrane is impermeable to acyl coenzyme A derivatives. The acyl group is transferred to carnitine to form carnitine derivatives which can pass through the membrane. The acyl coenzyme A derivative is then re-formed on the other side of the membrane.

Q46. What is active transport?

This is the process by which solutes can pass through membranes against their concentration gradient requiring energy. An active transport device which forces Na^+ out and K^+ in has been referred to as the "Sodium pump".

Q47. Define Isotopes.

Isotopes may be defined as atoms having the same atomic number but different atomic weights.

They are the subspecies of the same chemical element and occupy the same position in the periodic table, but have different physical properties, eg. H^1, H^2, H^3.

Stable isotopes : C^{12}, C^{13}. Radioactive isotopes : C^{11}, C^{14}.

Q48. What is radioactivity?

The process by which unstable nucleus of some atoms of some elements spontaneously emits photons of radiation is termed radioactivity.

Natural radioactive elements belong to one of the three different series : Uranium series, thorium series, actinium series.

Q49. What is meant by radioactive decay?

It is the spontaneous nuclear transformations that change an unstable radioactive nucleus to a more stable one with the simultaneous emission of ionizing radiations.

Q50. What is *radioisotopes*?

Radioactive isotopes differ from each other in radioactivity and mass number and are formed during the decay of a radioactive element.

Q51. What do you mean by Ci, r, rad, RBE, REM?

(a) **Ci (Curie)** : It is the international unit of radioactivity showing radioactivity of 1 g of radium. 1 Ci = 3.7×10^{10} nuclear disintegrations per second. 1 mCi = 10^{-3} Ci; 1 μCi = 10^{-6} Ci.

(b) **r (Roentogen)** : It is the amount of γ- or x-rays that deposits 8.33×10^{-6}j of energy per g. of air to produce 1.61×10^{12} ion pairs/g.

(c) **rad (Radiation absorbed dose)** : It is the quantity of γ- or x-rays that can deposit 100 ergs or 1×10^{-5}j/g in matter including biological tissues.

(d) **RBE (Relative Biological Equivalence)** : It is ratio between the doses (rad) of the radiation under investigation and of γ-rays producing identical biological effects. RBE is 1 for γ-rays.

(e) **REM (Roentgen Equivalent Mammals)** : It measures a radiation in terms of its effects on mammalian tissues. REM = RBE x rad.

Q52. State the effects of radiation.

(i) **Immediate effects :**

(a) Whole body exposure of 200-1,000 rads results in severe damage to hematopoietic system due to higher sensitivity. There is depletion of leukocytes and increased susceptibility to infection. Death occurs within 10-20 days.

(b) High radiation (1,000-5,000 rads) causes severe damage to mucosal epithelium, fluid loss, haemorrhage in G.I.T. resulting in nausea, vomiting, diarrhoea and death occurs within 3-5 days.

(c) Higher radiations (5,000-10,000 rads) cause meningitis, choroid plexilis and death occurs within 8-48 hours.

(ii) **Delayed effects :**

(a) Ionizing radiations can cause cancer. Rapidly dividing cells are prone to cancer-leukemia. There may be polycythemia vera, breast cancer and bone cancer.

(b) "In Utero" exposure leads to growth retardation, congenital malformation, neonatal death.

(c) Other effects are endocrine imbalance, nephrosclerosis, sterility, cataract etc.

(iii) **Genetic effects :**

The magnitude of the genetic defects depend on the stage of germ cell development, dose rate, dose fractionation and interval between exposure and conception.

Q53. Explain the term "Radiation Hazards".

Radioactivity exposure is hazardous to living organisms. The effects of radiation have come mainly from somatic and hereditary damages sustained by the atomic explosions.

Q54. Show the difference between radioactive isotope and stable isotope.

Radioactive isotope	Stable isotope
(a) Atomic number and atomic mass are continuously changing	Atomic number and atomic mass are constant.
(b) Spontaneous emission of radiation (α, β, γ rays) or neutrons.	No emission of any kind of radiation.
(c) New elements are produced as a result of disintegration.	No new elements are produced.
(d) Detected by external detectors.	Cannot be detected by external detectors.
(e) Emits radiation and can cause harmful effects on biological tissues.	Are not hazardous.
(f) Used for generation of power (energy) in nuclear reactors.	Not used.
(g) Used for destructive purposes i.e. for manufacture of nuclear weapons.	No such utility.
(h) Used for sterilization of medical instruments.	No such use.

Q55. What are the processes of detecting radioactive isotope?

Geiger-Muller Counter, Scintillators, Semi-conductor detectors, Autoradiography.

Q56. What is half-life?

The time taken for half the atoms in a quantity of a radioactive substance to disintegrate is known as the half-life.

$$T_{eff} = \frac{T_{\frac{1}{2}} \times T_b}{T_{\frac{1}{2}} + T_b}$$

Q57. What is meant by radiation protection?

(a) Minimum possible time should be spent near the radiation zone.

(b) Radioactive material should be handled from a maximum possible distance.

(c) Person should be shielded by lead.

Q58. What do you mean by Radioimmunoassay?

It is the technique used in the determination of microgram to picogram quantities of specific compounds such as peptide hormones, steroid hormones, prostaglandins, cyclic AMP, folate, cobalamin and heptitis B antigen in microlitre volumes of biological samples.

Q59. State the diagnostic importance of radioisotopes.

(a) ^{51}Cr—labelled erythrocytes are used for detecting internal hemorrhage.

(b) ^{131}I—labelled diodrast is used for renal function.

(c) ^{59}Fe—labelled erythrocytes are used for measuring circulating blood volume.

(d) CO^{60} is used for diagnosing pernicious anemia.

(e) I^{131} is used for determining thyroid function.

(f) Na^{24} has been used for diagnosing arterial diseases.

(g) CO^{60} is a powerful source of radiation for the treatment of various malignant disorders.

Q60. Mention the importance of Fe^{59} in the body.

The followings are determined with the help of Fe^{59} :

(a) The rate of disappearance of Fe^{59} from the plasma.

(b) The rate of absorption from the G.I. tract.

(c) The rate of "turnover" of plasma Fe indicates the extent of activity of the erythropoietic marrow.

(d) The rate of incorporation of Fe^{59} into the RBC which indicates the extent of effective erythropoiesis.

Q61. State the use of P^{32} in diseases.

P^{32} is used in the treatment of following diseases :

(a) Polycythemia vera.

(b) Multiple myeloma.

(c) Carcinoma of the breast and prostate.

(d) Chronic myeloid leukaemia.

(e) Primary haemorrhagic thrombocytosis.

Q62. Enumerate the uses of I^{131}.

(a) I^{131} has been used for the treatment of thyroid cancers.

(b) It has also been used for the treatment of thyrotoxicosis.

(c) I^{131} therapy of cardiac disease in euthyroid patients.

(d) I^{131} is used for the control of ectopic arrythmias.

Q63. How do radiations from radioactive isotope influence a cell?

(a) It causes structural changes in chromosomes.

(b) It causes chemical changes in genes.

(c) It inactivates enzymes of the cell.

Q64. How do radiations cause cancers?

(a) High rates of exposure to gamma radiations of 100 rads or more can cause cancer.

(b) Exposure to radiation produces cancers most frequently in the hemopoietic tissue of bone-maroow, and in some extrahemopoietic tissues such as bones, the thyroid, the skin and the lungs.

(c) High dose rates of total body irradiation can cause acute and chronic myeloid leukemias and acute lymphoid leukemia.

(d) Bone-seeking radioisotopes like ^{90}Sr, ^{41}Ca and ^{226}Ra are concentrated in the mineral structure of bone and damage both bones and the hemopoietic bone-marrow tissues by their radiations. Both leukemias and bone cancers may result.

(e) ^{241}Pu is concentrated in the periosteum and endosteum to produce similar damages.

(f) High incidence of lung cancers is available in pitchblende miners inhaling radioactive radon gas.

Q65. Mention the processes to be adopted for protection from radiation.

(a) Minimum time should be utilized near the radiation area.

(b) Radioactive materials should be handled from a maximum distance.

(c) Film badge should be used and the individual should be shielded by lead.

Q66. What do you mean by Schilling test?

(a) The absorption of vitamin B_{12} from the gut is studied by Schilling test.

(b) The radioisotopes CO^{60} or CO^{58} (half life = 7 1 days), CO^{57} (half life = 27 days) are used for labelling vitamin B_{12}.

(c) The normal urinary excretion is greater than 15 per cent of the test dose. The excretion is less than 5 per cent in case of the patients with pernicious anemia.

Q67. Explain Scanning.

An organ has the capacity to concentrate selectively a γ-emitting radioisotope when administered. An image of the distribution of the γ-emitter can be obtained by using a scanner over the organ which can give the following informations :

(a) Position, size, and boundaries of the organ.

(b) Shape of the organ.

(c) Presence or absence of any lesion. An impairment of function like cyst, abscess etc. is noticed as **void** (called as **cold area**). An abnormal accumulation of radio activity much in excess of the surrounding tissues appears as **dense** (called as **hot area**) showing the presence of tumors.

Q68. Mention the radioisotopes used for scanning different organs.

(a) **For brain scan :** Technetium (TC^{99m}) and indium (In^{133m}).

(b) **For lung scan :** I^{131}-human serum albumin macroaggregates.

(c) **For renal scan :** Hg^{203} or Hg^{197}-labelled chlormerodin.

(d) **For thyroid scan :** I^{131}. Recently, 99m technetium pertechnate (Tc^{99m}).

Q69. What is meant by extra corporeal blood irradiation?

(a) It is used in chronic leukemic patients.

(b) The blood is collected from the patient through his forearm artery and then it is circulated around a cesium^{137} source which emits powerful γ-rays. This irradiated blood is then returned to the same patient through his forearm vein.

Q70. Explain the importance of Boron^{10} neutron in brain tumors.

(a) Recently, Boron^{10} is used in the treatment of fatal brain tumor like glioblastoma multiforme.

(b) When Boron^{10} is injected IV in the patient, it is readily taken up by the tumor tissue in brain. After 10 minutes, the head of the patient is placed in a beam of slow neutrons. Boron^{10} in tumor tissue absorbs neutrons rapidly and is converted to Boron^{11} which disintegrates immediately to lithium isotope giving out α-particles. The tumor tissue is destroyed by the ionizing property of α-particles.

Q71. State the uses of I^{131} labelled Rose Bengal test.

(a) This test is used for the assessment of hepatic function.

(b) In case of the normal liver function, the dye becomes excreted through liver, neck count goes down but count over abdomen increases.

(c) In case of parenchymal liver diseases, neck count is high and practically no rise in count over the abdomen.

(d) If I^{131}-labelled Rose Bengal is administered IV, neck count goes down and count over the abdomen is practically nil.

Q72. Mention the principle of dilution studies for measuring volume and space.

A known amount of radioactive "tracer" is introduced into an unknown volume and thoroughly mixed for a known time. The concentration of the radioactive "tracer" is estimated. The total volume is then estimated by the formula given below :

$$V = \frac{N}{n}$$

where, V = Volume to be measured.

N = Total number of counts injected.

n = Number of counts per ml.

CHAPTER 2

CARBOHYDRATES

Q1. Define carbohydrate.

It may be defined as the polyhydric alcohols with aldehydes or ketones and their derivatives.

Q2. Classify carbohydrate.

(a) Monosaccharides (simple sugars) eg. glucose, fructose, galactose, mannose etc.

(b) Disaccharides (combination of two units of simple sugars) eg. maltose, lactose, sucrose.

(c) Oligosaccharides (combination of 3 to 10 units of simple sugars) eg. maltotriose.

(d) Polysaccharides (combination of more than 10 units of simple sugars) eg. starch, glycogen, cellulose, inulin etc.

Q3. How are monosaccharides classified?

Carbon atoms	Aldoses	Ketoses
Trioses	Glyceraldehyde or Glycerose	Dihydroxyacetone
Tetroses	Erythrose	Erythrulose
Pentoses	Ribose Xylose Arabinose	Ribulose Xylulose
Hexoses	Glucose Mannose Galactose	Fructose
Heptoses	Glucoheptose Galactoheptose	Sedoheptulose

Q4. What is *Asymmetric carbon atom*?

The central carbon atom No. 2 which is attached to the four different atoms or groups of atoms is called asymmetric carbon atom.

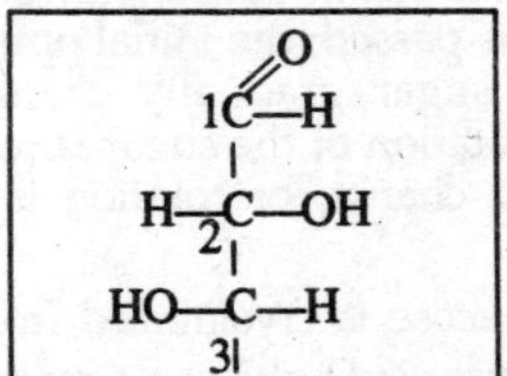

Q5. What do you mean by stereoisomers or geometric isomers?

Compounds that have the same structural formula but differ in spatial configuration are known as stereoisomers or geometric isomers. The presence of asymmetric carbon atoms allows the formation of isomers. The number of possible isomers depends on the number of asymmetric carbon atoms (n) and is equal to 2^n. Glucose with 4 asymmetric carbon atoms have 16 isomers.

Q6. Explain D and L isomerism.

When the –OH group around the carbon atom adjacent to the terminal primary alcohol carbon (eg. carbon atom 5 in glucose) is on the right the sugar is a member of D-series. But when it is on the left, it is a member of L-series.

Q7. What is optical isomerism?

The presence of asymmetric carbon atoms also confers optical activity on the compound. When a beam of plane polarised light is passed through a solution of an optical isomer it will be rotated either to the right, **dextrorotatory(+)** or to the left, **levorotatory (–)**.

When equal amounts of D and L isomers are present, the resulting mixture has no optical activity. The activities of each isomer cancel one another. Such a mixture is said to be **racemic** or DL-mixture.

Q8. Explain the term pyranose and furanose ring structures.

The six membered ring structure of monosaccharides is said to be pyranose ring structure and the five membered ring structure is said to be furanose ring structure.

Q9. What are α and β anomers?

When the –OH group around carbon No. 1 is on the right, it is α anomer. But when the –OH group is on the left, it is β anomer.

Q10. What do you mean by *mutarotation*?

When an aldohexose is first dissolved in water and the solution is put in optical path so that plane polarized light is passed, the initial optical rotation shown by the sugar gradually changes unitl a constant fixed rotation of the sugar is reached. This phenomenon of change of rotation is termed as mutarotation.

When D-glucose is crystallized from water at room temperature, α-D-glucose separates having the initial specific rotation of +112° which changes to a fixed rotation +52.5°.

When crystallization occurs from water at temperature above 98°C, a different β-glucose separates having initial specific rotation of +19° which changes to fixed rotation +52.5°.

α-D-glucose		β-D-glucose
+ 112° ⟶	+ 52.5°	⟵ + 19°

Q11. What are *epimers*?

Isomers formed as a result of interchange of the –OH and –H on carbon atoms 2, 3 and 4 of glucose are known as epimers. The most important epimers of glucose are mannose and galactose formed by epimerization at carbons 2 and 4, respectively. Epimerization takes place in the body by the enzyme **epimerase**.

Q12. What do you mean by Kiliani Synthesis?

A method for the synthesis of higher monosaccharide from a lower monosaccharide (eg. erythrose and threose from glycerose) was first proposed by Kiliani by the addition of HCN.

Q13. What are glycosides?

These are the compounds formed by the condensation reaction between a simple sugar and the hydroxyl group of a second compound or aglycone which may or may not be another simple sugar.

Q14. What are aminosugars?

Sugars containing an amino group are called amino sugars. Examples are D-glucosamine, D-mannosamine, and D-galactosamine. These amino sugars are formed by the replacement of –OH group around carbon No. 2 of a simple sugar.

Q15. What does happen on the oxidation of glucose?

On mild oxidation (by Br_2, H_2O) glucose is converted to gluconic acid by the oxidation of aldehyde group of it. On further oxidatin by HNO_3 gluconic acid is converted to saccharic acid by the oxidation of primary alcoholic group also and ultimately it is converted to glucuronic acid having the aldehyde group.

Q16. What happens when galactose is oxidized?

Galactose, on oxidation with HNO_3, produces the dicarboxylic mucic acid.

Q17. Give the reduction products of glucose, mannose, galactose and fructose.

Glucose	yields	Sorbitol
Mannose	yields	Mannitol
Galactose	yields	Dulcitol
Fructose	yields	Mannitol and Sorbitol

Q18. How are glucose, fructose and mannose interconverted?

Through common enediol form in presence of weak alkali such as $Ba(OH)_2$ or $Ca(OH)_2$.

Q19. Discuss the clinical importance of monosaccharides.

(a) Glyceraldehyde and dihydroxyacetone are the precursors of glycerol which form different types of lipids. Both glyceraldehyde and dihydroxyacetone in the form of phosphate ester become the intermediates of glycolysis and are ultimately involved in energy production.

(b) Erythrose in the form of phosphate ester becomes the intermediate of HMP shunt which is an alternative pathway for glucose oxidation.

(c) Ribose is a constituent of RNA and of certain coenzymes like FAD, NAD, coenzyme A. Deoxyribose is a constituent of DNA. L-xylulose is a metabolite of D-glucuronic acid and it is excreted in urine of humans affected by hereditary abnormality in metabolism causing **pentosuria**. L-fucose (methyl pentose) is involved in glycoproteins.

(d) D-glucose (dextrose, grape sugar) is the chief sugar present in blood and all tissues utilize glucose for energy purposes. It is mainly stored as glycogen in liver and muscle. It is a constituent of disacccharides and polysaccharides. Glucuronic acid is a constituent of hyaluronic acid. Mannose, the epimer of glucose, is a constituent of N-acetyl neuraminic acid. Galactose, the epimer of glucose, is involved in the formation of milk sugar lactose. Fructose (fruit sugar) is a constituent of sucrose and inulin. Fructose is abundant in seminal fluid and sperms utilize fructose for energy. Sedoheptulose in the form of phosphate ester is an intermediate of HMP shunt.

(e) Glycosides are found in many drugs. Cardiac glycosides are used in cardiac insufficiency. They are derivatives of digitalis.

(f) Glucosamine is a constituent of certain mucopolysaccharides. Galactosamine is a constituent of chondroitin sulphate present in cartilages, bones, tendons, and heart valves. Certain antibiotics such as erythromycin, carbomycin contain amino sugars.

Q20. Give example of disaccharides with their sources.

Maltose (Malt sugar) = 2 molecules of glucose. Present in malt.

Lactose (Milk sugar) = 1 mol. glucose + 1 mol. galactose. Present in milk.

Sucrose (table sugar) = 1 mol. glucose + 1 mol. fructose. Present in cane sugar.

Q21. What will be the oxidative product of lactose after prolonged boiling with HNO_3?

After boiling with HNO_3 lactose will be hydrolyzed to produce galactose which on oxidation will produce mucic acid.

Q22. Why is sucrose a non-reducing sugar?

Like maltose and lactose in the structure of sucrose there is no free aldehyde or ketone group which are the symbols for reducing sugars. The aldehyde group of glucose is combined with the ketone group of fructose to form the structure of sucrose. Hence it is a non-reducing sugar.

Q23. Why is sucrose said to be the "invert sugar"?

Sucrose is dextrorotatory and its specific rotation is +66.5°. The hydrolytic products are levorotatory having a specific rotation −19.5°. Because the laevorotation of fructose is greater than the dextrorotation of glucose. Since the hydrolytic products invert the rotation the resulting mixtures of glucose and fructose is said to be "invert sugar".

The process is said to be inversion.

Q24. Is honey invert sugar?

The greater sweetness of honey is due to the presence of fructose and fructose is levorotatory. Hence honey is an invert sugar.

Q25. State the clinical importance of disaccharides.

(a) Various baby foods contain maltose which is easily digested showing nutritional significance.

(b) Lactose is synthesized in the mammary gland from galactose and the presence of lactose in breast milk is the best source of energy for the new born body.

(c) Many organisms (eg. E. Coli, Str. Lactis) which are present in milk convert lactose of milk to lactic acid causing souring of milk.

(d) Sucrose if parenterally introduced can change the osmotic condition of the blood causing a flow of water from the tissue into the blood. Hence the clinicians use it in edema like cerebral edema.

(e) If the disaccharides are not hydrolyzed owing to the deficiency of suitable enzymes, diarrhea occurs.

Q26. What is the importance of oligosaccharide?

(a) Antibodies and coagulation factors contain oligosaccharides.

(b) The oligosaccharide units of glycoprotein are rich in information and are functionally important.

Q27. What do you mean by polysaccharides?

Polysaccharides are the polymers of monosaccharides.

Q28. Classify polysaccharides.

(a) Homopolysaccharides (Example : starch, glycogen, cellulose, dextrin, inulin).

(b) Heteropolysaccharides (Example : Mucopolysaccharides).

Q29. What are mucosubstances?

They have the property of forming slimy, viscous solution which occur in the body. They have structural and protective functions. They are divided into glycoproteins and mucopolysaccharides.

Q30. What are sialic acids?

(a) They are actually a family of compounds derived from neuraminic acid (pyruvic acid + mannosamine).

(b) They are widely distributed in vertebrate tissues.

(c) They are isolated from certain strains of bacteria.

(d) N-acetyl neuraminic acid is an example of sialic acid.

Q31. What are glycoproteins (mucoproteins)?

(a) They are protein-polysaccharide compounds occurring in the tissues, particularly in mucus secretions.

(b) They do not contain uronic acids although they contain acyl hexosamines. They also contain sialic acid and L-fucose.

(c) They form viscous solutions which function as lubricant and protective "screens" in the body.

(d) The continuously secreted mucus of the respiratory tract is a protection against invasion by bacteria.

(e) They are also found in α_1- and α_2-globulins of plasma.

Q32. Give examples of homopolysaccharides with their chemical nature.

Starch : A glucosan (formed by an α-glucosidic chain) consists of amylose (15–20%) having nonbranching helical structure and amylopectin (80–85%) having branched chains composed of 24–30 glucose residues united by 1→4 linkages in the chains and by 1→6 linkages at the branch points. Found in cereals, potatoes, legumes and other vegetables. The most important food source of carbohydrate.

Glycogen (Animal starch) : Storage polysaccharide of the animal body. More highly branched structure and has chains of 10–18–α–D-glucopyranose residues by α (1→4) glucosidic linkage with branching by α (1→6) glucosidic bonds. Stored in liver and muscle.

Cellulose : Consists of β-D-glucopyranose units linked by β(1→4) bonds to form long straight chains. Chief constituent of the framework of plants cannot be digested by humans because of the absence of a hydrolase that attacks the β linkage. Important source of "bulk" in the diet.

Dextrins : Hydrolytic breakdown of starch.

Inulin : A fructosan (formed by fructose only). Found in tubers and roots of dahlias and dandelions. Easily soluble in warm water and has been used for determination of the rate of glomerular filtration.

Q33. What are mucopolysaccharides?

These are acidic substances containing uronic acids and N-acelyted amino sugars which may in some compound be sulphated. The members of this group are as follows :

(a) Hyaluronic acid :
- (i) Occurs in synovial fluid, vitreous humor, skin, and bone.
- (ii) Acts as a lubricant in the body and has enormous capacity to hold water.
- (iii) The enzyme **hyaluronidase** in semen facilitates fertilization.

(b) Chondroitin sulphates :
- (i) Occur in the ground substances of connective tissues in the walls of large blood vessels and in tendons.
- (ii) Regulate flow and concentration of cations round the cell.

(c) Heparin :
- (i) Present in liver, lung, and the arterial wall.
- (ii) Used in medicine as anticoagulant.

(d) Blood groups :
- (i) Glucosamine and galactosamine are involved in blood groups.
- (ii) L-fucose is the constituent.
- (iii) The blood group substance of erythrocytes which are responsible for the immunologic reactions of blood are glycoprotein.

Q34. Discuss the clinical importance of polysaccharides.

(a) Starch is the chief carbohydrate and provides maximum glucose which yields energy.

(b) Glycogen is stored in liver and muscle and glycogen storage disease occurs in the deficiency of certain enzymes.

(c) Cellulose is stable insoluble compound and has no nutritional value in humans. It is of considerable human dietetic value since it stimulates peristalsis causing elimination of indigestible food residues. It is the best curing factor for persons suffering from constipation.

(d) Dextrins are widely used in infant feeding.

(e) Inulin is used in the determination of glomerular filtration rate (GFR). It is also used for estimation of body water volume.

(f) The enzyme **hyaluronidase** present in hyaluronic acid dissolves the viscid substances surrounding the ova to permit penetration of spermatozoa.

(g) Mucopolysaccharidosis is a disorder due to inherited enzyme defect in which there is mental retardation, skeletal changes, and corneal clouding.

Q35. What is citrate?

Citric acid + ionised citric acid.

Q36. Show the chemical composition of saliva.

(a) Organic substances : Mucin, glycoprotein, the enzyme ptyaline or salivary amylase.

(b) Inorganic substances : Chlorides, phosphates, bicarbonates of potassium and calcium, a trace of potassium.

Q37. Enumerate the digestion of carbohydrate.

(a) In the saliva :

$$\begin{matrix}\text{Starch}\\\text{Glycogen}\end{matrix} \xrightarrow[\text{P}^{\text{H}}\ 6.6\text{–}6.8]{\text{Salivary amylase}} \text{Maltose + Oligosaccharides + Maltotriose.}$$

(b) In the pancreas :

$$\begin{matrix}\text{Starch}\\\text{Glycogen}\end{matrix} \xrightarrow[\text{P}^{\text{H}}\ 7.1]{\text{Pancreatic amylase}} \text{Maltose + Oligosaccharides + Maltotriose.}$$

(c) In the small intestine :

$$\text{Sucrose} \xrightarrow[\text{P}^{\text{H}}\ 5.0\text{–}7.0]{\text{Sucrase}} \text{Glucose and fructose.}$$

$$\text{Lactose} \xrightarrow[\text{P}^{\text{H}}\ 5.4\text{–}6.0]{\text{Lactase}} \text{Glucose and galactose.}$$

$$\text{Maltose} \xrightarrow[\text{P}^{\text{H}}\ 5.8\text{–}6.2]{\text{Maltase}} \text{Glucose and glucose.}$$

Q38. State the absorption of carbohydrate.

(a) The products of carbohydrate digestion are absorbed from the jejunum into the blood of the portal venous system in the form of monosaccharides, chiefly as hexose and as pentose sugar (ribose).

(b) Fructose, pentose sugars and mannose are absorbed by simple diffusion.

(c) Glucose and galactose are absorbed by active transport mechanism because the OH on carbon 2 has got the same configuration in both the sugar. The active transport mechanism of glucose is as follows :

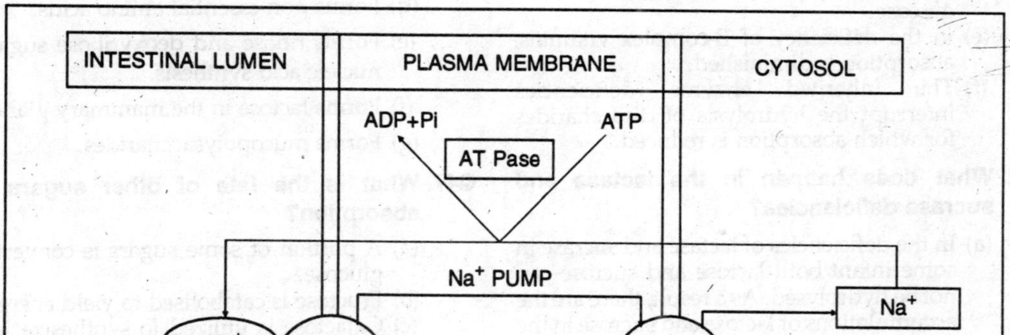

(d) Fructose is absorbed more slowly than glucose and galactose. Its absorption is proceeded by diffusion with the concentration gradient by means of an Na^+– independent transporter. Glucose facilitates fructose absorption. There is little fructose in blood.

Q39. Why is fructose not absorbed by active transport mechanism?

Because no. OH is present on carbon 2 of fructose and has not got the same configuration as in glucose and galactose regarding OH on carbon 2. This why its absorption is also slow.

Q40. Is the rate of absorption the same for all monosaccharides?

(a) Glucose and galactose are more rapidly absorbed than fructose by active transport mechanism.
(b) The phosphorylated sugars are more quickly absorbed within the cells of the intestinal mucosa.
(c) Fructose and Pentose sugars are slowly absorbed.

Q41. Evaluate the factors influencing rate of absorption of carbohydrate.

(a) Absorption suffers in case of unhealthy mucous membrane.
(b) Thyroid hormones increase absorption of hexoses and act directly on intestinal mucosa.
(c) Adrenocortical hormone deficiency decreases absorption of glucose and galactose due to the decreased concentration of sodium in body fluids.
(d) Insulin has got no effect on absorption of glucose.
(e) In the deficiency of B-complex vitamins, absorption is diminished.
(f) The inherited enzyme deficiencies interrupt the hydrolysis of disaccharides for which absorption is reduced.

Q42. What does happen in the lactase and sucrase deficiencies?

(a) In the deficiencies of lactase and sucrase in some infant both lactose and sucrose can not be hydrolysed. As a result, there are the accumulations of lactose and sucrose in the intestinal tract which hold water developing diarrhoea.
(b) The accumulated lactose and sucrose are fermented by intestinal bacteria producing gases which develop flatulence, distension and abdominal cramps.

Q43. What is disacchariduria?

(a) The deficiency of disaccharidase causes the increased excretion of disaccharides.
(b) The patients with intestinal damage (in sprue and celiac disease) excrete more than 300 mg of disaccharides.

Q44. What do you mean by metabolism?

It is the general term covering all the chemical changes occurring in the body.

Metabolism = Anabolism + Catabolism + Amphibolism

Anabolism : Synthesis of larger molecules from smaller ones requiring energy.

Catabolism : Breakdown of larger molecules into smaller ones with the release of energy.

Amphibolism : Acting as links between anabolism and catabolism eg. TCA cycle.

Q45. What is intermediary metabolism?

(a) Metabolism is generally interpreted to mean intermediary metabolism.
(b) The fate of dietary components after digestion and absorption constitutes intermediary metabolism.
(c) It includes all chemical processes within cells and tissues that are concerned with their building up and breaking down and in their functional operation.

Q46. What does happen to glucose after absorption?

(a) Utilized to maintain normal blood glucose level.
(b) Stored as glycogen in liver and muscle.
(c) Catabolised to form energy.
(d) Forms non-essential amino acids.
(e) Forms ribose and deoxyribose sugars for nucleic acid synthesis.
(f) Forms lactose in the mammary gland.
(g) Forms mucopolysaccharides.

Q47. What is the fate of other sugars after absorption?

(a) A portion of some sugars is converted to glucose.
(b) Fructose is catabolised to yield energy.
(c) Galactose is utilized to synthesize lactose in the mammary gland and to produce energy.
(d) Pentose sugars (ribose and deoxyribose) are used to form nucleic acids.

Q48. Define glycogenesis.

It is the process for the formation of glycogen in the cytoplasm of liver and muscle from glucose and other non-carbohydrates (eg. glycerol and amino acids).

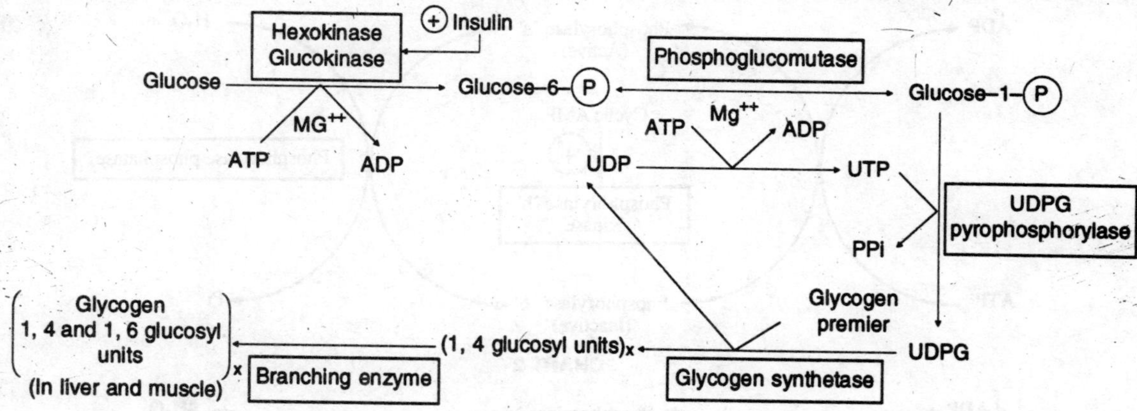

CHART 1

Q49. Mention the steps for glycogenesis.

See CHART 1 above.

Q50. What is UDPG?

(a) It is uridine diphosphate glucose. It is an intermediate product in the pathway of glycogenesis.

(b) It is required for the production of ascorbic acid (vitamin C) by the uronic acid pathway.

(c) It is essential for the synthesis of lactose from galactose in the mammary gland.

Q51. Enumerate the clinical importance of glycogen.

(a) Liver glycogen levels have got influence over certain forms of detoxication.

(b) Liver glycogen level depresses the rate of deamination of amino acids making them available for protein synthesis in the tissues.

(c) Liver glycogen also reduces the rate of ketone bodies formation.

(d) A high level of glycogen in the liver protects the liver cells against the harmful effects of many poisons and chemicals.

(e) The stored liver glycogen is utilized for the maintenance of blood glucose, particularly in between meals.

(f) Muscle glycogen acts as readily available source of intermediates of glycolysis for provision of energy within the muscle itself.

Q52. Name the substance for the stimulation and inhibition of glycogenesis.

Stimulators : Insulin, Glucocorticoids, glucose in higher concentration.

Inhibitors : Increased concentration of glycogen and cyclic AMP.

Q53. What is glycogenolysis?

Breakdown of glycogen to glucose in the cytoplasm of liver only. In the muscle, glycogen is broken down to lactic acid.

Q54. Why glucose is not formed in the muscle from glycogen?

Muscle is lacking in the enzyme glucose-6-phosphatase which is responsible for the conversion of glucose-6-(P) to glucose.

Q55. Mention the steps of glycogenolysis with their enzymes.

See CHART below.

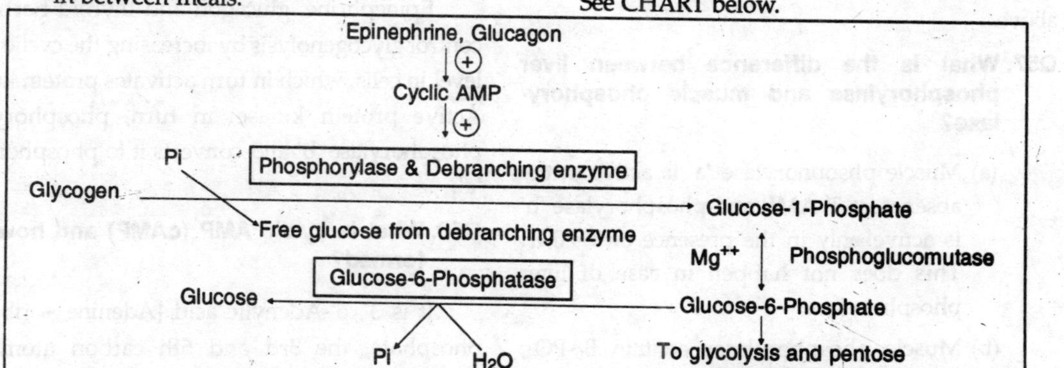

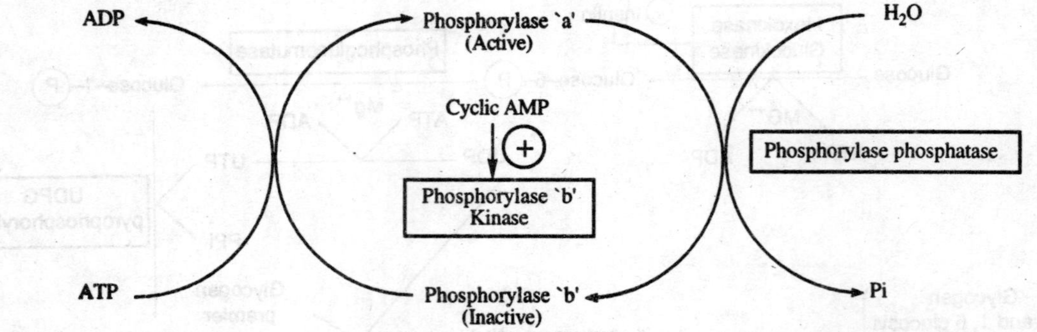

CHART 2

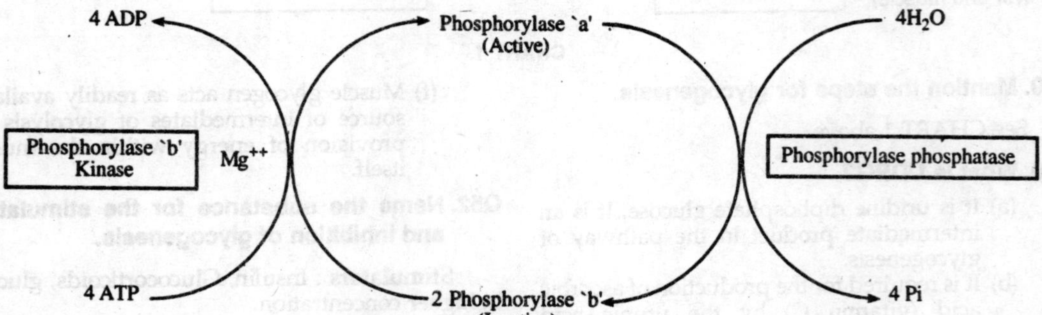

CHART 3

Q56. What do you know about phosphorylase?

(a) In liver, phosphorylase exists in two forms—phosphorylase 'a' (active) and phosphorylase 'b' (inactive). The conversion of one to the other is given in chart 2 :

(b) In muscle, phosphorylase is present in two forms—phosphorylase 'a' (active in the absence of 5′-AMP) and phosphorylase 'b' (active only in the presence of 5′-AMP).

Phosphorylase 'a' contains 4 mols of B_6-PO_4.

Phosphorylae 'b' contains 2 mols of B_6-PO_4.

The conversion of one to the other is as in chart 3 above.

Q57. What is the difference between liver phosphorylase and muscle phosphorylase?

(a) Muscle phsophorylase 'a' is active in the absence of 5′-AMP and phosphorylase 'b' is active only in the presence of 5′-AMP. This does not happen in case of liver phosphorylaes.

(b) Muscle phosphorylases contain B_6-PO_4; but not so in case of liver phosphorylases.

(c) No cleavage in the structure of liver phosphorylase.

(d) Muscle phosphorylase is not affected by glucagon.

(e) Active phosphorylase inhibits liver synthetase phosphatase.

(f) Muscle phosphorylase kinase has 4 types of subunits, α, β, γ and δ.

Q58. What is calmodulin?

It is Ca^{++} binding protein.

Q59. How is glycogenolysis regulated?

Epinephrine, glucagon and thyroid hormones control glycogenolysis by increasing the cyclic AMP level in cells, which in turn activates protein kinase. Active protein kinase, in turn, phosphorylates phosphorylase 'b' and converts it to phosphorylase 'a'.

Q60. What is cyclic AMP (cAMP) and how is it formed?

It is 3′, 5′-Adenylic acid [Adenine + ribose + phosphate, the 3rd and 5th carbon atoms are attached to phosphate].

Formation :

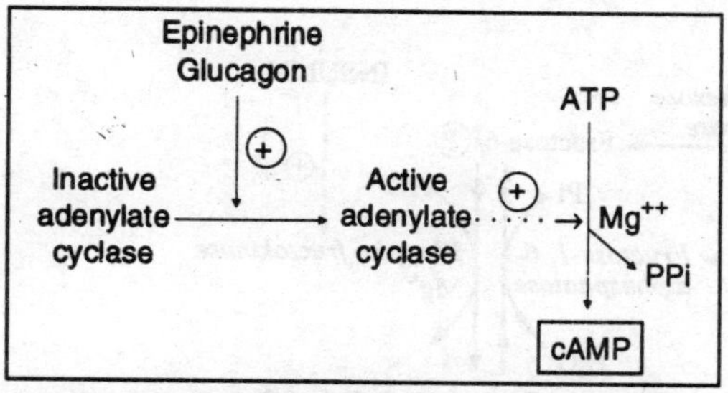

Q61. What is cyclic GMP and how is it formed?

(a) It is 3′, 5′-Guanylic acid [Guanine + ribose + phosphate, the 3rd and 5th carbon atoms are attached to phosphate].

(b) Prolactin causes elevation of cGMP in mouse mammary gland explants. May be mediated by PGF_2 and also raises the level of cGMP in the rat uterus.

(c) PGE_5 at high concentration appears to stimulate cGMP accumulation.

(d) PG endoperoxides may also stimulate the production of cGMP in platelets and recent evidence indicates that PGG_2 activates that soluble guanylate cyclase from platelets.

Formation :

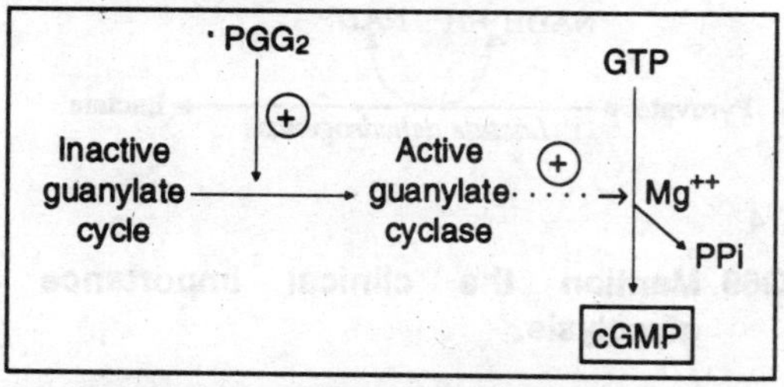

Q62. What is meant by "glycogen storage disease"?

It is a genetic one in which abnormal quantities of glycogen are deposited in the liver, kidney, heart and muscle. This happens due to the deficiency of enzyme. Seven types of glycogenesis from Type 1 to Type VII have been recognized.

Q63. What are the clinical features found in Type 1 glycogenesis (Von Gierke's Disease)?

(a) Owing to the deficiency of the enzyme glucose-6-phosphatase, very little glucose is derived from liver glycogen. Hypoglycemia occurs in children with this disease.

(b) Fat is then utilized as energy source which undergoes oxidation with the formation of more acetyl-CoA that ultimately forms ketone bodies and cholesterol.

(c) Persistent hypoglycemia inhibits insulin secretion which in turn inhibits protein synthesis causing dwarfism.

(d) Hypoglycemia also stimulate secretion of catecholamines and glucagon which cause glycogen breakdown of muscle forming lactic acid accumulation.

(e) Increased lactic acid of blood competes with uric acid excretion by kidneys leading to increased blood uric acid level which develops symptoms of gout.

Q64. What is glycolysis?

It is the process of breakdown of glucose and glycogen to pyruvate or lactate in the cytoplasm of liver and muscle under anerobic condition.

Q65. Enumerate the steps of glycolysis.

See Chart 4 next page.

Q66. Show the number of ATP formed from one molecule of glucose upto pyruvic acid and lactic acid by glycolysis.

From one molecule of glucose two molecules of pyruvic acid or lactic acid are formed starting from two molecules of glyceraldehyde-3-phosphate.

Upto pyruvic acid formation :

	Reaction	ATP formed
(a)	Glyceraldehyde-3-phosphate →1, 3-diphosphoglycerate	6
(b)	1, 3-diphosphoglycerate →3-phosphoglycerate	2
(c)	Phosphoenolpyruvate→Enolpyruvate	2
		10
	ATP consumed in	
(a)	Glucose→Glucose-6-phosphate	1
(b)	Fructose-6-phosphate →Fructose-1, 6-diphosphate	1
	Net ATP Synthesized	8

Upto lactic acid formation :

(a)	1, 3-diphosphoglycerate →3-phosphoglycerate	2
(b)	Phosphoenolpyruvate→Enolpyruvate	2
		4
	ATP consumed in	
(a)	Glucose→Glucose-6-phosphate	1
(b)	Fructose-6-phosphate →Fructose-1, 6-diphosphate	1
	Net ATP Synthesized	2

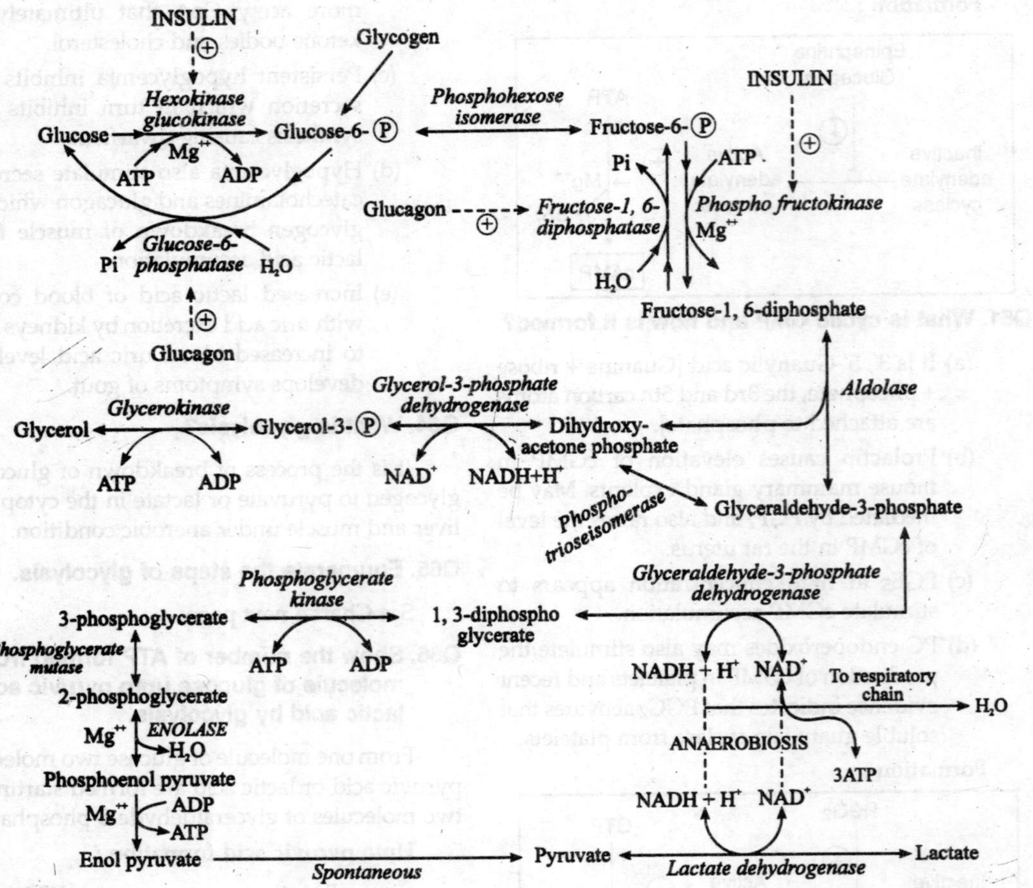

CHART 4

Q67. Name the inhibitors of glycolysis.

(a) **Iodoacetate** inhibits the enzyme glyceraldehyde-3- phosphate dehydrogenase.

(b) **Arsenite** inhibits the enzyme phosphoglycerate kinase.

(c) **Fluoride** inhibits the enzyme enolase.

Q68. State the differences between hexokinase and glucokinase.

Hexokinase	Glucokinase
(a) Found almost in all tissues	Found only in liver.
(b) More stable	More labile.
(c) Non-specific, can phosphorylate any of the hexoses	Specific, can phosphorylate glucose only.
(d) K_m is low; hence high affinity for glucose	K_m is high; hence low affinity for glucose.
(e) No change with glucose feeding	Increased by glucose feeding after fasting.
(f) Allosteric inhibition by glucose-6-phosphate	Not inhibited by glucose-6-phosphate.

Q69. Mention the clinical importance of glycolysis.

(a) Glycolysis provides ATP in skeletal muscle even in absence of O_2. Muscles can survive anoxic episodes.

(b) Heart muscle has poor glycolytic activity and poor survival under conditions of ischemia.

(c) The rate of glycolysis is very high in case of fast-growing cancer cells. So, more pyruvic acid is formed than TCA cycle can utilize. Therefore, more accumulation of pyruvic acid leads to the excessive formation of lactic acid producing lactic acidosis.

(d) Hemolytic anaemia is developed owing to the deficiency of hexokinase and pyruvate kinase enzymes of glycolysis.

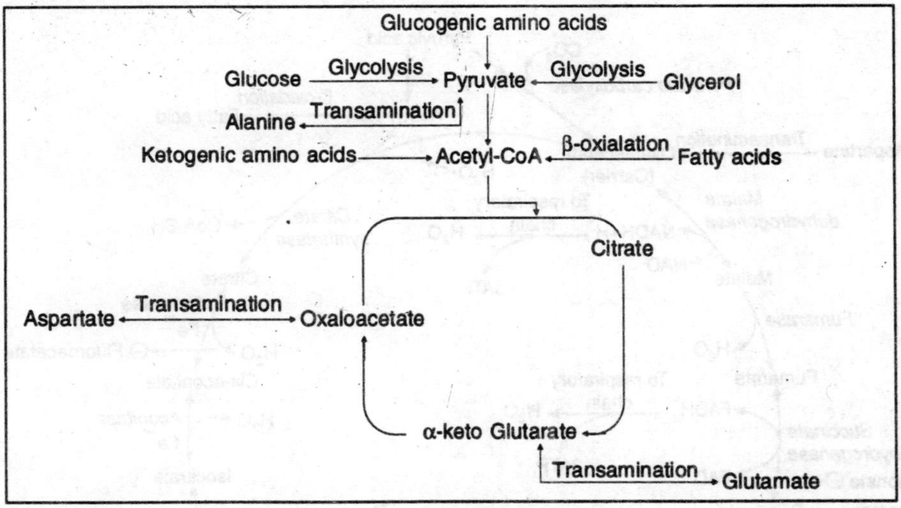

CHART 5

Q70. Name the enzyme, coenzyme and inhibitor in the conversion of pyruvate to acetyl-CoA.

Enzyme :

Pyruvate dehydrogenase complex consisting of

(a) Pyruvate dehydrogenase 29 mols.

(b) Dihydrolipoyl transacetylase 1 mol.

(c) Dihydrolipoyl dehydrogenease 8 mols.

Coenzymes :

NAD^+, CoA.SH, TPP, lipoic acid, FAD.

Inhibitors :

(a) Arsenite inhibits the enzyme pyruvate dehydrogenase.

(b) Dietary deficiency of thiamine allows pyruvate to accumulate.

Q71. Why is it called citric acid cycle or TCA cycle?

Since citric acid is first formed in the cycle the cycle is said to be citric acid cycle.

Since citric acid is a tricaroxylic acid it is said to be TCA cycle.

Q72. Why is citric acid cycle said to be amphobolic in nature?

Citric acid cycle provides metabolites for anabolic processes such as fatty acid and amino acid synthesis and gluconeogenesis. It also provides CO_2, H_2O and ATP.

Q73. Where and under what conditions does TCA cycle function?

(a) Mitochondria of the cell.

(b) Under aerobic conditions (Presence of O_2).

Q74. Why is TCA cycle called a common metabolic pathway?

TCA cycle is the only pathway for the complete oxidation of the foodstuffs—carbohydrates, fats, proteins yielding ATP, CO_2 and H_2O.

See Chart 5 above.

Q75. Enumerate the steps of TCA cycle.

See chart 6 next page.

Q76. Show the number of ATP formed in TCA cycle.

Reaction	ATP formed
(i) Pyruvate→Acetyl-CoA	3
(ii) Isocitrate→Oxalosuccinate	3
(iii) α-Keto glutarate→succinyl-CoA	3
(iv) Succinyl-CoA→Succinate	1
(v) Succinate→Fumarate	2
(vi) Malate→Oxaloacetate	3
Total	15

∴ 1 molecule of pyruvic acid being oxidized in the TCA cycle yields 15 ATP and 1 molecule of Acetyl-CoA being oxidized in the TCA cycle yields 12 ATP.

Q77. Show the number of ATP formed when one molecule of glucose is completely oxidized.

One molecule of glucose produces 2 molecules of pyruvic acid by glycolysis and these two molecules of pyruvic acid are completely oxidized in TCA cycle.

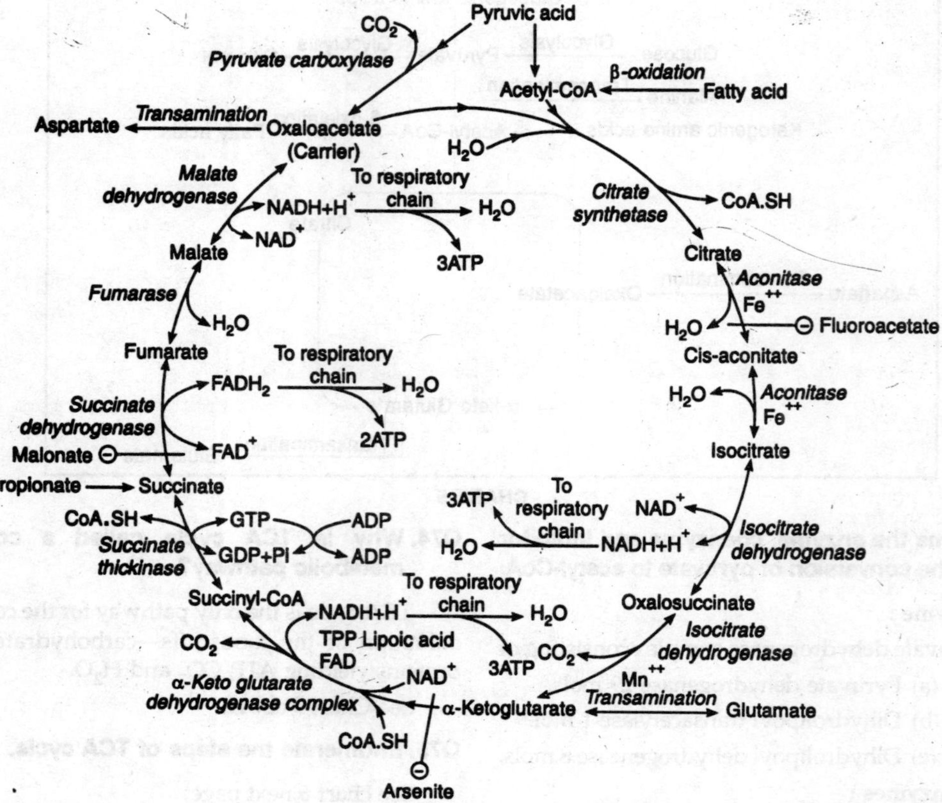

CHART 6

So, the number of ATP formed in TCA cycle from two molecule of pyruvic acid	= (15×2)= 30
and the number of ATP formed in glycolysis in the formation of 2 molecules of pyruvic acid	= 8
Total	= 38

Q78. Show the number of CO_2 and H_2O formed in TCA cycle.

Pyruvate→acetyl-CoA	1 molecule of CO_2
Oxalosuccinate→α-Ketoglutarate	1 molecule of CO_2
α-Ketoglutarate→Succinyl-CoA	1 molecule of CO_2
Total :	3 molecules of CO_2
The oxidation of reduced NAD and FAD by repiratory chain produces	5 H_2O
Utilized in TCA cycle	3 H_2O
Net	2 H_2O

Q79. Name the carrier substance in TCA cycle.

Oxaloacetate is the carrier in TCA cycle. It facilitates the conversion of a large quantity of acetyl units to CO_2. It may also be considered to play the catalytic role.

Q80. Mention the inhibitors in TCA cycle.

(a) Fluroacetate inhibits the enzyme aconitase to prevent the conversion of citrate.

(b) Arsenite inhibits α-ketoglutarate dehydrogenase for which α-ketoglutarate is accumulated.

(c) Malonate or oxaloacetate inhibits succinate dehydrogenase and succinate is accumulated.

Q81. How many APTS are formed in the conversion of α-ketoglutarate to succinate?

4 ATP.

Q82. What does happen to the ATPS formed from glucose oxidation?

(a) Provide energy as heat or other forms depending on the organism.

(b) Transfer energy to creatine producing creatine phosphate which is the immediate source of energy for muscle contraction.

(c) Provide energy for anabolic processes.

(d) Hydrolysed by ATPase of muscle with the liberation of energy for physical work.

Q83. What is the fate of acetyl-CoA in the body?

(a) Oxidized in the TCA cycle yielding energy.

(b) Synthesizes fatty acids.

(c) Synthesizes cholesterol and thus bile acids, vitamin D, and corticosteroid hormones.

(d) Forms ketone bodies.

Q84. Show the efficiency of the complete oxidation of glucose.

One molecule of glucose after complete oxidation yields 38 ATP.

∴ The total energy captured in ATP = 7,600 × 38 = 2,88,800 calories.

$$\text{Hence efficiency} = \frac{288800}{68,600} \times 100 = 42\%$$

Q85. Name the vitamins that play key role in TCA cycle.

(a) Thiamin (vitamin B_1) in the form of TPP, a coenzyme in the pyruvate dehydrogenase complex and in α-ketoglutarate dehydrogenase complex.

(b) Riboflavin (Vitamin B_2) in the form of FAD, a coenzyme in the pyruvate dehydrogenase complex, in α-ketoglutarate dehydrogenase complex, and in succinate dehydrogenase.

(c) Niacin in the form of NAD, a coenzyme for pyruvate dehydrogenase complex, α-ketoglutarate dehydrogenase complex, isocitrate dehydrogenase, malate dehydrogenase.

(d) Pantothenic acid in the form of coenzyme A, a coenzyme for pyruvate dehydrogenase complex and α-ketoglutarate dehydrogenase complex.

Q86. State the clinical importance of citric acid cycle.

(a) The chief function of citric acid cycle is to act as the final common pathway for the oxidation of carbohydrate lipids and protein.

(b) It plays a major role in gluconeogenesis, transamination, deamination and lipogenesis. Liver is the only tissue in which all occur to a significant extent.

(c) The enzymes of the citric acid cycle are located in the mitochondrial matrix which facilitates the transfer of reducing equivalents to the adjacent enzymes of the respiratory chain for the production of ATP.

(d) Intermediates of TCA cycle play a major role in the synthesis of heme formation, formation of non-essential amino acids, fatty acid synthesis, synthesis of cholesterol, formation of vitamin D, bile acids, bile salts, steroid hormones.

Q87. How the propionic acid converted to succinyl-CoA?

See Chart 7 below.

Q88. What is "Cori cycle" or "Lactic acid cycle"?

The conversion of muscle lactate to glucose in liver and its reentry into muscle is called "Cori cycle" or "Lactic acid cycle" or "Glucose-alanine cycle".

See Chart 8 next page.

Q89. What is HMP shunt or pentose phosphate pathway?

This is an alternate aerobic pathway for the oxidation of glucose in the cytoplasm of the cells of liver, lactating mammary gland and adipose tissue when anaerobic glycolysis is blocked by inhibitors. This pathway does not yield ATP and starts with 3 molecules of glucose-6-P.

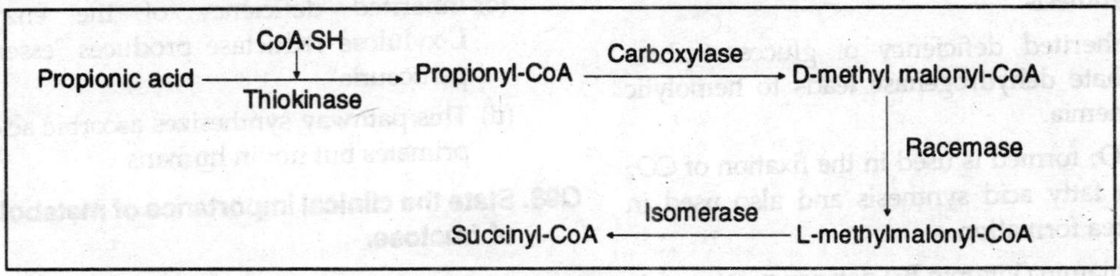

CHART 7

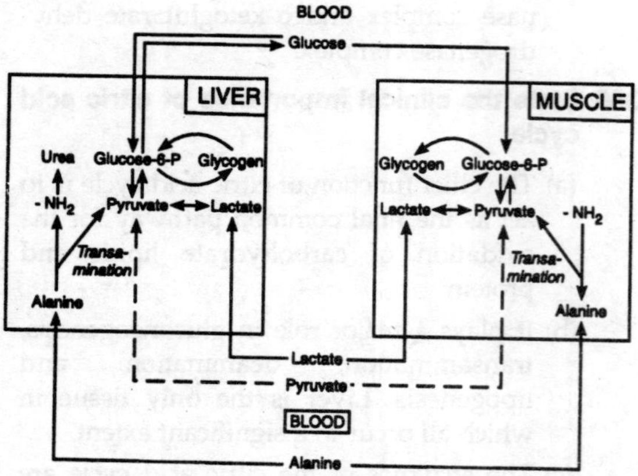

CHART 8

Q90. What are the main products of HMP shunt?

6 molecules of reduced NADP, 3 molecules of CO_2, 2 molecules of Fructose-6-phosphate, and the one molecule of glyceraldehyde-3-phosphate, ribose sugars.

Q91. State the role of thiamin in HMP shunt or pentose phosphate pathway.

Thiamin in the form of TPP acts as coenzyme for transketolase enzyme in the formation of sedoheptulose-7-phosphate.

Q92. What does happen in the hereditary deficiency of glucose-6-phosphate dehydrogenase?

(a) Red blood cells are hemolysed when the susceptible individuals are subjected to antimalarial drugs like primaquin and aspirin etc. Causing hemolytic anemia.

(b) The generation of reduced NADP is impaired.

Q93. Discuss the clinical importance of HMP shunt or pentose phosphate pathway.

(a) This pathway is not involved in energy production.

(b) It provides pentoses for nucleic acid synthesis.

(c) Inherited deficiency of glucose-6-phosphate dehydrogenase leads to hemolytic anemia.

(d) CO_2 formed is used in the fixation of CO_2 in fatty acid synthesis and also used in urea formation.

(e) Thiamin (Vitamin B_1) deficiency interrupts the pathway to be completed.

(f) It provides reduced NADP which is required for synthesis of fatty acids, cholesterol, steroids, sphingolipids, L-glutamate; for the conversion of phenylalanine to tyrosine, oxidized glutathione to reduced glutathione, methaemoglobin to Hb; and is also used in microsomal desaturation of fatty acid, uronic acid pathway.

Q94. Name the hormones involved in HMP shunt.

Insulin : It induces the synthesis of dehydrogenase enzymes for enhancing the activity of the pathway.

Thyroid hormones : They stimulate the activity of glucose-6-phosphate dehydrogenase of the pathway.

Q95. What do you mean by gluconeogenesis or neoglucogenesis?

It is the process by which glucose is formed from substances other than carbohydrates (fats and proteins).

See Chart 9 next page.

Q96. State Wernicke-Korsakoff syndrome.

(a) The enzyme transketolase cannot bind TPP affecting the transketolation reaction.

(b) The patient suffering from this disorder shows severe neuro-psychiatric symptoms.

(c) This is also characterized by deranged mental functions, depression, loss of memory, and mental confusion.

(d) There may be paralysis of eye movements.

Q97. Show the clinical importance of uronic acid pathway.

(a) It is an alternate pathway for oxidation of glucose without producing any ATP and starts with one molecule of glucose-6-phosphate.

(b) D-glucuronic acid produced in this pathway is utilized for detoxication.

(c) Inherited deficiency of the enzyme L-xylulose reductase produces "essential pentosuria".

(d) This pathway synthesizes ascorbic acid in primates but not in humans.

Q98. State the clinical importance of metabolism of fructose.

(a) Fructose is slowly absorbed, easily metabolized and a good source of energy.

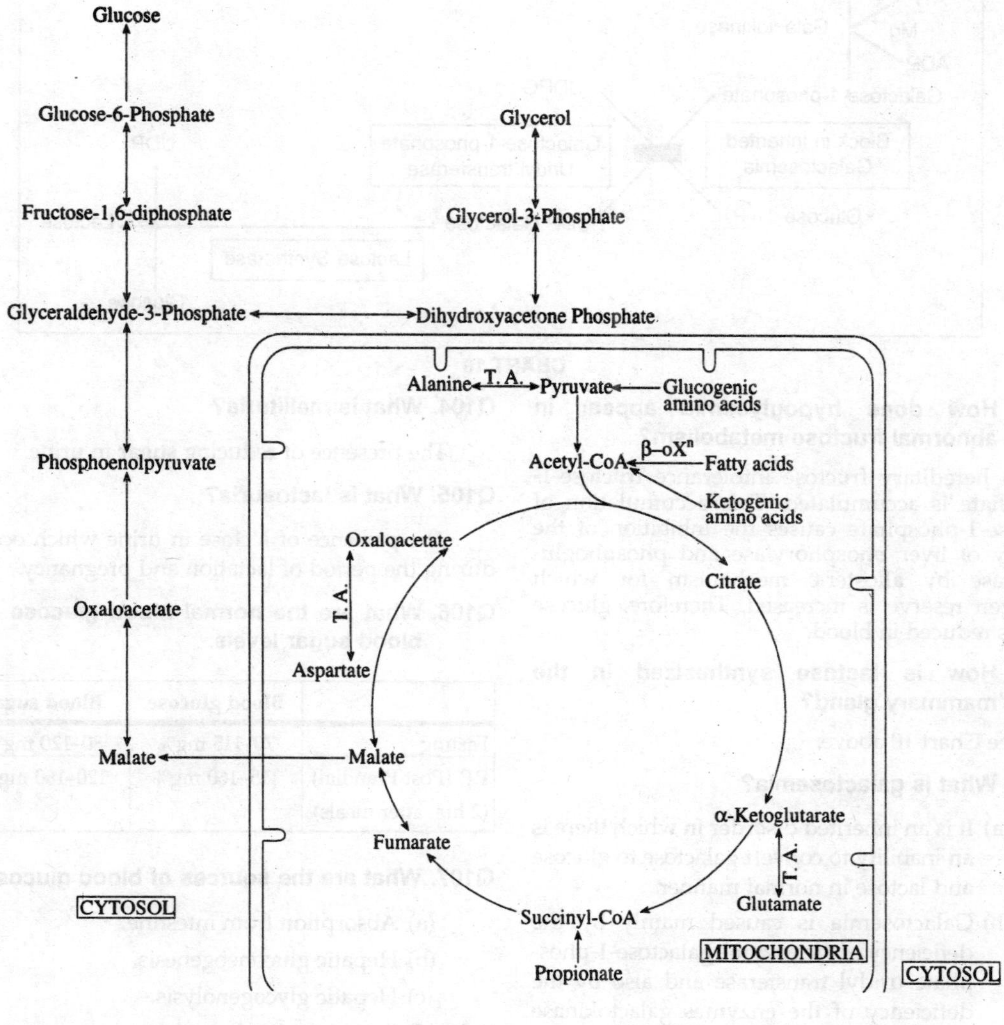

CHART 9

(b) Fructose is richly available in seminal fluid and spermatozoa uses fructose for energy.

(c) Dietary fructose in excessive amounts can cause increased synthesis of triglyceride.

(d) Fructose on parenteral administration for nutrition can cause depletion of adenine nucleotides in liver and liver cell necrosis.

(e) Fructose metabolism through sorbitol pathway causes the development of cataract in the lens of the eyes in diabetes mellitus.

(f) The absence of hepatic aldolase B leads to hereditary fructose intolerance causing rise in fructose level, hypoglycemia accompanied by nausea, vomitng and sweating.

Q99. What is essential fructosuria?

(a) A lack of hepatic fructokinase causes essential fructosuria.

(b) Diets low in fructose and sucrose are beneficial for this condition.

Q100. What do you mean by hereditary fructose intolerance?

(a) This condition appears due to the absence of hepatic aldolase B for which fructose-1-phosphate is not further metabolized.

(b) Diets low in fructose and sucrose are beneficial for this condition.

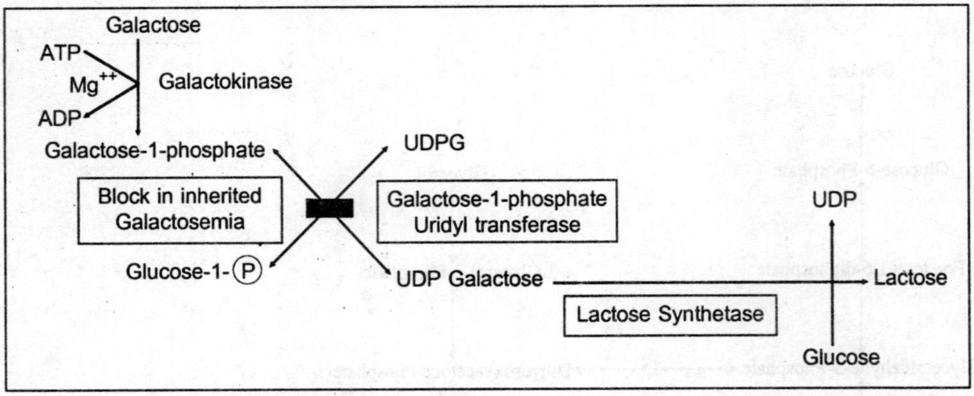

CHART 10

Q101. How does hypoglycemia appear in abnormal fructose metabolism?

In hereditary fructose intolerance fructose-1-phosphate is accumulated. This accumulation of fructose-1-phosphate causes the inhibition of the activity of liver phosphorylase and phosphoglucomutase by allosteric mechanism for which glycogen reserve is increased. Therefore, glucose level is reduced in blood.

Q102. How is lactose synthesized in the mammary gland?

See Chart 10 above.

Q103. What is galactosemia?

(a) It is an inherited disorder in which there is an inability to convert galactose to glucose and lactose in normal manner.

(b) Galactosemia is caused mainly by the deficiency of the enzyme galactose-1-phosphate uridyl transferase and also by the deficiency of the enzymes galactokinase and epimerase.

(c) The elevated level of blood galactose is reduced by aldose reductase in the eye to galactitol which accumulates causing cataract.

(d) Galactose cannot be converted to glucose due to the deficiency of the enzyme epimerase and increased galactose level increases insulin secretion. As a result, hypoglycemia occurs.

(e) The deficiency of inherited galactose-1-phosphate uridyl transferase affects the liver and red blood cells.

(f) The accumulated galactose-1-phosphate depletes the inorganic phosphate of liver causing liver failure and mental deterioration.

Q104. What is mellituria?

The presence of reducing sugar in urine.

Q105. What is lactosuria?

The presence of lactose in urine which occurs during the period of lactation and pregnancy.

Q106. What are the normal blood glucose and blood sugar levels.

	Blood glucose	Blood sugar
Fasting	70–115 mg%	80–120 mg%
P.P. (Post Prandial) (2 hrs. after meals)	115–160 mg%	120–160 mg%

Q107. What are the sources of blood glucose?

(a) Absorption from intestine.

(b) Hepatic gluconeogenesis.

(c) Hepatic glycogenolysis.

(d) From other carbohydrates.

Q108. What are the utilities of blood glucose?

(a) For oxidation in tissues to provide energy.

(b) For glycogenesis in liver, muscle, and other tissues.

(c) For lipogenesis in adipose tissues mainly.

(d) For synthesis of lactose, glycolipids, glycoproteins, nucleic acids, mucopolysaccharides etc.

(e) For excretion in the urine in case of high blood glucose level.

Q109. Mention the main role for the regulation of blood glucose.

Roles of liver, skeletal muscle, kidney, muscular exercise, and hormones.

Q110. What is the role of insulin on blood sugar?

(a) Insulin is bound to the cell wall by disulphide bonds and stimulates the uptake of glucose by some cells.

(b) It increases the permeability to the cells of muscles and adipose tissue.

(c) It inhibits growth hormone and ACTH and thus enhances hexokinase activity.

(d) It helps phosphorylation of glucose and thus assists glucose metabolism to proceed.

(e) It nelps in liver and muscle glycogenesis by stimulating hexokinase and glycogen synthetase.

(f) It helps fatty acid synthesis from acetyl-CoA.

Q111. What is hyperglycemia and hypoglycemia?

Increased blood glucose level above normal is called hyperglycemia; decreased blood glucose level below normal is called hypoglycemia.

Q112. What are the roles of growth hormones, ACTH, glucagon, epinephrine, glycocorticoids and thyroxine on blood glucose?

Growth hormone & ACTH :

(a) Decrease the uptake of glucose by muscle.

(b) Mobilize free fatty acids from adipose tissues which inhibit glucose utilization.

Glucagon and epinephrine :

(a) Stimulate glycogenolysis in liver accelerating phosphorylase.

(b) Cause glycolysis in muscle to form lactic acid which is again converted to glucose by "Cori cycle".

Glucocorticoids :

Stimulate gluconeogenesis by increasing hepatic uptake of amino acids.

Thyroxine :

Stimulates glycogenolysis in liver.

Q113. What do you mean by "renal threshold" for glucose?

In normal individuals, glycosuria occurs when the venous blood glucose exceeds 160-180 mg/100 ml. This level of the venous blood glucose is termed as the "renal threshold" for glucose.

Q114. What is meant by Diabetes Mellitus and how does it cause?

Diabetes = Lack of insulin; Mellitus = Sweet.

Cause : In the deficiency of insulin, liver glycogenolysis is increased and muscle glycogenesis is decreased. As a result, there is hyperglycemia which results in glycosuria.

Q115. What is "diabetes insipidus"?

A disease due to disorder of carbohydrate metabolism having extreme diuresis—upto 30 litres of urine per day.

Q116. What is "Diabetic Acidosis"?

A disease in which ketone bodies are associated with uncontrolled diabetes mellitus.

Q117. What should be the treatment for diabetes mellitus?

Subcutaneous or intramuscular injection of soluble insulin and protein with less carbohydrate diet.

Q118. What does happen when insulin is orally administered?

Since insulin is a protein it is digested to amino acids in the stomach. Hence it has no effect on blood glucose.

Q119. How is insulin destroyed in the body?

(a) The destruction of insulin in the liver is primarily due to the enzyme glutathione insulin transhydrogenase and this enzyme brings about reductive cleavage of the S—S bonds connecting the chains A and B. Reduced glutathione is the coenzyme and donates H atoms for the reduction. The A and B chain peptides are then degraded by proteolysis.

Q120. What is "Diabetic coma"?

In hypoglycemia, if blood sugar level reaches upto 30 mg/100 ml of blood.

In hyperglycemia, if blood sugar level reaches upto 550 mg/100 ml of blood.

Q121. What is "juvenile diabetes"?

The β-cells of pancreas are exhausted and become atrophied causing very low secretion of insulin forming diabetes mellitus which develops during the first 40 years of life, ketoacidosis and diabetic coma may develop if insulin treatement is delayed.

Q122. Express the changes in carbohydrate, fat, and protein metabolism in diabetes mellitus.

Carbohydrate metabolism :

(a) Diminished oxidation of glucose.

(b) Decreased glycogen formation.

(c) Decreased synthesis of fat.

Fat Metabolism :

(a) Fats are oxidized more to form more acetyl-CoA which forms more ketone bodies and cholesterol.

(b) Ultimately produces ketoacidosis and atherosclerosis if untreated.

Protein metabolism :

(a) More breakdown of tissue protein and the body is in negative nitrogen balance.

(b) More deamination of amino acids leads to more urea formation causing increased urea excretion in urine.

Q123. Name the conditions in which glucose tolerance is less.

Infections, obesity, hyperactivity of pituitary or adrenal cortex, atherosclerosis and cirrhoris of liver.

Q124. Evaluate the chief differences between hypoglycemic coma and diabetic coma.

Hypoglycemic coma	Diabetic coma
(a) Sudden and in good health	The patient suffers from ill-health for several days.
(b) Sweating, headache, tremor, semi consciousness, seeing double image.	Dehydration, abdominal pain, vomiting.
(c) Moist skin, full pulse, normal breathing, normal or raised blood pressure.	Dry skin, weak pulse, low blood pressure, air hunger.
(d) Ketone bodies and glucose absent in urine.	Ketone bodies and glucose present in urine.
(e) Blood sugar level comes down to 30 mg/100 ml.	Blood sugar level reaches to 550 mg/100 ml.

Q125. What do you mean by glucose tolerance?

The capacity of the body to utilize glucose.

Q126. How will you perform glucose tolerance test (GTT)?

(a) The individual is kept fasting for at least 12 hours after the last meal.

(b) Samples of venous blood and urine are collected in the fasting state to find out the blood and urine glucose level.

(c) Immediately after this 50 gms of glucose is administered to that subject. Samples of blood and urine are collected at an interval of 30 minutes for a period of 2 to 3 hours. The results of blood glucose analysis are plotted as a graph against time.

(d) Similar method is adopted in case of prediabetic, mild diabetic, and severe diabetic patients. The graph will show the glucose tolerance in each case.

Q127. What is glycosuria?

The condition in which abnormal quantities of glucose are excreted in urine is called glycosuria.

Q128. How many type of glycosuria are there?

Alimentary glycosuria, emotional glycosuria, renal glycosuria, glycosuria in pregnancy and lactation, diabetic glycosuria.

Q129. What do you mean by G : N or D : N ratio?

If the kidney becomes infected, the reabsorption of sugar becomes less causing hypoglycemia which stimulates secretion of epinephrine and glucagon. These two hormones cause glycogenolysis in the liver to maintain blood sugar level. But this sugar is also excreted out. The tissue protein then starts breaking down to supply glucose to maintain blood sugar level causing the tissue waste. The urine in such cases is tested for glucose and nitrogen. It is found that the ratio of glucose and nitrogen (G : N) or dextrose and nitrogen (D : N) is constant.

Q130. Show the pathway of conversion of carbohydrates into fats.

See Chart 11, next page.

Q131. What is the key enzyme of glycogenesis?

Glycogen synthase (Synthetase).

Q132. Name the key enzyme of glycogenolysis.

Phosphorylase.

Q133. Mention the key enzymes of glycolysis, TCA cycle, Gluconeogenesis, HMP shunt.

Glycolysis	→ Glucokinase, Phosphofructokinase-1, and Pyruvate kinase.
TCA cycle	→ Citrate synthase, Isocitrate dehydrogenase, and α-ketoglutarate dehydrogenase.
Gluconeogenesis	→ Pyruvate carboxylase, Phosphoenolpyruvate carboxykinase, fructose-1, 6-bisphosphatase, and glucose-6-phosphatase.
HMP shunt	→ Glucose-6-phosphate dehydrogenase.

Q134. What is the inhibitor of lactate dehydrogenase?

Oxamate.

Q135. What is favism?

Acute hemolytic anemia with hemoglobinuria may occur in glucose-6-phosphate dehydrogenase deficient persons who are sensitive to fava beans either on ingestion of uncooked or lightly cooked beans or inhalation of pollens.

Q136. Name some drugs which increase the formation of D-glucuronic acid by uronic acid pathway.

(a) Barbiturates,

(b) Amino pyrine,

(c) Antipyrine.

Q137. How is cataract formed?

The high concentration of sorbitol from glucose in the cells increase the osmotic pressure which is responsible for the development of cataracts of lens of the eye.

Q138. What do you mean by lag curve (Oxyhyperglycemic curve)?

Fasting blood glucose level is normal, but it rises rapidly in the next $\frac{1}{2}$ to 1 hour and exceeds the renal threshold showing the presence of glucose in urine. The return to normal level is rapid and complete.

Q139. Mention the conditions in which there is lag curve.

(a) In hyperthyroidism.

(b) After gastroenterostomy.

(c) During pregnancy.

(d) In early diabetes mellitus.

Q140. State the role of fructose 2, 6-bisphosphate in the regulation of glycolysis and gluconeogenesis in liver.

(a) It is the allosteric effector of phosphofructokinase-1 and inhibitor of fructose-1, 6-bisphosphatase in liver.

(b) It relieves inhibition of phosphofructokinase-1 by ATP and increases affinity for fructose-6-phosphate.

(c) It inhibits fructose-1, 6-bisphosphatase by increasing the Km for fructose 1, 6-bisphosphate.

Q141. What is substrate (Futile) cycle?

The control points in glycolysis and glycogen metabolism involve a cycle of phosphorylation and dephosphorylation catalyzed by the enzymes like glucokinase and glucose-6-phosphatase; phosphofructokinase-1 and fructose-1, 6-bisphosphatase, pyruvate kinase, pyruvate carboxylase and phosphoenol pyruvate carboxykinase; glycogen synthase and phosphorylase.

Q142. Is glucose excreted in normal health?

In normal health, a very small amount less then 0.5 gm of glucose is excreted in urine daily. But this small amount is not detected by Benedict's qualitative test.

Q143. What is the effect of parenteral administration of fructose?

It causes depletion of adenine nucleotides in liver producing liver cell necrosis.

Q144. What is the metabolic importance of fructose?

(a) Fructose is metabolized easily and hence a good source of energy.

(b) Seminal fluid is rich in fructose and spermatozoa uses fructose for energy.

(c) Excess dietary fructose leads to increased synthesis of triglyceride.

(d) In diabetes mellitus, fructose metabolism through **sorbitol pathway** causes the development of cataracts and neuropathy.

Q145. Name the test for the detection of carbohydrate and discuss its method. MOLISCH'S TEST.

Method : 2 ml. sugar solution + 2 drops of Molisch's reagent. Mixed thoroughly. Added 2 ml conc. H_2SO_4 by the side of the test tube. The test tube is carefully erected. Purple ring is formed at the junction of two liquids indicating the presence of carbohydrate.

Q146. What is the test for detecting polysaccharides? State its method.

IODINE TEST

Method : 3 mol. of provided solution + 3 drops of dilute solution of iodine.

Development of blue colour indicates starch.

Development of red colour indicates glycogen.

Development of purple colour indicates dextrins.

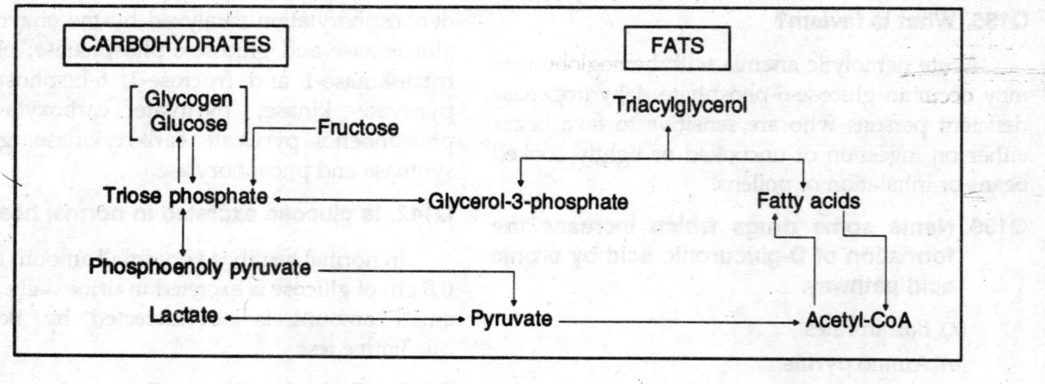

CHART 11

Q147. What are reducing sugars?

The sugars having free aldehyde or ketone groups in their structures which can reduce $CuSO_4$ to Cu_2O of benedict's (Qualitative) solution on boiling are called reducing sugars.

Examples : Glucose, fructose, mannose, galactose, maltose, lactose etc.

Q148. Why sucrose is not a reducing sugar?

Sucrose has no free aldehyde or ketone group in its structure.

Q149. Name the test for detecting reducing sugars and discuss its method.

BENEDICT'S TEST

Method : 5 ml. of Benedict's (Qualitative) solution + 8 drops of provided solution. Boiled for 3 minutes. The formation of light green, green, yellow, brick red precipitates indicates the presence of reducing sugars.

Light green ppt. ⟶ 0.1–0.5% reducing sugars.
Green ppt. ⟶ 0.5–1.0% reducing sugars.
Yellow ppt. ⟶ 1.0–2.0% reducing sugars.
Brick red ppt. ⟶ Above 2% reducing sugars.

Q150. Why does Benedict's test sometimes form white ppt?

Owing to the presence of high concentration of phosphates in urine.

Q151. Why is Benedict's test preferred to peform in the detection of glucose in urine?

(a) Benedict's test is less likely to give weakly positive results with concentrated urine due to the action of creatinine and uric acid.

(b) It is more sensitive to small quantities of glucose because of the weaker alkali it contains.

(c) The strong alkali (KOH) of the Fehling's solution can polymerize small quantities of glucose when vigorously boiled.

Q152. By which test monosaccharides are distinguished from disaccharides?

The copper acetate of Barfoed's solution is reduced by monosaccharides after boiling for 30 seconds.

The copper acetate of Barfoed's solution is reduced by disaccharides after boiling for several minutes.

Q153. What is the confirmative test for detecting sucrose? Discuss the method.

HYDROLYSIS TEST

Method : 5 mol of provided solution + 1 or 2 drops of thymol blue indicator + 2 drops of dil. HCl. Pink colour develops. Divided into two portions. One portion is boiled and cooled for hydrolysis. Both the portions are neutralized by 2% Na_2CO_3 solution. Benedict's test is performed for both. The hydrolyzed portion gives Benedict's positive while the unhydrolyzed portion is non-responsive.

Q154. Why galactose (although hexose aldose sugar) does not form the same type of osazone as does glucose or mannose?

The –OH group around carbon No. 4 is not the same as that of glucose, fructose or mannose. Hence it gives different types of crystals.

Q155. State the method for osazone test.

7 ml. of provided solution + 1 pinch of phenylhydrazine hydrochloride + 2 pinches of sodium acetate + 1 mol of glacial acetic acid. Boiled for 3 minutes. Filtered and the filtrate in a test tube is kept in a boiling water bath until crystals are formed.

Q156. What do you mean by osazone?

Formation of crystals.

CHAPTER 3

LIPIDS

Q1. Define lipids.

The lipids are a heterogeneous group of compounds related to fatty acids and are insoluble in water but soluble in organic solvents such as ether, chloroform, and benzene.

Q2. Differentiate between fats and oils.

Fats are solids at 20°C but oils are liquids at 20°C.

Q3. Evaluate the biomedical importance of lipids.

(a) Lipids are the important dietary constituents and serve as an efficient source of energy.

(b) Some deposits of lipids act as an insulating material in the body while lipids around kidney protect the organ.

(c) Since some vitamins A, D, E, K are fat soluble lipids are essential for these vitamins.

(d) The nervous system is rich in lipids and are essential for proper functioning.

(e) Lipids supply essential fatty acids which cannot be synthesized in the body and are essential for normal health and growth.

(f) The phosphatides of blood platelets are involved in the production of thromboplastin activity in the early stages of blood clotting.

(g) Lipoproteins and phospholipids are important constituents of many natural membranes.

(h) Lipoproteins are also carriers of triglycerides, cholesterol and phospholipids in the body.

(i) Acetic acid, a normal breakdown product of fatty acids, can be used by the body for the synthesis of cholesterol, vitamin D and corticosteroid hormones.

Q4. Classify lipids.

A. Simple lipids :

(i) Fats : Esters of fatty acids with glycerol.

(ii) Waxes : Esters of fatty acids with alcohols other than glycerol.

B. Compound lipids :

(i) Phospholipids : Neutral fat + Phosphoric acid + Nitrogen containing compounds or other substituents.

eg. Lecithin, Cephalin, Phosphatidyl serine, Phosphatidyl inositol, Plasmalogens, Sphingomyelins.

(ii) Glycolipids : Sphingosine (alcohol) + Fatty acids + Carbohydrate + Substance having nitrogen base.

eg. Cerebrosides and gangliosides.

(iii) Sulpholipids : Lipids having sulphate groups.

(iv) Aminolipids : (Proteolipids) eg. phosphatidyl ethanolamine, phosphatidyl serine.

(v) Lipoproteins : (Triglyceride, phospholipids, cholesterol, cholesteryl esters, free fatty acids) + Proteins.

C. Derived lipids :

Substances derived from above groups by hydrolysis. eg. cholesterol, fatty acids, glycerol, fatty aldehydes, steroids.

Q5. Why are glycerides, cholesterol and cholesteryl esters said to be neutral lipids?

Because these are uncharged.

Q6. What are steroids?

These are the compounds containing a special ring cyclopentanoperhydrophenanthrene nucleus (CPP-nucleus).

Examples : Cholesterol, bile acids, sex hormones.

Q7. What do you mean by saponification and saponification number?

Hydrolysis of fat by alkali is called saponification. The products are glycerol and the alkali salts of the fatty acids which are called soaps.

The number of milligrams of KOH required to neutralize the free fatty acids of 1 gram of fat.

Q8. What is Reichert-Miessl number?

The number of milliliters of 0.1 Normal KOH required to neutralize the soluble fatty acids from 5 grams of fat.

Q9. What do you mean by rancidity?

Natural fats are oxidized when exposed to air, light, moisture if warm developing an unpleasant odour and taste due to the formation of peroxides at the double bonds of unsaturated fatty acids. Vitamin E is an important natural antioxidant.

Q10. What is lecithin?

It is phosphatidyl choline. It is the valuable agent for the emulsification of fats and oils.

Q11. What is sphingomyelin?

(a) It is a phospholipid consisting of sphingosine (an amino alcohol having 18 carbon atoms), fatty acid, phosphoric acid, and choline.

(b) Its concentration is increased in Niemann-Pick disease.

Q12. Explain glycolipids.

Glycolipids contain sphingosine, fatty acid, and a carbohydrate moiety. It is classified into cerebrosides and gangliosides.

(a) **Cerebrosides** : Consist of sphingosine, a high molecular fatty acid, galactose.

These are differentiated by the type of fatty acid in the molecule. These are :

Kerasin : Containing lignoceric acid.

Cerebron : Containing a hydroxylignoceric acid (cerebronic acid).

Nervon : Containing nervonic acid (unsaturated homologue of lignoceric acid)

Oxynervon : Containing the hydroxy derivative of nervonic acid.

(b) **Gangliosides** : Sphingosine + higher fatty acid containing 18 carbon atoms + 1 molecule of glucose, or 2 molecules of galactose or 3 molecules of sialic acid or 1 molecule of N-acetyl-galactosamine.

Q13. What are lipoproteins?

These are formed by triacylglycerol (45%), phospholipids (35%), cholesterol and cholesteryl esters (15%), free fatty acids (less than 5%) and protein.

There are four groups of lipoproteins :

(a) Chylomicrons.

(b) Very low density lipoproteins (VLDL) or Pre- β-lipoproteins.

(c) Low density lipoproteins (LDL) or β-lipoproteins.

(d) High dessity lipoproteins (HDL) or α-lipoproteins.

Lipoproteins are separated by immuno-electrophoresis.

Q14. In which diseases the concentrations of VLDL and LDL are increased?

Atherosclerosis and coronary thrombosis.

Q15. What is iodine number?

The number of grams of iodine absorbed by 100 grams of fat. This is the measure of the degree of unsaturation of fat.

Q16. What is acid number?

The number of milligrams of KOH required to neutralize the free fatty acids of 1 gram of fat.

Q17. What is cardiolipin?

It is diphosphatidylglycerol formed from phosphatidic acid present in mitochondria.

Q18. State the importance of lipoproteins.

(a) Lipoproteins transport and deliver the lipids to tissues.

(b) They maintain structural integrity of cell surface, mitochondria and microsomes.

(c) The concentration of LDL is increased in severe diabetes mellitus, atherosclerosis etc.

(d) The concentrations of chylomicrons, VLDL, are increased in atherosclerosis and coronary thrombosis.

Q19. Enumerate the biomedical importance of phospholipids.

(a) They increase the rate of fatty acid oxidation.

(b) They help in the transport of lipids from intestine.

(c) They help blood clotting.

(d) They act as prosthetic group to certain enzyme.

(e) They are the carriers of inorganic ions across the membranes.

(f) They are involved in lipotropic action.

(g) They act as cofactor in lipoprotein lipase and triacylglycerol lipase.

Q19. What are saturated fatty acids?

These fatty acids have no double bonds in their structure.

Examples : Acetic acid, propionic acid, butyric acid, palmitic acid ($C_{15}H_{31}COOH$), Stearic acid ($C_{17}H_{35}COOH$), Lignoceric acid ($C_{22}H_{47}COOH$).

Q20. Name the monounsaturated fatty acids.

Palmitoleic acid and oleic acid.

Q21. Name the polyunsaturated fatty acids showing the number of double bonds.

(a)	Linoleic acid ($C_{17}H_{31}COOH$) with two double bonds.	(Δ^{9}, Δ^{12})	18 : 2
(b)	Linolenic acid ($C_{17}H_{29}COOH$) with three double bonds.	($\Delta^{9}, \Delta^{12}, \Delta^{15}$)	18 : 3
(c)	Arachidonic acid ($C_{17}H_{27}COOH$) with four double bonds.	($\Delta^{5}, \Delta^{8}, \Delta^{11}, \Delta^{14}$)	20 : 4
(d)	Clupanodonic acid ($C_{17}H_{25}COOH$) with five double bonds.	($\Delta^{7}, \Delta^{10}, \Delta^{13}, \Delta^{16}, \Delta^{19}$)	22 : 5
(e)	Cervonic acid ($C_{17}H_{23}COOH$) with six double bonds.	($\Delta^{4}, \Delta^{7}, \Delta^{10}, \Delta^{13}, \Delta^{16}, \Delta^{19}$)	22 : 6

Q22. Discuss the clinical importance of essential fatty acids (EFA).

(a) Lipids of gonads contain high concentrations of polyunsaturated fatty acid which are involved in reproductive function.

(b) Prostaglandins are synthesized from arachidonic acid.

(c) Leukotrienes are also formed from arachidonic acid.

(d) The deficiency of EFA causes swelling of mitochondria.

(e) Fats with high content of polyunsaturated fatty acids tend to lower serum cholesterol level.

(f) Clotting time is prolonged by the ingestion of fats rich in EFA.

(g) Fibrinolytic activity is increased by the ingestion of fats rich in EFA.

(h) Deficiency of EFA produces fatty liver.

(i) Linolenic acid is necessary in the diet for optimal vision.

(j) Deficiency of EFA causes skin lesion, cessation of growth, kidney damage, decrease resistance to stress, fatty liver, abnormalities of pregnancy and lactation in adult females, eczema like dermatitis.

Q23. Name the essential fatty acids.

Linoleic acid, Linolenic acid, arachidonic acid.

Q24. What is "Cholesterol"?

(a) It is a derived lipid and an alcohol having imperical formula $C_{27}H_{46}O$.

(b) It contains the cyclopentanoperhydrophenanthrene nucleus.

(c) The best source of cholesterol is the egg-yolk.

Q25. What is the normal level of cholesterol in the blood?

140–220 mg/100 ml of blood.

Q26. State the clinical importance of cholesterol.

(a) It is the essential constituent of cell.

(b) It increases the permeability of the cells.

(c) It transports fats to the liver in the form of cholesterol ester.

(d) It is involved in the formation of bile acids and bile salts, vitamin D, and corticosteroid hormones.

Q27. Express the biomedical importance of bile.

(a) Cholic acids permit the absorption of many valuable insoluble compounds.

(b) It helps digestion of fat by emulsification lowering the surface tension and thus facilitates the action of lipase on fat.

(c) Bile salts are activators of lipase.

(d) Bile salts are reabsorbed from the intestine and pass back to the liver where they stimulate further secretion of bile.

(e) Bile is an important source of alkali for neutralisation of hydrochloric acid entering the intestine from the stomach.

(f) Bile is a channel for the excretion of bile pigments, certain drugs, toxins, copper, iron, calcium and some inorganic salts.

(g) Fat is not digested in the absence of bile. This undigested fat forms a layer round particles of other foodstuffs and diminish the surface exposed to enzymes.

Q28. Mention the composition of bile.

Solids, water, bile acids and bile salts, cholesterol, mucin and pigments, fatty acids, and inorganic salts.

Q29. What are bile acids and bile salts?

Bile acids : Cholic acids, deoxycholic acid, chenodeoxycholic acid and lithocholic acid.

Bile salts : Sodium or Potassium glycocholate (Na or K + Cholic acid + glycine).

Sodium or Potassium taurocholate (Na or K + Cholic acid + taurine, a cysteine derivative).

Q30. What do you mean by eicosanoids?

(a) These are compounds derived from **eicosa**-(20-C) polyenoic fatty acids.

(b) They consist of **Prostanoids** and **leukotrienes.**

(c) Prostanoids include
   (i) Prostaglandins (PG)
   (ii) Prostacyclins (PGI)
   (iii) Thromboxanes (TX).

Q31. What are prostaglandins?

(a) They exist in every mammalian tissue.

(b) They act as local hormones.

(c) They have physiologic and pharmacologic activities.

(d) They are synthesized from arachidonic acid (contains 20-C) to form a cyclopentane ring.

(e) Three different eicosanoic fatty acids give rise to three groups of eicosanoids characterized by the number of double bonds in the side chains, eg. PG_1, PG_2, PG_3.

(f) PGE_2 has a keto group in position 9; whereas the "F" type has a hydroxyl group in this position.

Q32. What are leukotrienes?

(a) They are a third group of eicosanoid derivatives formed via the lipoxygenase pathway rather than cyclization of the fatty acid chain.

(b) They were first described in leukocytes.

(c) They are characterized by the presence of three conjugated double bonds.

(d) 5-lipoxygenases form leukotrienes. The first formed in leukotriene A_4 which is metabolized to leukotriene B_4 or leukotriene C_4.

Q33. How many prostaglandins have been isolated so far?

14 prostaglandins have so far been isolated from male reproductive tract.

Q34. What is "suicide enzyme"?

Cyclooxygenase is "suicide enzyme" because of self-catalyzed destruction.

Q35. State the clinical importance of prostaglandins.

(a) As little as 1 ng/ml of prostaglandins causes contraction of smooth muscle in animal.

(b) They cause vasodilatation and decrease blood pressure.

(c) They are used as contraceptives to prevent conception.

(d) They effectively induce labor.

(e) They are used to prevent gastric ulcers and control inflammation.

(f) They are the effective nasal decongestant.

(g) They increase cAMP in platelets, thyroids, fetal bone, and lung but lower cAMP in renal tubule cells and adipose tissue.

Q36. How are prostaglandins metabolized?

Prostaglandins are quickly metabolized by the enzyme 15-hydroxy prostaglandin dehydrogenase.

Q37. Mention the side-effects of prostaglandins.

(a) PGE_2 causes gastrointestinal smooth muscle to contract which leads to cramping and diarrhoea.

(b) PGE_2 irritates the mucosa lining of the throat causing pain and coughing.

Q38. What are the inhibitors of prostaglandins?

Aspirin, indomethacin, phenanzine.

Q39. Evaluate the clinical importance of prostacyclins.

(a) Prostacyclins act as an effective vasodilator and bring down blood pressure.

(b) They diminish gastric acid secretion.

(c) They increase renal blood flow causing renal vasodilatation and thus enhance glomerular filtration rate. They stimulate renin production from juxtra glomerular cells.

(d) They are destroyed by liver undergoing β- and ω-oxidation and ultimately excreted in urine.

Q40. State the biomedical importance of thromboxanes (TX).

(a) They produce vasoconstriction and increase blood pressure.

(b) They increase platelets aggregation.

(c) They contract smooth muscles on blood vessels, gastrointestinal tract, uterus, bronchi and bronchioles, but constriction is much weaker than PGE.

(d) They induce release of serotonin and Ca^{++} from platelet granules.

Q41. Mention the significance of leukotrienes clinically.

(a) They increase vascular permeability.

(b) Inhalation of leukotrienes causes bronchospasm. They are more potent as bronchoconstrictors in normal subjects as compared to histamine.

(c) They are the stimulators of mucus secretion from human airway tissue.

Q42. How is fat digested?

(a) **Lingual lipase** secreted by the dorsal surface of the tongue acts best at P^H 4.0 to 4.5. It is more active on triglyceride having shorter fatty acid chains. Milk fat is the best substrate for this enzyme. The released short chain fatty acids can be absorbed directly from the stomach wall and enter the portal vein.

(b) The presence of small amounts of gastric lipase has negligible activity on fats because of high acidity and non-emulsification of fats in the gastric juice. The gastric lipase is not capable of hydrolyzing fats containing long chain fatty acids.

(c) In the pancreas fat is emulsified by bile salts for which the surface tension of fat is reduced. The pancreatic lipase at the optimum P^H 8.0 hydrolyzes fat into glycerol and fatty acids.

(d) In the small intestine lecithin is hydrolyzed by lecithinase to form glycerol, fatty acids, phosphoric acid and choline.

Q43. How is fat absorbed?

(a) Glycerol being water-soluble is easily absorbed from the intestine.

(b) The fatty acid and monoglycerides are sparingly soluble in aqueous medium. So, these with the bile salts form a hydrotopic complex which can be absorbed from the intestinal lumen into the mucosal cells. The fatty acids being released from the bile salts combine with a glycerol moiety to resynthesize a neutral fat. The bile salts are then absorbed into the portal circulation and pass to the liver where they are re-excreted in the bile causing enterohepatic circulation.

Q44. Enumerate the main functions of the large intestine.

(a) Storage of faeces and their evacuation.

(b) Absorption of water but not of foodstuffs to any significant extent.

(c) Excretion of phosphate but not of iron or calcium.

Q45. What do you mean by lipotropic factor?

The substance which prevents the accumulation of abnormal quantities of fat in the liver is called a lipotropic factor.

Example : Choline is the most important one. Methionine, glycine, serine all of which contribute to the synthesis of choline.

Q46. What is transmethylation?

Substances which supply methyl groups act indirectly as lipotropic agents by contributing to the synthesis of choline.

Example : Methionine is the most important source.

Q47. Name the chief fat depots and show the characteristic of depot fat.

Subcutaneous fatty tissue, retroperitoneal fatty tissue, inter-muscular fatty tissue.

Characteristic :

(a) Neutral fat (Triglyceride) in the depot.

(b) Neutral fat and phospholipids in the tissues.

Q48. Mention the sources of fat in the body.

(a) The fat of the foodstuffs.

(b) The carbohydrate of the food.

(c) To some extent, the protein of the food.

Q49. What is lipogenesis?

The synthesis of fatty acids and triglycerides from acetyl-CoA in adipose tissue and to a lesser extent in other tissues including the liver.

Q50. How does alcoholism cause cirrhosis?

The intake ethanol is oxidized to acetaldehyde with the liberation of excess reduced NAD which competes with reducing equivalents from other substrates for the respiratory chain inhibiting their oxidation. This causes the reduced activity of the citric acid cycle. This, ultimately, leads to the decreased fatty acid oxidation and increased hepatic triacylglycerol synthesis. The more accumulation of triglyceride in the liver causes cirrhosis.

Q51. State the biosynthesis of triglyceride (Triacylglycerol).

See Chart 1 below.

Q52. Discuss the role of different hormones in lipolysis.

(a) Thyroid hormones stimulate epinephrine and norepinephrine hormones.

(b) Adenylate cyclase enzyme is stimulated by epinephrine, norepinephrine, ACTH, TSH, glucagon but depressed by insulin and prostaglandin E_1.

(c) Adenylate cyclase then converts ATP to cAMP. cAMP dependent portein kinase converts hormone-sensitive triacylglycerol lipase b (inactive) to hormone-sensitive triacylglycerol lipase a (active) which hydrolyzes triacylglycerol to diacylglycerol and FFA.

(d) Diacylglycerol lipase then hydrolyzes diacylglycerol to monoacylglycerol and FFA.

(e) Monoacylglycerol lipase hydrolyzes monoacylglycerol to glycerol and FFA.

Q53. Explain why some persons can "eat and not get fat".

(a) Brown adipose tissue is reduced or absent in obese persons.

(b) This tissue is characterized by a well-developed blood supply and a high content of mitochondria and cytochromes but low activity of ATP synthase. Metabolic emphasis is given on oxidation of glucose and fatty acids.

(c) Norepinephrine liberated from sympathetic nerve endings is important in increasing lipolysis in the tissue. Oxidation and phosphorylation are not coupled in mitochondria of this tissue. Thus, oxidation produces much heat and little free energy is trapped in ATP.

(d) Since lipolysis is prominent in such case fat formation is less.

Q54. How are fats oxidized?

(a) Fats are hydrolyzed to glycerol and fatty acid by lipase. Glycerol is oxidized by glycolysis.

(b) Fatty acids undergo α-, β-, and ω-oxidation.

(c) The unsaturated fatty acids and the saturated fatty acids with even number of carbon atoms undergo β- oxidation; while the fatty acids with odd number of carbon

Glycerol + ATP —(Glycerokinase)→ Glycerol-3-Phosphate + ADP

Glycerol-3-Phosphate + 2 Acyl-CoA —(Glycerophosphate acyl transferase)→ 1, 2-diacylglycerolphosphate.

↓ Phosphatase

1, 2-diacylglycerol + Phosphoric acid

1, 2-diacylglycerol + Acyl-CoA —(Diacylglycerol acyl transferase)→ Triacylglycerol (Triglyceride)

CHART 1

atoms undergo β-oxidation until propionyl-CoA is formed. However, β-oxidation is the most important pathway.

Q55. What do you mean by α–, β–, and ω-oxidation?

α- oxidation : This involves decarboxylation of the COOH group after hydroxylation and the formation of fatty acid containing an odd number of carbon atoms which undergo repeated β- oxidation. It removes one carbon at a time from the carboxyl end of the molecule. It does not require CoA intermediates and does not lead to the generation of high-energy phosphates.

β- oxidation : β-carbon atom of fatty acid is oxidized to –COOH by the removal of 2 carbon atoms at a time in the form of acetyl-CoA.

ω- oxidation : The oxidation takes place by hydroxylase enzyme involving cytochome P-450 in microsomes. The $-CH_3$ group is converted to a $-CH_2OH$ group which is ultimately oxidized to –COOH group forming a dicarboxylic acid which is then converted to smaller dicarboxylic acids by β- oxidation.

Q56. Why is triacylglycerol (triglyceride) advantageous over carbohydrate or protein?

(a) The caloric value of triglyceride is over twice as great as that of carbohydrate or protein.

(b) It exists in the most concentrated form in which potential energy can be stored.

(c) Fatty acids provide more metabolic water upon oxidation than other metabolic fuel which is advantageous to mammals occupying dry environments.

Q57. What is Zellweger's syndrome?

(a) It is a rare inherited disorder.

(b) The inherited absence of peroxisomes in all tissues and its enzymes fail to oxidize long-chain fatty acids in peroxisomes resulting in accumulation of fatty acids (C_{26}–C_{38}) chain length in brain tissue and other tissues like liver or kidney.

Q58. State the role of carnitine in fatty acid metabolism.

(a) Carnitine (β- hydroxytrimethylammonium butyrate) is widely distributed and particularly abundant in muscle. It is protein in nature.

(b) It stimulates the oxidation of long chain fatty acids in mitochondria.

(c) Long chain acyl-CoA as such can not penetrate mitochondrial membrane and carnitine itself also cannot penetrate mitochondrial membrane. The combined form acyl carnitine is influenced by carnitine palmityl acyl transferase present in mitochondrial membrane and facilitate acyl CoA to enter into the mitochondria from undergoing β- oxidation.

(d) Another enzyme carnitine acetyl acyl transferase present in mitochondrial membrane causes the transfer of short chain acyl groups between CoA and cornitine.

Q59. What does happen in the deficiency of carnitine?

(a) Accumulation of lipids.

(b) Impaired ketogenesis in the presence of increased plasma FFA.

(c) Hypoglycemia.

(d) Muscular weakness.

Q60. How many ATPs are formed by one turn of β- oxidation of fatty acid and where does β- oxidation take place?

5 ATP; Mitochondria of the cell.

Q61. How many molecules of acetyl-CoA is formed by β- oxidation of one molecule of stearic acid?

9 molecules of acetyl-CoA.

Q62. How many ATPs are formed by complete β-oxidation of one molecule of stearic acid?

(a) Stearic acid contains 18 carbon atoms. It undergoes 8 turns of β-oxidation and 9 molecules of acetyl-CoA are formed.

(b) Acetyl-CoA enters TCA cycle for complete oxidation.

(c) Two ATP is used for the activation of stearic acid.

∴ The total ATP = $[(5 \times 8) + (12 \times 9)] - 2 = 146$.

Q63. How does the oxidation of fatty acids containing odd number of carbon atoms take place?

$$\text{Fatty acid} \xrightarrow{\beta\text{-oxidation}} \text{Propionyl-CoA} \rightarrow \text{Succinicacid} \rightarrow \text{TCA cycle}$$

(with odd number of carbon atoms)

Q64. How is carbohydrate converted to fat?

(a) Glucose is converted to pyruvate by glycolysis in the cytosol.

(b) Pyruvic acid is oxidized in the mitochondria to acetyl- CoA which then forms citric acid being combined with oxaloacetate.

(c) Citrate enters into the cytosol and splitted to acetyl-CoA and oxaloacetate by ATP citrate lyase.

(d) Acetyl-CoA in the cytosol then yields palmitate by biosynthetic process.

Q65. What are the main constituents required for fatty acid synthesis?

Acetyl-CoA, CO_2, multienzyme complex, reduced NADP.

Fatty acids are synthesized best in the cytosol of the cell.

Q66. What are ketone bodies or acetone bodies or ketones?

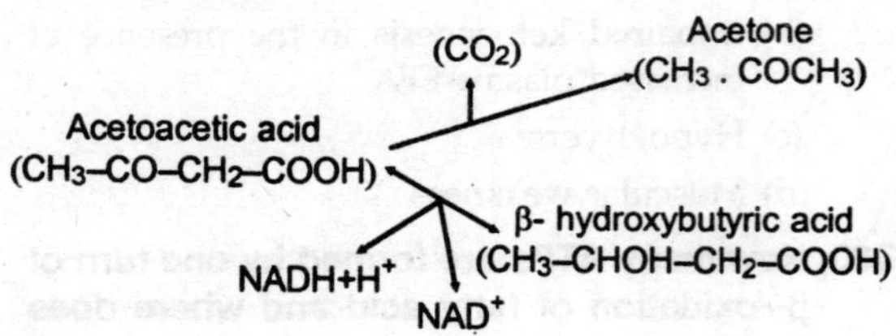

Q67. What do you mean by ketogenesis and ketolysis.

The formation of ketone bodies in the liver is termed ketogenesis.

The breakdown of ketone bodies in the extra-hepatic tissues is termed ketolysis.

Q68. What is ketosis?

The condition in which there is a high concentration of ketone bodies in tissues and blood is termed ketosis.

Q69. What is the normal concentration of ketone bodies in the blood and in the urine?

0.2 – 2.0 mg/100 ml of blood.

Less than 1 mg/day in the urine.

Q70. What are the conditions for ketosis?

Severe diabetes mellitus, prolonged starvation, glycogen storage diseases, toxemia of pregnancy, infective hepatic diseases, continued fever.

Q71. How are ketone bodies formed?

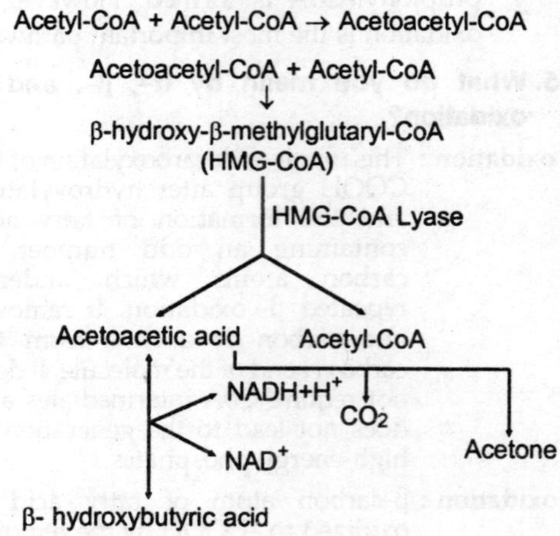

Q72. How are ketone bodies utilized in the extra hepatic tissues?

First reaction :

Acetoacetate + Succinyl-CoA $\xrightarrow{\text{Transferase}}$ Acetoacetyl-CoA + Succinic acid

Second reaction :

Acetoacetate + ATP + CoA . SH $\xrightarrow{\text{Thiolase}}$ Acetoacetyl-CoA + AMP + PPi

Acetoacetyl-CoA $\xrightarrow{\text{Thiolase}}$ 2 Acetyl-CoA ⟶ TCA cycle

Q73. How does ketoacidosis occur?

Acetoacetic acid and β-hydroxybutyric acid are moderately strong acids and are buffered in blood or tissues. Their excretion in large quantities causes some loss of buffer cation which depletes the alkali reserve developing ketoacidosis.

Q74. How is ketosis prevented?

(a) Incase of **diabetic ketosis,** carbohydrate diet, intramuscular injection of insulin and antiketogenic substance (aspartic acid) should be administered.

(b) In case of **prolonged starvation ketosis,** carbohydrate diet and antiketogenic substance (aspartate) should be given.

(c) Intravenous injection of isotonic solution of sodium salts should be given to restore the electrolytes and the fluids of the body.

Q75. How is cholesterol synthesis controlled by regulation of HMG-CoA reductase?

(a) During fasting cholesterol synthesis is reduced.

(b) Cholesterol inhibits HMG-CoA reductase in liver by feed back mechanism.

(c) Insulin or thyroid hormone increases HMG-CoA reductase activity, whereas glucagon or corticoids decrease it.

(d) The enzyme exists in both active and inactive forms that may be modified by phosphorylation-dephosphorylation mechanism.

Q76. How is cholesterol synthesized in the body?

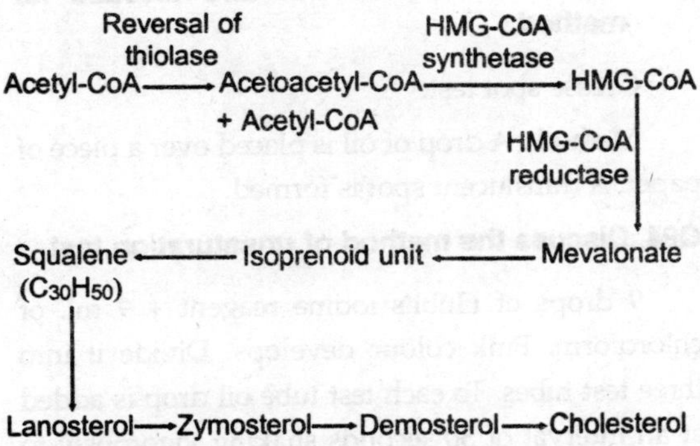

Q77. Explain the effect of drugs on cholesterol biosynthesis.

(a) Sitosterol : A hypercholesterolemic agent which blocks the esterification of cholesterol in the G.I.T. and thus reduces cholesterol absorption.

(b) Choloxin and neomycin : They cause the increased fecal excretion of cholesterol and bile acids.

(c) Clofibrate : It inhibits hepatic cholesterol synthesis.

(d) Estrogens and nicotinic acid : They are the hypocholesterolemic drugs.

Q78. Under which conditions does hypercholesterolemia occur?

Uncontrolled diabetes mellitus, obstructive jaundice, glomerulonephritis, impairment of liver, hypothyroidism, and nephrosis.

Q79. Name the conditions in which hypocholesterdemia occurs.

Anemia, hepatic diseases, infection, hyperthyroidism, carcinoma, and acute pancreatitis.

Q80. What do you mean by atherosclerosis?

This is caused by the deposition of cholesteryl ester and other lipids in the connective tissue of the arterial walls. This disease is accompanied by high blood pressure, obesity, lack of exercise.

Q81. What are the factors involved in higher levels of free fatty acids in blood?

Emotional stress, nicotine from cigarette smoking, and coffee drinking.

Q82. How is serum cholesterol level decreased?

The polyunsaturated fatty acids form cholesteryl esters of polyunsaturated fatty acids which are rapidly metabolized by the liver and other tissues which may enhance excretion.

Q83. What is the normal level of cholesterol in the blood?

140-220 mg/100 ml. of blood.

This level increases with ages and during pregnancy.

Q84. What is the importance of cholesterol in the body?

(a) It helps in the permeability of cells.

(b) It protects red cells from being easily hemolyzed.

(c) It has defensive action.

(d) It transports fats to liver in the form of cholesterol ester for oxidation.

(e) It helps to form bile acids and bile salts, 7-dehydrocholesterol and vitamin D_3, corticosteroid hormones, androgens, estrogens and progesterone.

Q85. Name the factors affecting cholesterol level in the blood.

(a) Dietary fats (butter fat, hydrogenated fat) containing higher saturated fatty acids cause increased serum cholesterol.

(b) Excessive amounts of sucrose consumption cause an increase in serum cholesterol level.

(c) Intake of excess calories causes an increased serum cholesterol level.

(d) Increased protein intake does not change the blood cholesterol level. Low protein intake causes reduction in plasma cholesterol level.

(e) Large amounts of nicotinic acids cause lowering of plasma cholesterol level.

(f) Physical exercise lowers the serum cholesterol level.

(g) Increased fibre contents of diets cause excretion of cholesterol and bile acids.

Q86. In what form cholesterol is available in the blood?

(a) Free cholesterol.

(b) Cholesterol ester (65 to 75 P.C. of total cholesterol).

Q87. How is the activity of HMG-CoA influenced?

(a) The activity of HMG-CoA is decreased during fasting.

(b) Its activity is inhibited by cholesterol feeding.

(c) Its activity is not reduced in diabetes mellitus.

Q88. What do you mean by chylomicrons?

(a) These consist of a triglyceride code covered with phospholipid and protein.

(b) These are found in chyle which is formed only in the lymphatic system draining the intestine.

Q89. What is "fixation of CO_2"?

Addition of CO_2 eg. CO_2 is added to acetyl-CoA to form malonyl- CoA in fatty acid synthesis.

Q90. State the main functions of essential fatty acids.

(a) These retard atherosclerosis being esterified and emulsified with cholesterol.

(b) These effect the prolongation of clotting time and increase the fibrinolytic activity.

(c) The fatty livers are cured by its introduction in the diet.

Q91. What is phosphorylation?

The liberation of phosphate from ATP is an energy-producing reaction and the energy is utilized for chemical transfers but not as heat. ATP is reformed from ADP and inorganic phosphate.

Q92. What is Gaucher's disease?

This disease is caused by the accumulation of cerebrosides in reticuloendothelial cells due to the deficiency of a specific gluco-cerebroside cleavage enzyme (gluco-cerebrosidase).

Q93. Name the test for fat and discuss its method.

Grease spot test.

Method : A drop of oil is placed over a piece of paper. A translucent spot is formed.

Q94. Discuss the method of unsaturation test.

9 drops of Hubl's iodine reagent + 9 ml. of chloroform. Pink colour develops. Divide it into three test tubes. To each test tube oil drop is added at an interval of 30 seconds shaking vigorously to discharge the pink colour. The number of drops is counted for each oil in each test tube. The less the number of drops required, the more is the unsaturation of fat.

Q95. Name the tests for the detection of cholesterol and discuss the methods.

(a) **Salkowski's test :** A little cholesterol is dissolved in 2 ml. chloroform.
An equal volume of conc. H_2SO_4 is added to it. Shaked gently. The upper layer of chloroform turns red and the sulphuric acid layer assumes a yellow colour with green fluorescence.

(b) **Libermann-Burchard reaction :** A crystal of cholesterol is dissolved in 2 ml. chloroform in a dry test tube. 10 drops of acetic anhydride and 2 drops of conc. H_2SO_4 are added. Mixed well. A red rose colour develops which quickly changes through blue to green.

Q96. What do you mean by lipidosis?

It means the abnormal concentrations of lipoproteins in blood or of specific lipids in tissues.

Q97. What is "xanthomatosis"?

It means lipid accumulation in tissues in association with large "foam cells". The lipid is mostly cholesterol. Xanthomata develops in hyperlipoproteinemia, diabetes mellitus, and glycogen storage disease.

Q98. Mention the clinical importance of glycorol.

It is nontoxic and reduces cerebral edema with improvement in cerebrospinal fluid.

Q99. What is Norum's disease?

(a) It is an inherited disorder caused by the deficiency of the enzyme lecithin-cholesterol acyl transferase. The esterification of cholesterol in plasma does not take place.

(b) This disease is characterized by rise in free cholesterol and fall in cholesterol ester.

Q100. Name the different lipases in the body.

(a) Lingual lipase : In saliva.
(b) Gastric lipase : In gastric juice.
(c) Pancreatic lipase : In Pancreatic juice.
(d) Intestinal lipase : In intestinal epithelial cells.
(e) Adepolytic lipase : In adipose tissue.
(f) Serum lipase : In serum.

Q101. Why are vegetable fats preserved for longer periods than animal fats?

Vegetable fats contain vitamin E, phenols, hydroquinones, and tannins etc. which are anti-oxidants that prevent development of rancidity.

Q102. Explain the role of hormones in liopgenesis.

(a) Glucagon inhibits fatty acids synthesis by inhibiting the key enzyme acetyl-CoA carboxylase.

(b) Insulin increases fatty acid synthesis by decreasing lipolysis, activating protein phosphatase, activating citrate lyase, and enhancing formation of acetyl-CoA by stimulating glycolysis.

Q103. Why are young ladies protected against myocardial infarction?

Young ladies have higher levels of HDL which is the effect of increased estrogen level in young ladies. This raised HDL level protects them from myocardial infarction.

Q104. Name the factors involved in causing coronary heart disease.

(a) High blood pressure.
(b) Smoking.
(c) Lack of exercise.
(d) Male gender.
(e) Obesity.
(f) Drinking soft water.

Q105. What are the factors leading to higher levels of free fatty acids?

(a) Emotional stress,
(b) Cigarette smoking,
(c) Coffee drinking,
(d) Partaking of a few large meals.

Q106. Indicate the reason that premenopausal women are protected against many of the undesirable factors.

These women have higher concentration of HDL than do men and postmenopausal women.

Q107. Why is red wine particularly beneficial?

Red wine contains antioxidants.

Q108. How does aspirin prevent thrombus formation?

(a) Aspirin irriversibly acetylates the platelet cyclooxygenase system and inhibits TXA_2 formation.

(b) It also inhibits PGI_2 synthesis in vascular endothelial cells. The vascular endothelial cells regenerate cyclooxygenase very soon.

(c) The net result appers that the formation of PGI_2 prevents platelet aggregation and thrombus formation.

Q109. What is Tay-Sachs disease (GM_2 gangliosidosis)?

(a) It is an inherited disorder of ganglioside.

(b) In the absence of the enzyme hexosaminidase A, GM_2 can not be degraded. Rather, GM_2 accumulates in the brain causing widespread injury to ganglion cells in brain and retina.

CHAPTER 4

PROTEINS

Q1. Define α-amino acid.

It is the hydrolytic product of protein. It has at least one amino group and a carboxyl group. The amino group is attached to the alpha (α) carbon atom of this acid.

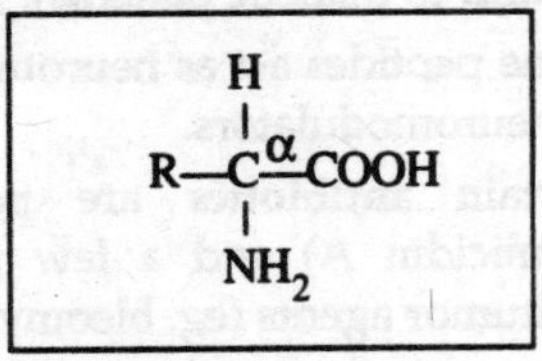

Q2. What are proteins?

These are organic substances of high molecular weight formed by a number of amino acids united by a peptide linkage.

Q3. State the biological significance of proteins.

(a) These are the "building blocks" for the synthesis of the structural constituent of the body.

(b) These synthesize hormones, enzymes, antibiotics, and glutathione.

(c) These supply lost nitrogen, sulphur, and energy etc.

(d) They have the defensive action against infections by means of protein antibodies.

(e) They are involved in blood clotting.

Q4. Enumerate the clinical importance of amino acids.

(a) Besides their roles in protein, L-amino acids and their derivatives are involved in nerve transmission, regulation of cell growth and the biosynthesis of porphyrins, purines, pyrimidine, and urea formation.

(b) Both D- and L-amino acids are present in polypeptide antibiotics elaborated by microorganisms.

(c) Genetic defects in the metabolism of amino acids result in severe illness. Several rare genetic diseases of amino acid catabolism, if left untreated, result in mental retardation and early death.

(d) The transport defects of specific amino acids result in the excretion in the urine of increased amounts of one or more amino acids causing aminoacidurias.

Q5. What do you mean by the isoelectric P^H (P^I) of an amino acid?

The isoelectric P^H is the P^H midway between PK values on either side of the isoionic species.

$$\therefore\ P^I = \frac{PK_1 + PK_2}{2}$$

$$\text{eg. For alanine, } P^I = \frac{2.35 + 9.69}{2} = 6.02$$

Q6. What are ampholytes or zwitterion?

The amino acids forming proteins contain both acidic and basic groups and react with both alkalies and acids to form salts. These amino acids are amphoteric and are called ampholytes or zwitterion.

Q7. Why are the amino acids ornithine, citrulline, arginosuccinic acid not found in the classification of amino acids?

Because they are not involved in protein formation but they fulfil important roles in metabolic processes.

Q8. Which amino acids are involved in protein formation?

(a) Only naturally occurring 20 L-α-amino acids are involved in protein formation.

(b) Over 20 D-amino acids occur naturally. Some of them occur in certain bacterial cell walls and a variety of D-amino acids in antibiotics.

Q9. Mention the difference between polypeptides and proteins.

The sequence of amino acids linked by the peptide bond are called polypeptides. They do not contain amino acids more than 100.

Proteins contain amino acids more than 100.

Q10. Classify amino acids.

(a) **Neutral amino acids** : Glycine, alanine, valine, leucine, isoleucine.

(b) **Aromatic amino acids** : Phenylalanine, tyrosine, tryptophan.

(c) **Sulphur containing amino acids** : Methionine, cysteine.

(d) **Acidic amino acids** : Aspartic acid, glutamic acid.

(e) **Basic amino acids** : Arginine, histidine, lysine.

(f) **Imino acid** : Proline.

Q11. What are essential amino acids? Name them.

The amino acids which are necessary for tissue synthesis and growth are said to be essential amino acids. They are not synthesized in the body and must be supplied in the diet. Methionine, tryptophan, threonine, leucine, isoleucine, valine, phenylalanine, lysine, and histidine.

Q12. Classify proteins.

(a) **Simple proteins** : Albumin, globulin, prolamines, protamines, histones.

(b) **Conjugated proteins** : Lipoproteins, phosphoproteins, nucleoproteins, chromoproteins, metalloproteins.

(c) **Derived proteins** : Proteoses, peptides, peptones.

Q13. What do you mean by the denaturation of protein?

Disruption of the structural pattern of the protein due to the cleavage of non-covalent bonds by heat, ultraviolet light, high pressure, organic solvents, acids or alkalies.

Q14. Name the agents for denaturation.

Heat, strong acids, high salt concentration or heavy metals etc.

Q15. What is the importance of denaturation?

(a) For the estimation of glucose, uric acid, and drugs etc of blood the protein of the blood is removed as precipitate being formed by denaturation.

(b) Denaturation is used to know the enzyme catalyzed reaction of an extract at the loss of enzyme activity.

Q16. Enumerate the clinical importance of peptides.

(a) Many major hormones are peptides and may be given to patients to overcome deficiency states (eg. administration of **insulin** to diabetic patients).

(b) Some peptides act as neurotransmitters or as neuromodulators.

(c) Certain **antibiotics** are peptides (eg. gramicidin A) and a few peptides are antitumor agents (eg. bleomycin).

(d) The recombinant DNA technology has facilitated the manufacture of good amounts of peptide hormones. The same technology is adopted for the synthesis of other peptides (eg. certain viral peptides and proteins) for the use in **vaccines.**

Q17. State the structural features of proteins.

(a) **Primary structure** : The sequence of amino acids and location of any disulfide bond. It is encoded in the genes.

(b) **Secondary structure** : It describes the folding of polypeptide chains into multiply H-bonded motifs such as the α-helix and the β- pleated sheet.

(c) **Tertiary structure** : Combination of these motifs (eg. β-α-β).

(d) **Quarternary structure** : It consists of two or more polypeptide chains.

Q18. What is electrophoresis?

The method by which different components of the mixture of amino acids are analysed under electric field.

Q19. What is Ehlers-Danlos syndrome?

It is the disease associated with defects in collagen maturation.

Q20. Name the amino acid absent in α–helix.

Proline.

Q21. State the clinical importance of protein precipitation.

Protein free filtrate is used for estimation of blood sugar, blood urea etc.

Q22. How will you differentiate flocculation from coagulation of proteins?

(a) Flocculation means maximum precipitation of protein and soluble in extremes of pH.

(b) Flocculations on further heating produces dense white coagulation and insoluble in extremes of pH.

Q23. Mention the levels of plasma proteins and albumin where edema can occur.

Plasma total protein level : Below 5.0 gms per cent.

Plasma albumin level : Below 2.5 gms per cent.

Q24. What is oroso-mucoid? State its function.

(a) Orosomucoid is α_1-acid glycoprotein. It is the reliable indicator of acute inflammation.

(b) It functions as a transport protein for progesterone.

Q25. What do you mean by α_1-fetoprotein?

(a) It is an α_1-globulin present in high concentration in fetal blood during pregnancy.

(b) Normal adult blood has less than 1μg/100 ml.

(c) Human blood containing increased amount of α_1-fetoprotein indicates hepato cellular carcinoma or teratoblastomas.

Q26. What is hepatoglobin?

(a) It is an α_2-globulin synthesized in liver.

(b) It binds Hb forming Hp-Hb complex which circulates in the blood and is finally destroyed by RE cells.

Q27. State the digestion of protein.

(a) **In the stomach :**

$$\text{Protein} \xrightarrow[\text{P}^{\text{H}}\ 1.0-2.0]{\text{Pepsin}} \text{Proteoses and peptones.}$$

$$\text{Casein of milk} \xrightarrow[\text{Calcium, P}^{\text{H}}\ 4.0]{\text{Rennin}} \text{Coagulates milk.}$$

(b) **In the pancreas :**

$$\text{Proteoses and peptones} \xrightarrow[\text{P}^{\text{H}}\ 7.9]{\text{Trypsin}} \text{Polypeptides and dipeptides.}$$

$$\text{Proteoses and peptones} \xrightarrow[\text{P}^{\text{H}}\ 8.0]{\text{Chymotrypsin}} \text{Polypeptides and dipeptides, more coagulated milk}$$

$$\text{Polypeptides at the free carboxyl end of the chain} \xrightarrow[\text{P}^{\text{H}}\ 7.9]{\text{Carboxypeptidase}} \text{Lower peptides. Free amino acids.}$$

(c) **In the small intestine :**

$$\text{Polypeptides at the free amino end of the chain} \xrightarrow{\text{Aminopeptidase}} \text{Lower peptides. Free amino acids.}$$

$$\text{Dipeptides} \xrightarrow{\text{Dipeptidase}} \text{Amino acids.}$$

Q28. Discuss the absorption of protein.

(a) Under normal conditions, the dietary proteins are completely digested to amino acids and these end products are then rapidly absorbed from the intestine into the portal blood.

(b) The natural (L) isomers—but not the (D) isomers—of an amino acid is actively transported across the intestine from the mucosa to the serosa requiring vitamin B_6 as pyridoxal phosphate. 2, 4-dinitrophenol inhibits this transport.

(c) Amino acids are transported through the brush border by means of carriers, many follow Na^+ -dependent mechanisms.

(d) Very small amounts of peptides and even of intact protein molecules may be absorbed in jejunum.

Q29. What does happen in celiac disease in children?

Proteins of larger molecular size than amino acids are absorbed from the intestine under certain conditions.

Q30. What is body proteins pool?

Proteins constantly move from the plasma to the tissues and from the tissues to the plasma. This continuous exchange of proteins and amino acids occurs even in inert tissues like tendon.

Q31. What does it mean by negative nitrogen balance?

The total amount of nitrogen lost from the body through feces and urine exceeds the amount of nitrogen taken as food. This happens in advanced cancer, following surgery, failure to ingest adequate high-quality protein.

Q32. What is transamination.

This is the process by which the amino group of one amino acid is transferred to a keto acid for which the amino acid is converted to a keto acid and the keto acid is converted to an amino acid being catalyzed by transaminase involving pyridoxal phosphate (B_6-PO_4) as coenzyme. eg.

$$\text{Glutamic acid + Pyruvic acid} \xrightleftharpoons[(B_6\text{-}PO_4)]{\text{Glutamate Pyruvate transaminase}} \alpha\text{-Keto glutarate + Alanine.}$$

This occurs in most tissues and abundant in heart.

Q33. What is transamidation?

This is the process by which an amidine group $\left(\begin{array}{c} NH_2 \\ | \\ C=NH \end{array}\right)$ is transferred from one amino acid to the other. eg. Arginine and glycine to form glycocyamine and guanidoacetic acid.

Q34. What do you mean by oxidative deamination?

This is the process by which there is oxidation by the removal of hydrogen atoms and removal of amino group in the form of NH_3 from an imino acid forming keto acid.

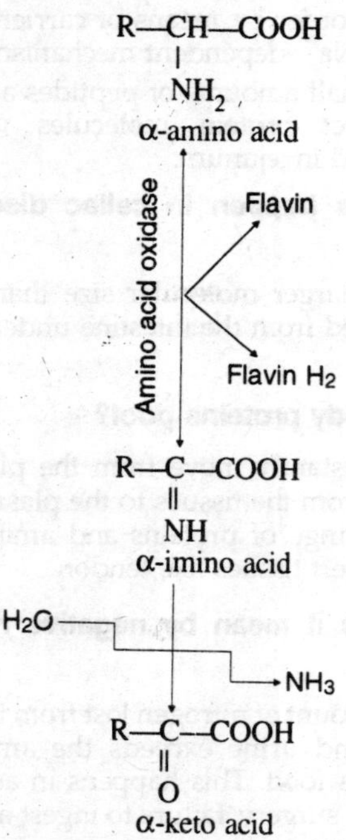

This happens in liver and kidney.

Q35. What is decarboxylation?

This is the process by which carbon dioxide (CO_2) is eliminated from an amino acid to form amines by the enzyme amino acid decarboxylase involving pyridoxal phosphate (B_6-PO_4) as coenzyme.

Q36. What is taurine?

Cysteic acid formed from cysteine is decarboxylated to give taurine which is used in conjugation of bile acids.

Q37. Why does protein of diet sometimes produce allergy?

Under certain conditions, food proteins are absorbed as such without undergoing digestion producing allergy because dietary proteins behave as an antigen.

Q38. Is nitrogen stored in the body?

Yes, the indication of it is that the subject grows or increased size of his muscles by continued exercise.

Q39. What are Ethereal sulphates?

(a) Esters of phenols with indoxyl sulphate.

(b) These are excreted in the urine.

Q40. What do you mean by body protein turnover?

(a)

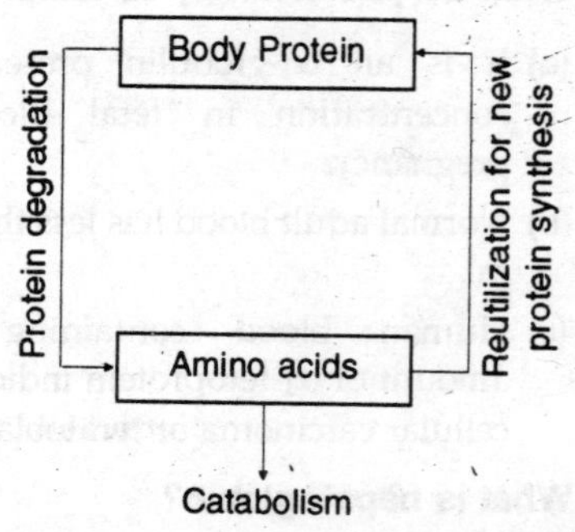

(b) Each day, humans turn over 1-2% of their total body protein, mainly muscle protein.

(c) 75-80% of the liberated amino acids are reutilized for new protein synthesis.

(d) The nitrogen of the remaining 20-25% forms urea.

(e) The carbon skeletons are then degraded to amphibolic intermediates.

Q41. Name the enzymes occupying the central positions in amino acid biosynthesis.

(a) Glutamate dehydrogenase, glutamine synthetase, and transaminases.

(b) Their combined effect is to catalyze transformation of inorganic ammonium ion into the organic α-amino nitrogen of various amino acids.

Q42. How is glutamine formed in the body?

Glutamine is formed from glutamate being catalyzed by glutamine synthetase, (a mitochondrial enzyme present in renal tissues in highest concentration) accompanied by ATP and NH_3.

Q43. How is tyrosine formed from phenylalanine in the body?

(a) Tyrosine is formed from phenylalanine by the reaction catalyzed by phenylalanine hydroxylase. The reaction is irreversible.

(b) The **phenylalanine hydroxylase complex** is a mixed function oxygenase present in mammalian liver.

(c) The reaction involves incorporation of one atom of molecular oxygen into the para position of phenylalanine while the other atom is reduced forming water.

(d) The reducing power (supplied ultimately by NADPH) is immediately provided as **tetrahydrobiopterin** (a pteridine which resembles folic acid).

Q44. Which is the only ketogenic amino acid?

Leucine.

Q45. How is glycine biosynthesized?

Liver cytosol contains glycine transaminases that catalyze the synthesis of glycine from glyoxalate and glutamate or alanine. This is the strongly favoured glycine synthesis other than two additional important mammalian routes.

Q46. What is the benefit of excess amino acid intake?

Excess amino acids are not stored. The amino acids not immediately incorporated into new protein are rapidly degraded. Consumption of excess amino acids serves no purpose that cannot be equally served at a lower cost by carbohydrates and lipids.

Q47. State the fate of α-amino nitrogen.

The α-amino nitrogen from amino acids is excreted in the form of one of the three end products—ammonia, uric acid or urea.

(a) Teleostean fish, which are **ammonotelic,** excrete nitrogen as **ammonia**.

(b) Land animals convert nitrogen either to **uric acid** (Uricotelic organisms) or to **urea** (Ureotelic organisms).

(c) Humans are **ureotelic** and excrete the highly water soluble compound urea.

(d) The high blood urea levels in patients with renal disease are a consequence of **malaise.**

Q48. How does ammonia intoxication become life threatening?

(a) Ammonia formed by bacteria is absorbed into the portal venous blood which contains higher levels of ammonia.

(b) Since the liver normally promptly removes this ammonia from the portal blood, peripheral blood is virtually ammonia-free. This is essential.

(c) In case the portal blood bypasses the liver, ammonia may rise to toxic levels in the systemic blood.

(d) The **normal level** of ammonia in blood is 10-20 mg/dL. Even traces of ammonia are toxic to the central nervous system. The symptoms of **ammonia intoxication** include tremor, slurred speech, blurred vision, and in severe cases, coma and death.

Q49. How does ammonia maintain acid-base balance?

(a) Excretion into the urine of ammonia produced by renal tubular cells facilitates regulation of acid-base balance and cation conservation.

(b) Ammonia, derived from glutamine by renal glutaminase, is increased in metabolic acidosis and decreased in alkalosis.

Q50. How is ammonia removed from liver and brain?

(a) By the formation of urea in the liver and to a small extent in the brain.

(b) By the formation of glutamine from glutamate by glutamine synthetase in the brain.

Q51. How is urea biosynthesized?

See Chart 1, next page.

Site : Liver is the principal site for urea formation. Urea is also formed in brain to a small extent.

Enzyme activator : N-acetyl glutamate.

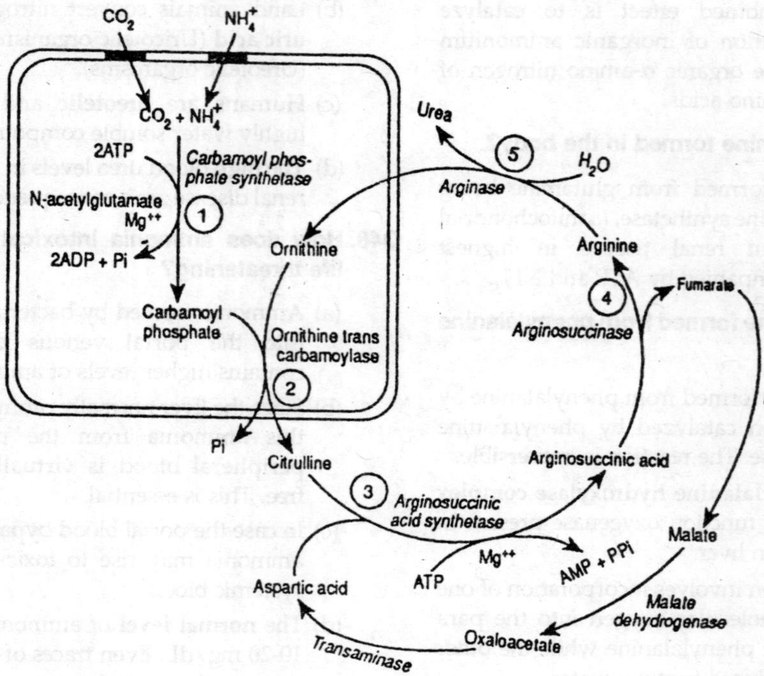

CHART 1

Required amino acids : Ornithine, citrulline, glutamate, aspartate, arginine, arginosuccinate.

Normal level in blood : 15-45 mg/100 ml.

Normally daily output through urine : 20-30 gms.

Q52. What is the test for urea in urine?

Biuret test : Positive.

Specific urease test :

Method :

(a) 2 ml urine in one test tube and 2 ml water in another test tube.

(b) A drop of phenol red is added to each of the test tubes.

(c) 2 per cent Na_2CO_3 is added drop by drop till the deep pink colour develops.

(d) 1 or 2 drops acetic acid is added to each test tube till the colour is just yellow.

(e) A pinch of soyabean powder is added to each and the test tubes are rotated between the palms or the temperature is raised to 60°C, the optimum for the enzyme urease.

(f) The development of pink colour in the test tube containing urine indicates the presence of urea.

Q53. Is urea a protein although urea and protein both give biuret test positive?

No. Urea is not a protein. Because it is not formed by amino acids by peptide linkages although it has a peptide link which is responsible for biuret test positive.

Q54. What are the metabolic disorders in the urea cycle?

Hyperammonemia type 11, citrullinemia, hyperargininemia, arginosuccinicaciduria, Hyperammonemia type 1.

Q55. How is Figlu formed?

Deamination of histidine produces **urocanate**, which on addition of H_2O plus an internal oxidation-reduction catalyzed by **urocanase**, is converted to 4-imidazolone-5-propionate which on hydrolysis forms N-formiminoglutamate (Figlu).

Q56. Why is Figlu clinically important?

(a) The formimino group of figlu is transferred to tetrahydrofolate forming glutamate. This reaction is totally blocked in the deficiency of folic acid and figlu is excreted in urine. Excretion of figlu following a test dose of histidine provides a diagnostic test for folic acid deficiency.

Q57. What is glycinuria?

Glycinuria results from a defect in renal tubular reabsorption. This is characterized by urinary excretion of 0.6 – 1 gm of glycine per day and a tendency to form oxalate renal stones.

Q58. What is cystonosis (cystine storage disease)?

It is a rare lysosomal disorder characterized by defective carrier mediated transport of cystine, cystine crystals are deposited in tissues and organs, particularly the reticuloendothelial system. This is usually accompanied by aminoaciduria.

Q59. Name the amino acid which forms a number of hormones in the body. What are the hormones so formed?

Phenylalanine.

Epinephrine and norepinephrine, Thyroxine.

Q60. What is melanin?

(a) It is a black pigment present in the skin, hair, and retina of the eye.

(b) This is formed from phenylalanine by the enzyme tyrosinase.

Q61. What do you mean by albinism?

It is the condition in which tyrosinase is lacking resulting the non-formation of melanin for which the hair, skin, and retina of the eye are not black pigmented.

Q62. What is DOPA?

It is dihydroxyphenylalanine formed from tyrosine which is further metabolised to form melanin. It is excreted out in case of tyrosinosis.

Q63. How is epinephrine formed?

Phenylalanine → Tyrosine → DOPA → Dopamine → Norepinephrine → Epinephrine.

Q64. How is Thyroxine formed?

2[Tyrosine] → 2[Di-iodo tyrosine] → Thyroxine.

Q65. Name the vitamin which is formed from tryptophan and the quantity of it.

Niacin (Nicotinic acid). One mg. of niacin is formed from 60 mg. of tryptophan.

Q66. Name the vitamins involved in the biosynthesis of Niacin.

Vitamin B_2 and Vitamin B_6.

Q67. What is serotonin and how is it formed?

Serotonin is a vasoconstrictor and a stimulant of the central nervous system. It occurs in intestinal mucosa where it enhances peristalsis.

Tryptophan —Hydroxylation→ 5-hydroxytryptophan

↓ Decarboxylation

5-hydroxytryptamine (Serotonin)

Q68. How is niacin formed from tryptophan?

Tryptophan → Kynurenine → 3- hydroxyanthranilic acid → Niacin.

Q69. What is melatonin?

It is the hormone formed from serotonin on acetylation and methylation. Melatonin is nothing but N-acetyl-5-methoxytryptamine. It is the hormone of pineal gland and inhibits the action of MSH (Melanocyte stimulating hormone).

Q70. What is active methionine?

Methionine is activated by ATP to form S-adenosylmethionine (active methionine). It is a methyl donor.

Q71. What is glutathione?

It is a tripeptide consisting of glycine, cysteine, and glutamic acid. It acts as a coenzyme in the formation of prostaglandin, formic acid, lactic acid.

Q72. What is the fate of arginine?

(a) Arginine is the formamidine donor for **Creatine** synthesis in primates and for **Streptomycin** synthesis in stretomyces.

(b) Other fates include conversion to putrescine, spermine, and spermidine and in invertebrate muscle to **arginine phosphate.**

Q73. What is GABA?

It is γ-aminobutyrate which is formed by decarboxylation of L-glutamate being catalyzed by the pyridoxal phosphate dependent enzyme L-glutamate decarboxylase present in the tissues of central nervous system, principally the gray matter.

Q74. Name the amino acids involved in creatine synthesis.

Glycine, arginine, and methionine.

Q75. How is creatine biosynthesized?

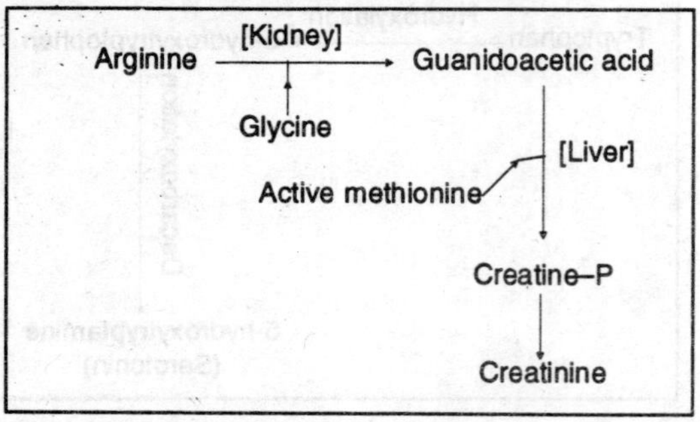

Q76. Is creatine excreted in urine?

Traces of creatine are normally present in urine. Creatinine, the anhydride of creatine, is mainly excreted in urine.

Q77. What is creatinine coefficient?

It is the number of milligrams of creatinine nitrogen excreted per kilogram of body weight in 24 hours.

Q78. State the clinical importance of creatine.

(a) Excessive creatinuria occurs in starvation, fabrile and wasting diseases, diabetes mellitus, and in certain myopathies.

(b) Excessive creatinuria has also been observed in hyperthyroidism, during the menstrual period and in pregnancy.

Q79. What does happen to the N-part of an amino acid?

α-NH_2 group → NH_3 → Urea → Excreted in urine.

Q80. Explain nonoxidative deamination.

Certain amino acids like serine, histidine, and cysteine are deaminated nonoxidatively by specific enzymes to form NH_3. But they do not fulfil major role in NH_3 formation.

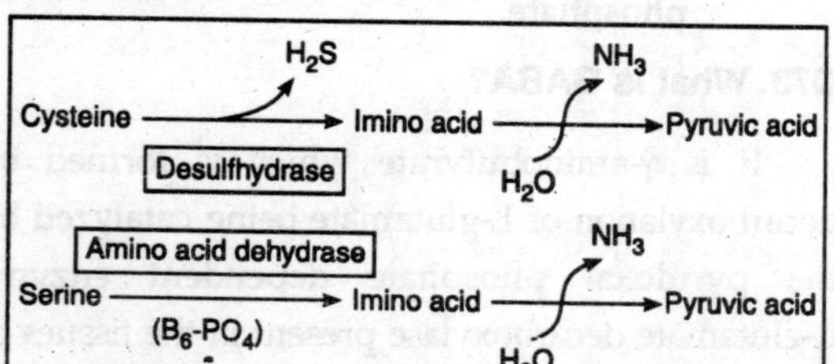

Q81. Why is NH_3 toxic?

(a) Glutamine formed from glutamate by the increased NH_3 concentration reduces "brain cell pool" of glutamatic acid which decreases formation of inhibitory neurotransmitter "GABA".

(b) α-ketoglutarate, the intermediate of TCA cycle, is aminated by high concentration of NH_3. As a result, the concentration of α-ketoglutarate is decreased for which the oxidation of carbohydrate, fat, and protein is highly affected.

(c) Increased level of glutamine enhances the overflow of glutamine from brain which allows entry of tryptophan by the same "transporter". Hence tryptophan concentration in brain cells increases which favours increased synthesis of serotonin.

Q82. Enumerate the important functions of glutamine in the body.

(a) Formation of guanylate (GMP).

(b) The amide N of glutamine is transferred to keto group of fructose 6-phosphate to form glucosamine phosphate.

(c) In conservation of Na^+ in kidney.

(d) Excess NH_3 entering brain is detoxicated by glutamine formation.

(e) Involved in conjugation reaction in detoxication.

(f) Certain tumors require glutamine / asparagine for their growth.

Q83. State the amines produced by putrefaction by intestinal bacteria.

(a) Cadaverine from lysine.

(b) Putrescine from ornithine.

(c) Agmatine from arginine.

Q84. What are polyamines?

(a) Spermidine and spermine are the polyamines.

(b) Increased polyamine excretion is claimed to be characteristic of malignant diseases.

(c) Polyamines are increased in leukemias and carcinoma of ovaries, lungs, colon, rectum, prostate, kidney, bladder, testes.

(d) Spermidine is the best "marker" of tumor cells destruction.

Q85. What will you suspect if a lady reports that her baby's napkin soiled with urine turns black on exposure to air?

Alkaptonuria.

Q86. What is ochronosis?

In prolonged cases of alkaptonuria, homogentisic acid derivatives are deposited in cartilages of ears and nose leading to generalized pigmentation of connective tissues. This is also deposited in joints leading to arthritis. This condition is said to be ochronosis.

Q87. Give some examples where active methionine acts as methyl donor.

(a) Norepinephrine $\xrightarrow{-CH_3}$ Epinephrine.

(b) Guanido acetic acid $\xrightarrow{-CH_3}$ Creatine.

(c) Ethanolamine $\xrightarrow{-CH_3}$ Choline.

(d) N-acetyl serotonin $\xrightarrow{-CH_3}$ Melatonin.

(e) Nicotinamide $\xrightarrow{-CH_3}$ N-methyl nicotinamide.

Q88. Enumerate the clinical features in carcinoid syndrome.

(a) Chronic diarrhea.

(b) Respiratory distress and bronchospasm.

(c) Right sided heart failure.

(d) Cyanotic appearance.

(e) Cutaneous vasomotor episodes of "flushing".

Q89. Why is it called Hartnup's disease?

The disease is named after the Hartnup family where it as discovered.

Q90. What are the causes of organic proteinuria?

(a) **Pre-renal causes :** Cardiac diseases, liver diseases, malignancies.

(b) **Renal causes :** Acute and chronic glomerulonephritis, nephrotic syndrome, TB of kidney, and carcinoma of kidney.

(c) **Postrenal causes :** Inflammatory degenerative and traumatic lesions of ureter, bladder, prostate, and urethra.

Q91. What do you mean by Tamm-Horsfall protein?

(a) It is a glycoprotein having molecular weight about 30,000.

(b) It is secreted by the mucous glands of the normal genitourinary tract.

(c) It serves to protect the epithelia of the genito uninary tract.

Q92. Are amino acids excreted in urine?

(a) Small amounts of free and combined amino acids are excreted in urine.

(b) The free amino acids normally amount to 1.4 mg/kg. for men and 2.3 mg/kg for women.

Q93. What is proteinuria or albuminuria?

In nephritis, albumin is excreted in urine due to the escape of it through kidney.

Q94. Name the test for protein. Discuss its method.

Biuret test

Method : 2 ml provided solution + 2 ml. 10% NaOH + 1 drop 10% $CuSO_4$ solution in a test tube. Formation of violet colour indicates the presence of protein.

Q95. Name the test for α-amino acid. Discuss its method.

Ninhydrin test

Method : 1 ml provided solution + 1 ml. ninhydrin solution in a test tube. Heated. Formation of violet colour indicates the presence of α-amino acid.

Q96. What are the tests for albumin and globulin? Discuss the method.

Full saturation test for albumin.

Half saturation test for globulin.

Full saturation test : Solid ammonium sulphate $[(NH_4)_2SO_4]$ is added to 5 ml provided solution by vigorous shaking until saturated. Formation of gelatinous ppt. in the solution indicates the presence of albumin.

Half saturation test : 5 ml saturated ammonium sulphate solution is added to 5 ml. Provided solution. Shaked vigorously. Formation of gelatinous ppt. indicates the presence of globulin.

Heat coagulation test for albumin : Two-third (2/3) of the test tube is filled with the provided solution and made acidic if alkaline by acetic acid. The upper portion is heated. A gelatinous ppt. is formed which is not dissolved by the addition of 10% acetic acid.

Q97. Name the other colour reaction test for amino acids.

(a) **Millon's test :** 3 ml. provided solution + 1 ml. $HgNO_3$ in HNO_3. Heated. Formation of red colour indicates the presence of phenylalanine and tyrosine.

(b) **Nitroprusside test :** 1 ml. sodium nitroprusside in dil. NH_4OH is added to 3 ml. Provided solution in a test tube. Formation of red colour indicates the presence of cysteine.

(c) **Hopkin's cole test :** $\frac{1}{2}$ ml. glyoxylic acid in 36N.H_2SO_4 is added to 3 ml. Provided solution in a test tube. Formation of purple colour indicates the presence of tryptophan.

(d) **Xanthoproteic test :** A few drops of conc. HNO_3 is added to 3 ml. Provided solution in a test tube. Boiled. Formation of yellow colour indicates the presence of tryptophan, phenylalanine, and tyrosine.

(e) **Sakagauchi test :** A few drops of alpha-Naphthol and sodium hypochlorite is added to 3 ml. Provided solution. Formation of red colour indicates the presence of arginine.

(f) **Pauly test :** A few drops of diazotized sulfanilic acid in alkaline solution is added to 3 ml. Provided solution. Formation of red colour indicates the presence of histidine.

Q98. What is code word or codon?

(a) The bases A, G, T and C found in DNA are organized into 3-letter code words called **codons.** The collection of these codons makes up the **genetic code.**

(b) A linear array of codons (**a gene**) specifies the synthesis of various RNA molecules, most of which are involved in protein synthesis.

(c) The **genetic code** explains the way in which protein defects may cause genetic disease and for the diagnosis and treatment of these disorders.

(d) It has been found that 61 of the 64 triplets do code for the 20 amino acids required for the synthesis of protein.

Q99. What do you mean by "chain-terminating triplets"?

Three of the 64 triplets do not code for any amino acid. They function rather as a signal (so-called "nonsense triplets") to terminate a polypeptide chain at that point. Owing to this function these triplets are now termed "chain-terminating triplets".

Q100. What is translation?

The cell possesses the machinery necessary to translate information accurately and efficiently from the nucleotide sequence of an mRNA into the sequence of amino acids of the corresponding specific protein. Clarification of this process is termed **translation.**

Q101. What is Wobble?

The degeneracy of the genetic code resides mostly in the last nucleotide of the codon triplet suggesting that the base pairing between this last nucleotide and the corresponding nucleotide of the anticodon is not strict. This is called **wobble.**

Q102. What is transcription?

DNA has got an important role in directing the synthesis of messenger RNA (mRNA) referred to as **transcription.**

Q103. What is "block copolymers"?

These are synthetic polynucleotides consisting of repeating sequence of two or three bases.

Q104. What is anticodon?

It consists of a triplet and relates to the codon for the amino acid to be carried.

Q105. What is "missense mutant"?

A codon for one amino acid is changed into that for another. This is referred to as missense mutant.

Q106. What are polysomes or ergosomes?

Many of the ribosomes can assemble to translate simultaneously a single mRNA molecule and form polysomes or ergosomes or polyribosome.

Q107. What do you mean by structural genes and operator genes?

(a) These are genes of chromosomes.

(b) The structural gene DNA directs the synthesis of specific enzyme and other proteins through the mRNA template.

(c) The structural genes, even though present in a given cell, cannot initiate protein synthesis except when suitable conditions prevail within the cell.

(d) The operator gene exerts control over the adjacent structural genes located on the same chromosomes, since formation of mRNA can be effected only at a point on DNA where the operator occurs.

Q108. What is operon?

One or more structural genes together with their operator gene constitute the genetic structure known as operon.

Q109. What do you mean by Repressor?

A third factor within the cell is known as the regulator gene and controls the activity of the operator gene. This regulator gene does so by its activity to induce the synthesis of protein macromolecules (probably RNA proteins) called **repressors.**

Q110. What is derepression?

The operon is active because the repressor system is itself inactivated. The phenomenon is said to be derepression.

Q111. Which amino acids have a single codon?

Tryptophan, methionine.

Q112. State the protein biosynthesis.

Protein biosynthesis can be described by activation, initiation, elongation, termination :

(a) Activation of amino acids by ATP by activating enzymes.

(b) The activated amino acids are transferred to tRNA. The amino acyl tRNA interacts with GTP and IF-2 to form a complex which along with IF-1 attaches 60S rebosomal subunits releasing IF-1, IF-2, IF-3 and hydrolysing GTP.

(c) EF-1 forms a complex with amino acyl -tRNA. This complex then allows the aminoacyl-tRNA to enter the A site. The alpha amino group of aminoacyl-tRNA at A site attacks the carboxyl group of the peptidyl-tRNA occupying the P site. Thus, peptide chain is elongated.

(d) After multiple cycles of elongation the terminating codon of mRNA appears in the A site. The termination signal in the A site causes the termination of the long peptide chain.

Q113. Name the inhibitons of proteins biosynthesis.

(a) **Puromycin** causes the premature release of the polypeptide.

(b) **Diphtheria toxins** catalyze the ADP ribosylation of EF-2 in mammalian cells and inactivates EF-2.

Q114. What are the types of gene regulation?

Positive regulation : When the expression of genetic information is quantitatively increased by the presence of a specific regulatory element.

Negative regulation : When the expression of genetic information is diminished by the presence of a specific regulatory element.

Q115. What is cistron?

It is the genetic unit coding for the structure of the subunit of a protein molecule, acting as it does as the smallest unit of genetic expression.

Q116. What is constitutive mutation?

The expression of some genes is **constitutive** means they are expressed at a reasonably constant rate and not known to be subject to regulation. As a result of mutation, some inducible gene products become constitutively expressed. A mutation resulting in constitutive expression is called a constitutive mutation.

Q117. What is lac operon?

(a) β-galactosidase hydrolyzes the β-galactoside lactose to galactose and glucose.

(b) The structural gene for β-galactosidase (the lac Z gene) is clustered with the genes responsible for the permeation of galactose into the cell (Y) and for galactoside acetylase (A).

(c) The structural genes for these three (3) enzymes are physically associated to constitute the lac operon.

Q118. How is protein translation controlled?

(a) The 4F complex is particularly important in controlling the rate of protein translation.

(b) 4F is a complex consisting of 4E which binds to the m^7G cap structure at the 5′ end of the mRNA and 4G. In addition to binding 4E, 4G binds to eIF-3 which links the complex to the 40S ribosomal subunit. It also binds 4A and 4B.

Q119. State the role of hormones in protein synthesis.

(a) Insulin enhances the increased incorporation of amino acids into protein at the level of translation by RNA.

(b) Growth hormone stimulates the synthesis of DNA for the elevated level of protein synthesis.

Q120. What is the action of diptheria toxin on protein synthesis?

Diptheria toxin binds with EF-2 and prevent its capacity to carry out translocation.

CHAPTER 5

INTERRELATIONSHIP IN THE METABOLISM OF PROTEIN, FAT AND CARBOHYDRATE

INTERRELATIONSHIP IN THE METABOLISM OF PROTEIN, FAT & CARBOHYDRATE

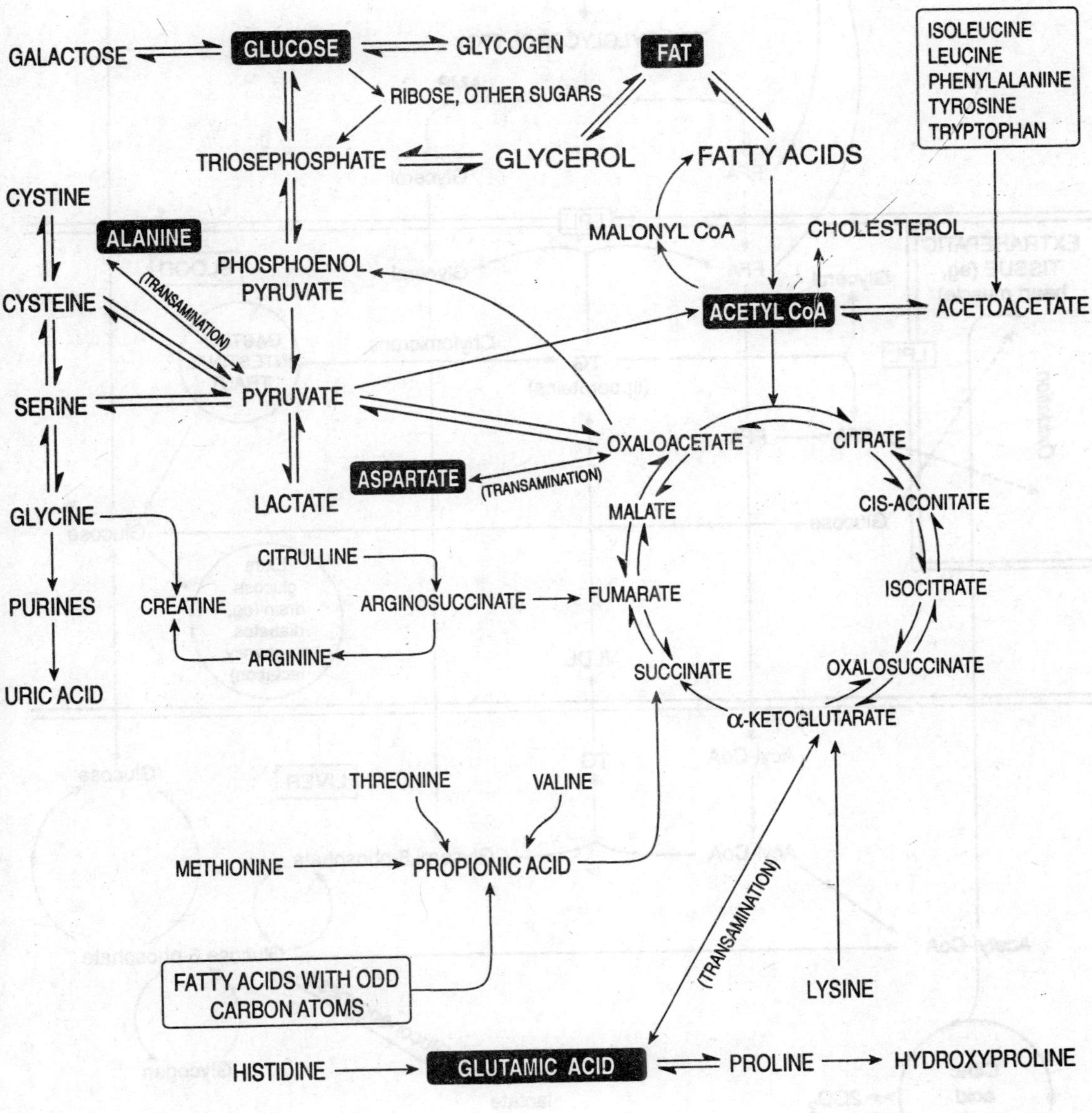

Metabolic interrelationship between adipose tissue, the liver, and extrahepatic tissues. (LPL, lipoprotein lipase; FFA, free fatty acid; VLDL, very low density lipoproteins.)

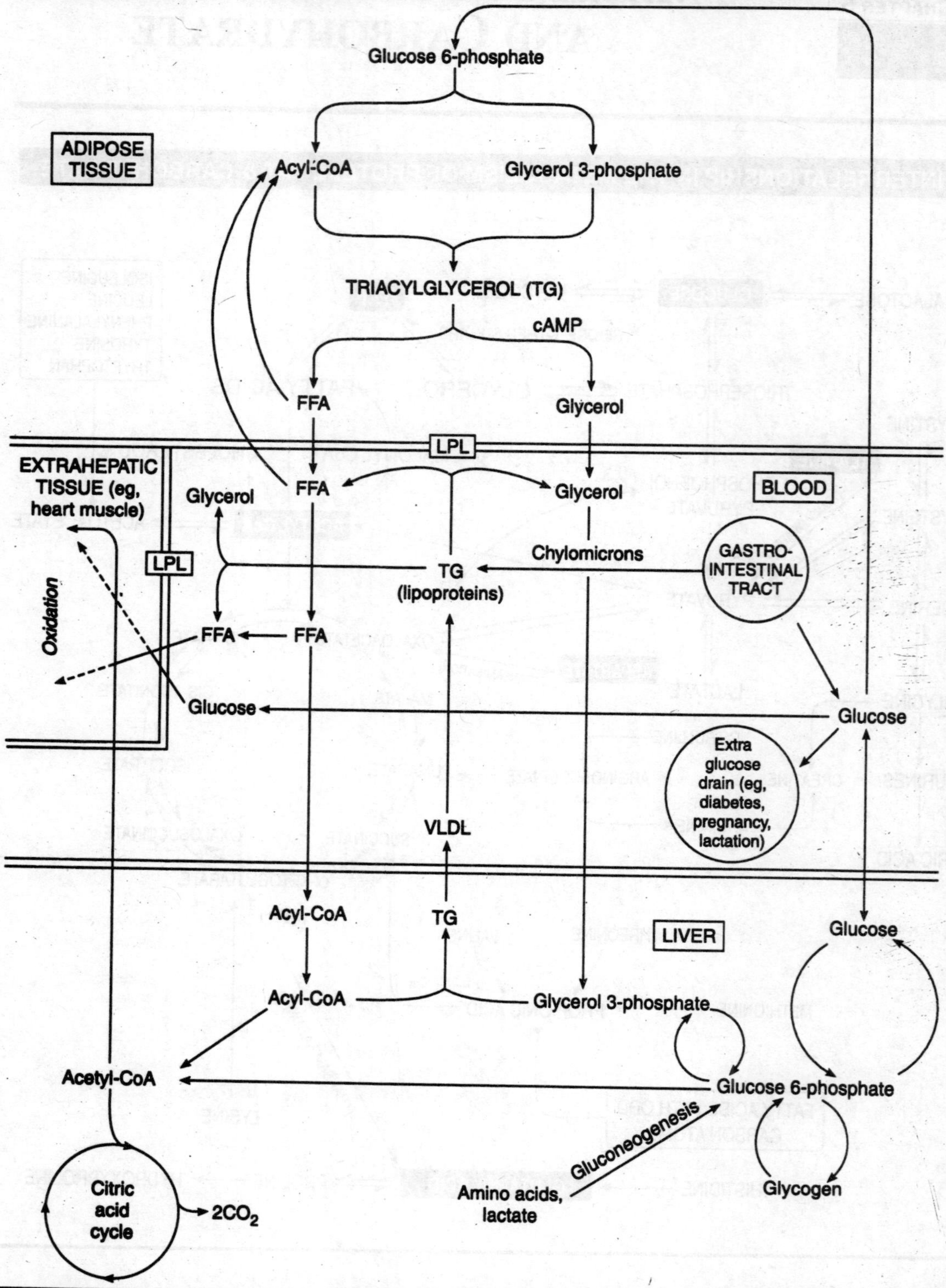

Table 5.1 : Summary of the major regulators of metabolic pathways

Pathway	Major Regulatory Enzymes	Activator	Inhibitor	Effector Hormone	Remarks
Citric acid cycle	Citrate synthase		ATP, long-chain acyl-CoA		Regulated mainly by the need for ATP and therefore by the supply of NAD^+
Glycolysis	Phosphofructo-kinase	AMP, fructose 2, 6-bisphosphate in liver, fructose 1, 6-bisphosphate in muscle	Citrate (fatty acids, ketone bodies), ATP, cAMP	Glucagon ↓	Induced by insulin
Pyruvate oxidation	Pyruvate dehydrogenase	CoA, NAD, ADP, pyruvate	Acetyl-CoA, NADH, ATP (fatty acids, ketone bodies)	Insulin ↑ (in adipose tissue)	Also important in regulating the citric acid cycle
Gluconeogenesis	Pyruvate carboxylase	Acetyl-CoA	ADP	Glucagon?	Induced by glucocorticoids, glucagon, cAMP. Repressed by insulin
	Phosphoenol pyruvate carboxykinase	cAMP?			
	Fructose-1, 6-bisphosphatase	cAMP	AMP, fructose 2, 6-bisphosphate in liver, fructose 1, 6-bisphosphate in muscle	Glucagon	
Glycogenesis	Glycogen synthase		Phosphorylase (in liver). cAMP, Ca^{2+} (in muscle)	Insulin ↑ Glucagon (liver) ↓ Epinephrine ↓	Induced by insulin
Glycogenolysis	Phosphorylase	cAMP, Ca^{2+} (muscle)		Insulin ↓ Glucagon (liver) ↑ Epinephrine ↑	
Pentose phosphate pathway	Glucose-6-phosphate dehydrogenase	$NADP^+$	NADPH		Induced by insulin
Lipogenesis	Acetyl-CoA carboxlase	Citrate	Long-chain acyl-CoA, cAMP	Insulin ↑ Glucagon (liver) ↓	Induced by insulin
Cholesterol synthesis	HMG-CoA reductase		Cholesterol, cAMP, mevalonate, bile acids	Insulin ↑ Glucagon (liver) ↓	Inhibited by certain drgus, eg, lovastatin

Table 5.2 : Summary of the major and unique feature of metabolismof the principal organs

Organ	Major Function	Major Pathways	Main Substrates	Major Products	Specialist Enzymes
Liver	Service for the other organs and tissues	Most represented, including gluconeogenesis; β-oxidation; ketogenesis; lipoprotein formation; urea, uric acid, and bile acid formation; cholesterol synthesis; lipogenesis[1]	Free fatty acids, glucose (well fed), lactate, glycerol, fructose, amino acids (Ethanol)	Glucose, VLDL (triacylglycerol), HDL, ketone bodies, urea, uric acid, bile acids, plasma proteins (Acetate)	Glucokinase, glucose-6-phosphatase, glycerol kinase, phosphoenolypyruvate carboxykinase fructokinase, arginase, HMG-CoA synthase and lyase, 7α-hydroxylase (Alcohol dehydrogenase)
Brain	Coordination of the nervous system	Glycolysis, amino acid metabolism	Glucose, amino acid, ketone bodies (in starvation) Polyunsaturated fatty acids in neonate	Lactate	
Heart	Pumping of blood	Aerobic pathways, eg, β-oxidation and citric acid cycle	Free fatty acids, lactate, ketone bodies, VLDL and chylomicron triacylglycerol, some glucose		Lipoprotein lipase. Respiratory chain well developed.
Adipose tissue	Storage and breakdown of triacylglycerol.	Esterification of fatty acids and lipolysis; lipogenesis[1]	Glucose, lipoprotein triacylglycerol	Free fatty acids, glycerol	Lipoprotein lipase, hormonesensitive lipase
Muscle Fast twitch Slow twitch	Rapid movement Sustained movement	Glycolysis Aerobic pathways, eg, β-oxidation and citric acid cycle	Glucose Ketone bodies, triacylglycerol in VLDL and chylomicrons, free fatty acids	Lactate	Lipoprotein lipase. Respiratory chain well developed.
Kidney	Excretion and gluconeogenesis	Gluconeogenesis	Fee fatty acids, lactate, glycerol	Glucose	Glycerol kinase, phosphoenolpyruvate carboxykinase
Erythrocytes	Transport of O_2	Glycolysis, pentose phosphate pathway. No mitochondria and therefore no β-oxidation or citric acid cycle	Glucose	Lactate	(Hemoglobin)

[1]In many species but not very active in humans

CHAPTER 6

NUCLEOPROTEINS

Q1. What do you mean by nucleoprotein?

It is a conjugated protein composed of simple protein (protamines or histones) with nucleic acid.

Q2. What is nucleic acid?

(a) It is the polynucleotide.

(b) The singly esterified 5′-phosphate of a nucleotide can esterify a second alcohol function group (–OH) forming a **diester**. This second –OH group resides on the pentose of another nucleotide at third position of pentose. This results in a **dinucleotide** in which the pentose moieties are linked by a 3′ → 5′ phosphodiester bond. The 3′ → 5′ phosphodiester bond forms the backbone of polynucleotides.

(c) Examples of nucleic acids are RNA and DNA.

Q3. Explain nucleotides.

(a) A nucleotide is composed of purine or pyrimidine bases, pentose sugars (ribose or deoxyribose) and phosphoric acid.

(b) Nucleotides form a portion of coenzymes such as FAD, NAD^+, $NADP^+$, coenzyme A, and S-adenosyl methionine.

(c) Nucleotides also serve regulatory functions.

(d) ADP levels regulate the rate of mitochondrial oxidative phosphorylation.

(e) Specific nucleotides act as allosteric regulators of enzyme activity, and cyclic AMP and cyclic GMP serve "second messenger" functions.

Q4. What is nucleoside?

(a) It is composed of purine or pyrimidine bases, and pentose sugars (ribose or deoxyribose).

(b) The synthetic analogs of nucleosides are involved in inhibiting the growth of cancer cells or of certain viruses.

Q5. Name the major and minor bases of nucleic acids.

Major bases : Adenine, guanine, cytosine, thymine, uracil.

Minor bases : 5-methylcytosine present in both bacterial and human DNA N^6-methyladenine, N^6, N^6-dimethyladenine, and N^7-methylguanine present in mammalian RNAs.

Bases present in the free state : Hypoxanthine (6-oxypurine), xanthine (2, 6-dioxypurine).

Q6. Name the methylated purines of plants of pharmacologic properties.

Caffeine of coffee : 1, 3, 7 –trimethylxanthine.

Theophylline of tea : 1, 3 – dimethylxanthine.

Theobromine of cocoa : 3, 7 – dimethylxanthine.

Q7. What is the nucleoside which replaces ribose by arabinose?

Cytarabine (the synthetic analog of nucloside) is used in the chemotherapy of cancer and viral infections.

Q8. What is the importance of 5-iododeoxyuridine?

5-iododeoxyuridine (the synthetic analog of nucleoside) is effective in the treatment of herpetic keratitis, an infection of the cornea by herpes virus.

Q9. How are nucleic acids digested?

In the small intestine.

Nucleic acid $\xrightarrow{\text{Polynucleotidase}}$ Nucleotides

Nucleotides $\xrightarrow{\text{Nucleotidase}}$ Nucleosides + Phosphoric acid.

Nucleosides $\xrightarrow{\text{Nucleosidase}}$ Purine or pyrimidine bases + pentose sugars (ribose or deoxyribose).

Q10. How does purine deficiency occur?

Purine deficiency states in humans are due to the deficiencies of folic acid and sometimes to a deficiency of vitamin B_{12}.

Q11. What are the sources of the nitrogen and carbon atoms of the purine ring?

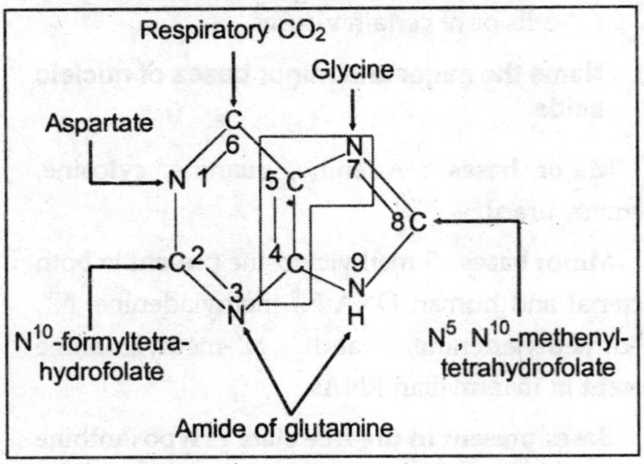

Q12. What are the sources of the nitrogen and carbon atoms of the pyrimidine ring?

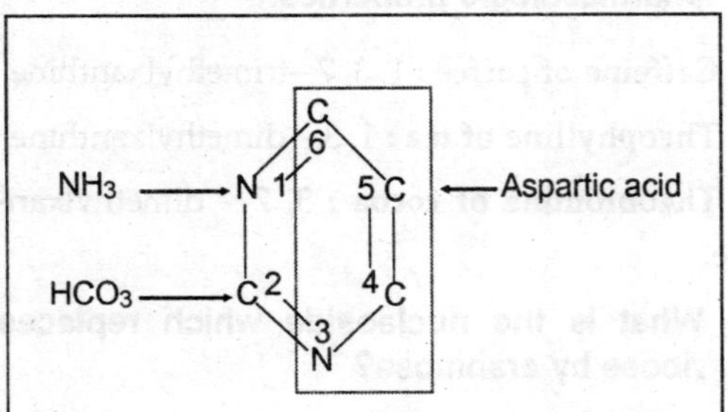

Q13. How does pyrimidine biosynthesis begin?

Pyrimidine biosynthesis begins with the formation from glutamine, ATP, CO_2, of carbamoyl phosphate.

Q14. Why are mammals said to be prototrophic?

Because they are capable of synthesizing purine and pyrimidine nucleotides and are not dependent on exogenous sources of these important compounds.

Q15. How are purine and pyrimidine catabolized in man?

(a) **Purine catabolism :**

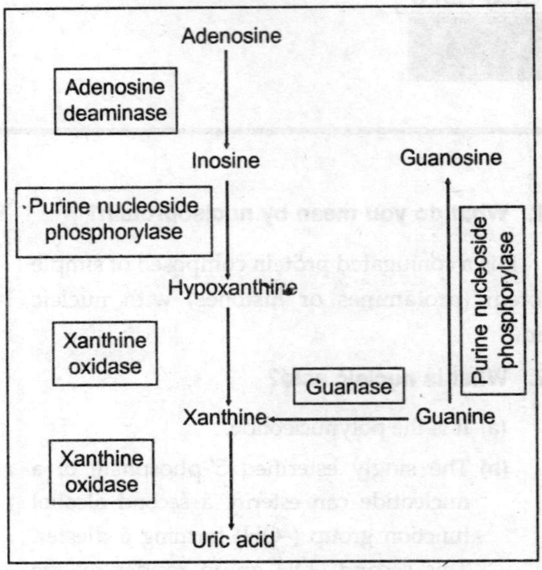

(b) **Pyrimidine catabolism :** Very little is known about the catabolism of pyrimidines. It is concluded that very little free pyrimidine is formed during digestion and that amount is not at all absorbed. However, the end products may be CO_2, NH_3, β-alanine, β-aminoisobutyrate.

Q16. What is the end product of purine catabolism of humans, amphibians, birds, and reptiles?

Uric acid. Since the enzyme uricase is lacking in them uric acid cannot be converted to allantoin.

Q17. What is urate?

Urate is the salt of uric acid. It is far more water soluble than uric acid.

Q18. What is the normal uric acid level of blood?

Males : 2.5 – 7.0 mg/100 ml.

Females : 2.5 – 5.5 mg/100 ml.

Q19. How much uric acid is excreted daily in urine in humans?

600 – 700 mg/day.

Q20. Name the competitive inhibitor of urate excretion and absorption.

High doses of aspirin.

Q21. How is uric acid catabolised in lower primates and other mammals?

$$\text{Uric acid} \xrightarrow{\text{Uricase}} \text{Allantoin.}$$

Q22. How does the output of uric acid increase?

(a) On a high protein diet.

(b) Some uricosuric agents such as salicylates, cinchophen and carinamide increase urinary elimination of uric acid by inhibiting its reabsorption in the renal tubules.

(c) ACTH and adrenocortical oxysteroids also increase the urinary excretion by inhibiting renal tubular reabsorption.

Q23. When does serum uric acid level increse?

In nephritis, eclampsia, sickle cell anemia, thalassemia, hemolytic anemia.

Q24. What are the products of uric acid breakdown in human?

Uric acid is broken down to urea and NH_3 by the intestinal bacteria.

Q25. What is hypouricemia?

(a) It occurs due to the deficiency of xanthine oxidase for which there is increased excretion of hypoxanthine and xanthine due either to genetic defect or to severe liver damage.

(b) In severe xanthine oxidase deficiency, patients may exhibit **xanthinuria** and **xanthine lithiasis.**

Q26. What do you mean by tophi?

The deposition of crystals of sodium urate in soft tissues particularly in joints.

Q27. What is "miscible uric acid pool"?

(a) Uric acid is mainly in urine and small amount in sweat and saliva.

(b) A portion of the uric acid is destroyed by intestinal bacteria. This intestinal uricolysis gives rise to urea and ammonia which are absorbed and excreted by the kidneys.

(c) Under conditions of normal production and removal, the body contains a "readily miscible uric acid pool".

Q28. What is Gout?

It is the disease in which sodium urate crystals are deposited in the soft tissues, particularly in joints due to the over-production of uric acid. Pain in the joints is the principal sign and the pain starts from the toes.

Q29. What should be the treatment for gout?

(a) The effective drug is the **allopurinol** which can reduce the formation of uric acid by inhibiting the enzyme xanthine oxidase responsible for uric acid formation.

(b) Bed rest.

(c) Avoidance of foods of high purine content (eg. meat, fish, peas, beans, sweet breads, liver, kidney,)

(d) Instruction to take foods of low purine diet (eg, fruits, milk, cheese, eggs, sweets, sugar, vegetables).

Q30. Name the test for the detection of uric acid in urine and discuss its method.

Murexide test.

Method : 5 ml urine is saturated fully with solid ammonium chloride in a test tube. A few drops of strong ammonium hydroxide is added to it. A gelatinous precipitate of ammonium urate is formed. The supernatant is filtered.

The residue on the filter paper is taken in a porcelain dish. 2 drops of conc. HNO_3 is added and heated to dryness. A reddish yellow residue remains. Cooled and added a few drops of very dilute ammonium hydroxide. The development of purplish red colour is due to the formation of murexide showing the presence of uric acid.

Q31. State the clinical importance of nucleic acid.

(a) Nucleic acids are able to transmit genetic information.

(b) They undergo mutation.

(c) RNA helps protein synthesis.

(d) In cell division, the nucleic acid chain is duplicated preserving in each daughter cell the information contained in the parent cell.

Q32. Show the distinction between RNA and DNA.

RNA	DNA
(a) Formed from adenine and guanine, cytosine and uracil, with ribose and phosphoric acid.	(a) Formed from adenine and guanine, cytosine and thymine, with deoxyribose and phosphoric acid.
(b) Present in cell cytoplasm and nucleolus.	(b) Present in nucleus, probably in the chromosomes.
(c) Carries no genetic information and undergoes no mutation.	(c) Carries genetic information from one generation of cell to the next and undegoes mutation.
(d) Concerned with protein synthesis.	(d) Concerned with protein synthesis.
(e) Contains 60 to 6,000 nucleotides, molecules are unbranched, structure is less rigid.	(e) Contains 1,600 to 9,000 nucleotides, molecules are long and thread like having a length of about 250 times greater than their breadth, structure is highly complex.

Q33. What do you mean by transcription and replication?

The genetic information stored in the nucleotide sequence of DNA serves two purposes :

(a) **Transcription** is the term which provides information for the synthesis of all protein molecules of the cell and organism through RNA.

(b) **Replication** is the term which provides the information inherited by daughter cells. Replication of the DNA molecule occurs in a semconservative manner.

Q34. How many of types of RNA are there?

Three types : Ribosomal RNA (rRNA), Transfer RNA (tRNA), Messenger RNA (mRNA).

Q35. What are the functions of RNA?

(a) Since the ribosomes are the site of protein synthesis, ribosomal RNA (rRNA) may play the vital roles in binding messenger RNA (mRNA).

(b) Soluble RNA (sRNA) or transfer RNA (tRNA) molecules combine specifically with the amino acids for peptide bond formation.

(c) Messenger RNA (mRNA) associates with ribosomes and controls protein synthesis.

Q36. What are mutations?

(a) Mutations are due to a change in the base sequence of DNA. They may result from the faulty replication, movement, or repair of DNA and occur with a frequency of about one in every 10^6 cell division. An abnormal gene product can be the result of mutations.

(b) The rate of mutation is increased by viruses, chemicals, ultraviolet light, and ionizing radiation.

(c) Somatic cells are often affected by mutations and are passed on to successive generations of cells within an organism.

(d) Most cancers are due to horizontal transmission of induced mutation.

Q37. What is chromatin?

(a) **Chromatin** is made up of very long double-stranded **DNA molecules** and having a nearly equal mass of small basic proteins **histones** and a smaller amount of nonhistone proteins and a small quantity of RNA.

(b) Chromatin studies have demonstrated dense spherical particles called **nucleosomes** which are 10 nm in diameter and connected by DNA filaments.

(c) HI histones are the least tightly bound to chromatin.

Q38. What do you mean by S phase?

In animal cells including human cells, the replication of the DNA genome occurs only at a specified time during the life span of the cell. This period is said to be synthetic or S phase.

Q39. What are the main principles for RNA synthesis?

(a) RNA synthesis involves initiation, elongation, and termination.

(b) RNA molecule formation is initiated at 5′ end and then follows with the release of factor when the elongation of the RNA molecule continues from 5′ to its 3′ end antiparallel to its template.

(c) Synthesis of RNA molecule is terminated by a signal which is recognized by the σ factor.

Q40. What are sigma (σ) factors?

(a) Sigma factors form a heterogeneous group of proteins.

(b) Most bacterial sigma factors are single polypeptides.

(c) All the related sigma proteins are acidic with an excess of negative charge.

Q41. Name the inhibitors of DNA and RNA synthesis.

(a) Actinomycin D inhibits RNA and DNA chain elongation.

(b) Rifampicin (Refampin) inhibits RNA initiation.

(c) Anthramycin inhibits DNA and RNA synthesis.

Q42. What do you mean by DNA replication?

DNA is a double helix. One strand of DNA is a complement of the other. Upon unwinding of the double helix, each strand acts as a template for the formation of new strand. This process is said to be DNA replication.

Q43. Explain conservative replication, dispersive replication, theta replication and semiconservative replication.

Conservative replication : The double helical DNA of the parent cell is just copied as an intact DNA. The two new strands are different from those of the parent DNA.

Dispersive replication : The parent DNA breaks into fragments and these fragments are amalgamated with the new ones.

Theta replication : Most prokaryotic DNA replicates when it is in a circular form. Such replication is said to be theta replication.

Semiconservative replication :

(a) The parent molecule unwinds at the location of replication, the remaining portion becomes intact.

(b) Each unwind strand acts as a "template" and a new daughter strand is formed.

(c) The daughter molecule is released. The one strand is only new, the other is from the parent. Hence, the mode of replication is semiconservative.

Q44. Mention the types of DNA polymerase in Prokaryotes and Eukaryotes.

In Prokaryotes :

(a) DNA polymerase 1 (Pol 1)

(b) DNA polymerase 11 (Pol. 11)

(c) DNA polymerase 111 (Pol 111)

In Eukaryotes :

(a) Polymerase α (maxi polymerase)

(b) Polymerase β (mini polymerase)

(c) Polymerase γ.

Q45. Explain Okazaki fragments.

(a) The formation of short RNA strands is required for the synthesis of DNA.

(b) In eukaryotes, the short RNA strand is about 10 nucleotides in length.

(c) The RNA and short DNA (about 150 bases) combination is known as **Okazaki** fragments.

Q46. State the role of DNA ligase.

(a) This is an enzyme polypeptide in nature having molecular weight 77,000.

(b) It joints the ends of two pieces of DNA strands by synthesizing a phosphodiester bond between 3′-OH at the end of one chain and 5′-phosphate at the end of the other.

(c) It joins segments of DNA in recombinant DNA.

(d) It helps to link the ends of linear DNA duplex to form circular DNA.

(e) It repairs single stranded nicks in duplex DNA.

Q47. How does DNA damage occur during DNA replication?

(a) By misincorporation of deoxynucleotides.

(b) By spontaneous deamination of bases during normal genetic functions.

(c) By X-radiation that causes "nicks" in DNA.

(d) By UV irradiation that may cause thymine dimer formation.

(e) By different chemicals that interact with DNA.

Q48. What are the methods for DNA repair?

(a) Repair of thymidine dimers and spontaneous deamination of cytosine to uracil.

(b) Photo reactivation.

(c) Recombinational repair.

Q49. What do you know about xeroderma pigmentosum?

(a) It is an inherited disorder in which DNA repair mechanisms are defective due to the deficiency of the enzymes **nicking endonuclease.**

(b) Increased cutaneous sensitivity to UV rays of sunlight produces blisters on skin.

(c) There is dry keratosis, hyperpigmentation, and atrophy of skin.

(d) Corneal ulcers may be produced.

(e) It is usually a fatal condition.

(f) Death may occur due to the formation of squamous cell carcinoma of skin.

Q50. What is ataxia telangectasia?

(a) It is a familial disorder which shows increased sensitivity to x-rays and UV rays.

(b) It is manifested by progressive cerebellar ataxia.

(c) There is frequent sino-pulmonary infections.

Q51. What do you mean by recombinant DNA?

It is the combinations of portions of DNA from one molecule to another in the same cell or cell of one species to another.

Q52. How is artificial DNA recombinant formed?

Artificial recombinant DNA technology is also called **genetic engineering**. This is the artificial modifications of the genetic constitution of a living cell by introduction of foreign DNA through experimental techniques.

Q53. Name the enzymes involved in genetic engineering.

Restriction endonucleases, exonucleases, endonucleases, reverse transcriptase, DNA polymerase, and DNA ligase.

Q54. What do you mean by vehicle DNA?

DNA which acts as the carrier is said to be vehicle DNA.

Q55. What do you know about cDNA (Complementary DNA)?

(a) It is synthesized on RNA template by reverse transcriptase.

(b) The single-stranded DNA is used as a template and a cDNA is synthesized by DNA polymerase.

(c) The cDNA can be linked to a vehicle DNA.

Q56. What is plasmid DNA?

(a) Plasmids are extra chromosomal DNAs that occur in closed circular froms in bacteria.

(b) Each plasmid carries its own replicator region and retains its characters after recombining with foreign DNA.

(c) It may carry additional information such as antibiotic resistance factor.

Q57. What is passenger DNA?

(a) the foreign DNA which is passively transferred from one cell to the other is known as passenger DNA.

(b) Synthetic DNA and cDNA are used as passenger DNA.

Q58. Explain DNA cloning.

Foreign DNA may be transcribed in a new host during the replication of vehicles in the bacteria. The passenger DNA also replicates alongwith the vehicles and forms a single copy from which multiple copies of specific DNA fragments are obtained. This is known as DNA cloning.

Q59. What is cosmid?

(a) It is a hybrid of phages and plasmids with the addition of "cos" site.

(b) It is used as a vector in recombinant DNA technology.

Q60. What are the applications of recombinant DNA technology?

(a) In production of hormones like insulin, growth hormone.

(b) In diagnosis of inherited disorders like sickle cell anemia, thalassemia etc.

(c) In gene therapy.

Q61. What are exons and introns?

Exons : They are the long sequences of DNA that do not supply the genetic information and are finally translated into the amino acid sequence of a protein molecule.

Introns : The sequences formed by long sequences of DNA with the amino acid sequence interrupt the coding region of structural genes. These intervening sequences are said to be introns.

Q62. What are splicesomes?

These consist of the primary transcript, five small nuclear RNAs and more than 50 proteins.

Q63. What do you know about Nortern Blot test?

The analysis of RNA is done by this test. The following are the steps to perform it.

(a) **Electrophoresis :** The total cellular RNA or isolated mRNA is subjected to agarose gel electrophoresis in the presence of denaturing agent to remove secondary structure contraints.

(b) **Blotting :** The gel is blotted in the same way as given in southern Blot test with the difference that the paper has been treated chemically so that it covalently binds RNA.

(c) **Hybridization :** A labelled probe is used. This allows visualization of the RNA species that are complementary to the probe.

Q64. Explain Southern Blot test.

Chromosomal DNA is analysed by this test. The following are the steps of it :

(a) **Cleavage :** DNA is cleaved by restriction endonuclease at specific sites.

(b) **Electrophoresis :** The DNA fragments are separated by agrose-gel electrophoresis.

(c) **Blotting :** The separated DNA fragments are transferred to a sheet of nitrocellulose by a flow of buffer. These fragments bind to the nitrocellulose forming a replica of the pattern of DNA fragments.

(d) **Hybridization :** Hybridization of fragments with a labelled probe. The probe is hybridized to filter. The probe may be a purified RNA, cDNA, or a segment of cloned DNA.

(e) **Autoradiography :** The pattern of bands that contain the DNA fragments are visualized by means of radiation from the probe.

Q65. State Western Blot test.

Proteins are analysed by this test. The steps are given below to perform it :

(a) **Electropharesis :** The elec rophoresis of the whole protein is done and transferred to a nitrocellulose membrane and fixed.

(b) **Probing :** The fixed protein is probed with radioactive antibody.

(c) **Autoradiography :** The patterns of bands are visualized by means of radiation from probe.

CHAPTER 7

DIGESTION AND ABSORPTION

Q1. What is digestion?

It is the process by which naturally occurring foodstuffs are disintegrated into assimilable forms.

The minerals and vitamins of the foodstuffs are made more assimilable during digestive reactions.

Q2. State the clinical aspects of digestion and absorption.

(a) Secretion of more gastric HCl causes **ulceration** or no secretion of HCl causes **achlorhydria**.

(b) Defects of bile secretion lead to **gallstones**.

(c) Exocrine pancreatic insufficiency in **cystic fibrosis** causes **steatorrhea.**

(d) Malabsorption of vitamin B_{12} and folic acid causes **anemia**.

(e) Malabsorption of vitamin D leads to **rickets** and **osteomalacia**.

(f) Lactase deficiency causes **milk intolerance.**

(g) Malabsorption of neutral amino acids leads to **Hartnup disease.**

Q3. What is saliva?

(a) Saliva, secreted by the salivary gland, consists of about 99.5% water.

(b) It acts as a lubricant for mastication and for swallowing.

(c) It is a vehicle for the excretion of certain drugs eg, ethanol and morphine, of inorganic ions such as K^+, Ca^{++}, HCO_3^-, SCN^- (thiocyanate), iodine, and of immunoglobulins (IgA).

(d) The P^H of saliva is about 6.8.

Q4. What is the quantity of saliva secreted per day?

500 – 1,500 ml.

Q5. How is salivary secretion excited and controlled?

(a) The presence of food in the mouth or awareness of the approach of food becomes the stimulus for the secretion of saliva through reflex action.

(b) Efferent nerves are the parasympathetic nerves to the secreting glands.

Q6. What is the function of salivary amylase?

(a) It is capable of bringing about the hydrolysis of strach and glycogen to maltose.

(b) It is readily inactivated at P^H 4.0 or less, so the digestive action on food in the mouth is ceased in the acid environment of the stomach.

Q7. What is a lingual lipase?

(a) It is secreted by the dorsal surface of the tongue (Ebner's glands).

(b) It is not so significant in humans.

Q8. Name the drugs that affect the salivary secretion?

(a) Acetylcholine, epinephrine, ephedrine, pilocarpine cause salivary secretion.

(b) Atropine has the reverse effect.

Q9. What are the secretary products in the lumen of the stomach?

Chief cells — Pepsin.

Oxyntic cells or Parietal cells — HCl.

Epithelial cells — Mucin.

Q10. What is the quantity and P^H of gastric juice?

Quantity : 2 to 3 litres per day.

P^H : About 1.0.

Q11. State the composition of gastric juice.

(a) It is a clear, pale yellow fluid.

(b) It consists of water (97–99%), HCl (0.2–0.5%), total solids (0.5%).

(c) Total solids : 1 organic [Rennin (in infants), pepsin, lipase (trace), glycoprotein, mucin] 2 Inorganic [H^+, Cl^-, Na^+, K^+, traces of Ca^{++}, Mg^{++}, phosphate,, sulphate].

Q12. How is gastric secretion controlled?

(a) A reflex phase depending on the nervous action to the stomach.

(b) An indirect action of the products of digestion in the stomach.

(c) An intestinal phase due to the action of the products of digestion in the duodenum.

Q13. What is meant by psychic or appetite juice?

The sight or even the smell of food causes a free flow of gatric juice, especially in a hungry animal.

Q14. How is HCl formed?

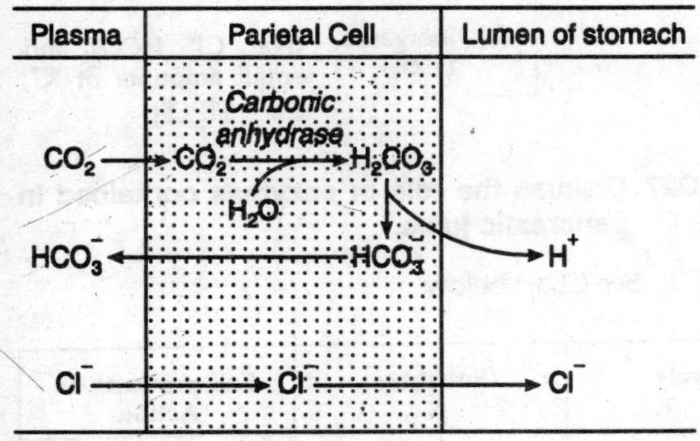

Q15. What is the mechanism for the production of flow of gastric juice?

(a) Gastrin is liberated from the pyloric mucosa by food or the products of protein digestion. This hormone passes into the blood and is carried to the gastric glands which are stimulated to secrete.

(b) Gastrin is also liberated, although in small amounts, when the breakdown products of protein reach the duodenum.

Q16. What is the role of rennin?

(a) **Rennin**, the important enzyme in the digestive processes of infants, prevents the rapid passage of milk from the stomach.

(b) In the presence of calcium, Rennin changes the casein of milk to a paracasein which is then acted on by **pepsin**.

(c) Rennin is absent from the stomach of adults. It is used in the making of cheese.

Q17. State the function of gastric lipase.

(a) The stomach secretes **gastric lipase** which in human is the main preduodenal lipase.

(b) The heat of the stomach helps in liquidizing the bulk if dietary lipids and emulsification is aided by persistaltic contractions.

(c) **Lingual** and **gastric lipases** hydrolyze triacylglycerol containing short—medium, and unsaturated long chain fatty acids to form mainly free fatty acids and 1, 2 – diacylglycerols.

(d) The enzymes are destroyed at low P^H but are active after feeding because of the buffering action of dietary proteins in the stomach.

(e) Preduodenal lipases become important when the activity of the pancreatic lipase is low and milk fat needs to be digested.

Q18. What do you mean by free and total acidity of gastric juice?

Free acidity means HCl.

Total acidity = Free acidity + conjugated acidity (HCl combined with protein + weak acids like lactic acid)

Q19. What is meant by hyperchlorhydria, hypochlorhydria, and achlorhydria?

Hyperchlorhydria : Increased level of HCl due to an increase in the volume of the gastric juice secreted.

Hypochlorhydria : Reduced level of HCl due to decrease in the volume of the gastric juice secreted.

Achlorhydria : The absence of HCl in the gastric juice.

Q20. Explain the changes occurring in the food in the stomach.

(a) The action of ptyalin continues until stopped by the penetration of HCl into the food mass.

(b) Proteins are broken down into shorter form by pepsin.

(c) Sucrose is partially hydrolysed by HCl.

(d) Micro-organisms in the food are destroyed or prevented from growing by HCl.

Q21. Why is the stomach not digested by itself?

(a) The stomach wall is protected by mucous which can inhibit pepsin and also acts as a buffer.

(b) The gastric epithelium also possesses anti-enzymes.

Q22. What are the inhibitors of gastric secretion?

(a) Fat inhibits gastric digestion by liberating enterogastrone from the duodenum. This substance inhibits both the motility and the secretion of the stomach.

(b) The release of gastrin is reduced when the gastric contents become highly acidic.

Q23. What is the nature and P^H of pancreatic juice?

Nature : Distinctly alkaline fluid.

P^H : 7.5 – 8.0 or higher.

Q24. Mention the quantity of pancreatic juice produced in man per day.

Usually between 1,200 ml. and 1,500 ml.

Q25. How is the secretion of pancreatic juice regulated?

(a) The acid of gastric juice enter the upper part of the duodenum and stimulates the secretion of the hormone secretin which increases the flow of pancreatic juice.

(b) Cholescystokinin also stimulates the secretion of enzymes of the juice.

Q26. State the composition of pancreatic jucie.

Solids (1.8%) →

(a) Organic (0.6%) [Proteins, trypsinogen, chymotrypsinogen, elastase, carboxypeptidase, pancreatic amylase, Ribonuclease, deoxyribonuclease, cholesteryl ester hydrolase, phospholipase A_2]

(b) Inorganic (1.2%) [Na^+, Cl^-, HCO_3^-, and small amounts of K^+, Ca^{++}, $HPO_4^=$]

Q27. Discuss the role of enzymes contained in pancreatic juice.

See Chart below.

Method of activation and optimal conditions for activity	Enzyme	Substrate	End products or Action
Trysinogen —(Enterokinase of intestine, P^H 5.2 – 6.0)→ Trypsin	Trypsin	Protein. Peptides	Polypeptides. Dipeptides.
Chymotrypsinogen —(Trypsion, P^H 8.0)→ Chymotrypsin	Chymotrypsin	Protein. Peptides.	Same as trypsin. More coagulating power for milk.
Proelastase —(Trypsin)→ Elastase	Elastase	Protein. Peptides.	Polypeptides. Dipeptides.
Procarboxypeptidase —(Trypsin)→ Carboxypeptidase	Carboxypeptidase	Polypeptides at the free carboxyl end of the chain.	Lower peptides. Free amino acids.
P^H 7.1	Pancreatic amylase	Starch. Glycogen	Maltose plus 1 : 6 glucosides plus maltotriose.
Activated by bile salts, Phospholipids, colipase. P^H 8.0	Lipase	Primary ester linkages of triacyl-glycerol	Fatty acids. 2-mono-acylglycerols, glycerol.
Activated by bile salts	Cholesteryl ester hydrolase	Cholesteryl esters.	Free cholesterol plus fatty acids.
Prophospholipase A_2 —(Trypsin, Ca^{++})→ Phospholipase A_2	Phospholipase A_2	Phospholipids	Fatty acids. Lysophospholipids.
—	Ribonuclease.	Ribonucleic acid	Nucleotides.
—	Deoxyribonuclease	Deoxyribonucleic acid	Nucleotides.

Q28. How is the pancreatic secretion produced after a meal?

(a) The pancreas first secretes reflexly.

(b) When the acid contents of the stomach reach the duodenum both secretin and pancreozymin are set free and are absorbed into the circulation and then stimulate the pancreatic cell.

Q29. Show the composition of hepatic and of gallbladder bile.

	Hepatic bile (as secreted)		Bladder bile.
	Per cent of total bile	Per cent of total solids	Per cent of total bile
Water	97.00	—	85.92
Solids	2.52	—	14.08
Bile acids	1.93	36.9	9.14
Mucin and Pigments	0.53	21.3	2.98
Cholesterol	0.06	2.4	0.26
Esterified and non-esterified fatty acids	0.14	5.6	0.32
Inorganic salts	0.84	33.3	0.65
Specific gravity	1.01	—	1.04
p^H	7.1 – 7.3	—	6.9 – 7.7

Q30. State the Properties of bile.

(a) **Emulsification :**

(i) The bile salts have the ability to lower surface tension. This enables them to emulsify fats in the intestine and to dissolve fatty acids and water insoluble soaps.

(ii) The presence of bile in the intestine helps the digestion and absorption of fats as well as the fat-soluble vitamins A, D, E and K.

(iii) When fat digestion is impaired, fat covers the other foodstuffs and prevents enzymes from attacking them. Under these conditions, the activity of the intestinal bacteria causes putrefaction and formation of gas.

(b) **Neutralization of acid :** The bile neutralizes the acid chyme from the stomach and prepares it for digestion in the intestine.

(c) **Excretion :** Bile is an important vehicle for bile acids and cholesterol excretion. It also removes many drugs, toxins, bile pigments, various inorganic substances such as copper, zinc and mercury.

(d) **Bile pigment metabolism :** It helps in the metabolism of bile pigments.

Q31. What is the quantity of bile secreted per day?

300 ml – 1,000 ml.

Q32. What is the enzyme present in bile?

Alkaline phosphatase.

Q33. Why is the colour of bile yellow?

Due to the bilirubin present in bile.

Q34. Name the bile salts.

Sodium or potassium glycocholate, sodium or potassium taurocholate.

Q35. What do you mean by entorohepatic circulation?

Bile salts are absorbed in the intestine and pass into the blood, are carried into the portal system back to liver and are resecreted in the bile. This is known as enterohepatic circulation.

Q36. What are the glands involved for the secretion of intestinal juice?

Glands of Brunner and Lieberkuhn.

Q37. Discuss the role of enzymes contained in intestinal juice.

Enzymes	Optimal conditions for activity	Substrate	End products or Action
Aminopeptidase	—	Polypeptides at the free amino end of the chain	Lower peptides. Free amino acids.
Dipeptidase	—	Dipeptides	Amino acids.
Sucrose	p^H 5.0 – 7.0	Sucrose	Fructose, Glucose
Maltase	p^H 5.8 – 6.2	Maltose	Glucose
Lactase	p^H 5.4 – 6.0	Lactose	Glucose, Galactose.
Trehalase	—	Trehelose	Glucose
Phosphatase	p^H 8.6	Organic phosphates	Free phosphate
Isomaltase or 1 : 6 glucosidase	—	1 : 6 glucosides	Glucose
Polynucleotidase	—	Nucleic acid	Nucleotides.
Nucleosidase	—	Purine or pyrimidine nucleosides	Purine or pyrimidine bases. Pentose phosphate.

Q38. How are gall stones formed?

Various factors such as infection cause the supersaturated bile to precipitate the excess cholesterol as crystals. The crystals grow to form stones unless the newly formed crystals are rapidly excreted into the intestine with the bile.

Q39. Why is renal calculi formed?

(a) Prevalence of hot climate for more than six months in a year and the consumption of large amounts of tea (rich in oxalate).

(b) Consumption of diets rich in calcium, phosphate, and purines.

(c) Low urine volume and more excretion of uric acid, calcium and phosphate.

(d) Urine infection and stagnation.

Q40. What does happen with the food residues in the large intestine?

Since large amount of water is absorbed in the large intestine the semiliquid intestinal contents gradually become more solid. During this period, the bacteria produce various gases such as CO_2, methane, hydrogen, nitrogen, H_2S as well as acetic, lactic, propionic, and butyric acids. Lecithin is decomposed to choline and neurine.

Q41. Give a few examples of the decarboxylating action of intestinal bacteria on some amino acids.

Histidine ⟶ Histamine.
Tyrosine ⟶ Tyramine.
Arginine ⟶ Agmatine.
Lysine ⟶ Cadaverine.

Q42. What is the effect of the drug neomycin on ammonia intoxication?

The oral administration of the drug neomycin reduces the concentration of ammonia delivered from the intestine to the blood owing to its antibacterial action.

Q43. How will you justify that the intestinal bacteria are also beneficial?

(a) The intestinal bacteria synthesize essential amino acids and certain vitamins, particularly vitamin K and some members of B-complex (eg, biotin).

(b) In herbivora, the cellulose diet is decomposed by the intestinal bacteria and made available for absorption.

Q44. Show the site of absorption of nutrients.

Site	Nutrient
Jejunum	Glucose and other monosaccharides, some disaccharides. Monoacylglycerols, fatty acids, glycerol, cholesterol. Amino acids, peptides, vitamins, folate. Electrolytes, iron, calcium, water.
Ileum	Bile acids. Vitamin B_{12} Electrolytes. Water.

Q45. What is lactose intolerance? Mention its symptoms.

The deficiency of the enzyme lactase causes the diminished absorption of lactose resulting in intolerance to lactose of milk.

Symptoms : Abdominal cramps, diarrhea, intestinal irritation.

Q46. What is sucrase deficiency?

There is an inherited deficiency of sucrase and isomaltase. These two enzymes occur as a complex enzyme.

Q47. What is disacchariduria?

(a) This is a condition in which there is an increase in the excretion of disaccharides due to disaccharidase deficiencies.

(b) As much as 300 mg or more of disaccharide may be excreted in the urine of these people.

Q48. What does cause fructose-sorbitol malabsorption?

It causes watery feces, abdominal discomforts, and gas production owing to bacterial fermentation.

Q49. What is chyluria?

This is an abnormality in which the patient excretes milky urine because of the presence of an abnormal connection between the urinary tract and the lymphatic drainage system of the intestine, a so called chylous fistula.

Q50. Show the disturbances due to malabsorption.

Symptoms	Substances malabsorbed
Anemia	Iron, Vitamin B_{12}, Folate.
Edema	Products of protein digestion.
Tetany	Calcium, Magnesium, Vitamin D.

Symptoms	Substances malabsorbed
Osteoporosis	Calcium, Products of protein digestion, vitamin D.
Milk intolerance	Lactose.
Bleeding, bruising	Vitamin K.
Steatorrhea (fatty stools)	Lipids and fat-soluble vitamins.
Hartnup disease (defect in intestinal neutral amino acid carrier)	Neutral amino acids.

Q51. Enumerate the localization and major action of gastrointestinal hormones.

Hormones	Localization	Major action
Gastrin	Gastric antrum, duodenum	Gastric acid and pepsin secretion.
CCK	Duodenum, jejunum	Pancreatic amylase secretion.
Secretin	Duodenum, jejunum	Pancreatic bicarbonate secretion.
GIP	Small bowel	Enhances glucose-mediated insulin release, inhibits gastric acid secretion.
VIP	Pancreas	Smooth muscle relaxation, stimulates pancreatic bicarbonate secretion.
Motilin	Small bowel	Initiates interdigestive intestinal motility.
Substance P	Entire gastrointestinal tract	Physiologic action uncertain.
Somastatin	Stomach, duodenum, pancreas	Numerous inhibitory effects.
PP	Pancreas	Inhibits pancreatic bicarbonate and protein secretion.
BLI	Stomach, duodenum	Stimulates release of gastrin and CCK.

CHAPTER 8

ENZYMES

Q1. Define enzymes.

Enzymes are soluble, colloidal organic catalysts formed by living cells, protein in nature, specific in action, inactive at 0°C, and destroyed by moist heat at 100°C.

Q2. State the clinical importance of enzymes.

(a) Enzymes are the biocatalysts which regulate the rates at which all physiologic processes take place and it has got the central role in health and diseases.

(b) In liver cirrhosis, the important key enzymes, which are responsible for urea formation, are impaired. As a result, there is ammonia intoxication and ultimately hepatic coma.

(c) In the impairment of enzymes, disgestion of foodstuffs is stopped causing malabsorption and metabolic disturbance. The net result is no energy production and no growth of body.

(d) Measurement of increased or decreased concentrations of certain enzymes in blood serum provides valuable diagnostic and prognostic information to the physicians.

Q3. What are zymase and zymogen?

The extracellular enzyme when secreted ready for action is said to be zymase. eg, salivary amylase. The extracellular enzyme when secreted in inactive form and activated by other agent secreted by other cell is said to be zymogen eg, pepsinogen, trypsinogen etc.

Q4. What are intracellular and extracellular enzymes?

Enzymes which are used in the cells which make them are said to be intracellular enzymes. eg, metabolic enzymes.

Enzymes which are produced by other cell and are used to the other part of the body are said to be extracellular enzymes. eg, enzymes of digestive juices.

Q5. What is a substrate?

The substance on which an enzyme acts is called the substrate,. eg, lactose, maltose, sucrose etc.

Q6. What do you mean by active groups and active centres?

The amino, carboxyl, and sulphydryl groups of the side chains of amino acids are available for linkage between peptide chains. Some of these groups together with imidazole ring and the alcoholic group of serine are responsible for the union between substrate and enzyme. These groups are termed active groups.

The region of the protein surface at which these active groups are located are called active centres.

Q7. Classify enzymes.

(a) **Oxidoreductases :** Oxidases eg, tyrosinase; Anaerobic dehydrogenases eg, malic dehydrogenase; Hydroperoxidases eg, peroxidases.

(b) **Tansferases :** Transphosphorylases eg, hexokinases; transaminases eg, GOT, GPT.

(c) **Hydrolases :** Polysaccharases eg, amylase; Esterases eg, lipases.

(d) **Isomerases :** Glucose phosphate isomerase, triose phosphate isomerase.

(e) **Lyases :** Aldolase, decarboxylase.

(f) **Ligases :** Glutamine synthetase, succinate thiokinase, acetyl-CoA carboxylase.

Q8. What is meant by enzyme unit?

Under appropriate conditions, the measured rate of the reaction is proportional to the quantity of enzyme present. Results are expressed in enzyme units. Enzyme units are expressed in micromoles (μ mol), nanomoles (n mol), picomoles (p mol).

Q9. What are coenzymes, cosubstrates? Give examples.

Coenzymes are heat-stable, low molecular-weight organic compounds required for the activity of enzymes and prosthetic group of enzymes. The coenzymes are said to be cosubstrates. Examples : ATP, ADP, NAD, NADP, etc.

Q10. What are cofactors?

The metallic ions which are involved in enzyme activity. eg, Ca^{++}, Mg^{++}, Mn^{++}, Fe^{++}, etc.

Q11. Name some group transferring, hydrogen transferring, and nutritional coenzymes.

Group transferring coenzymes : ATP, B_6-PO_4, CoA etc.

Hydrogen transferring coenzymes : NAD^+, $NADP^+$, FAD, CoQ etc.

Nutritional coenzymes : TPP, NAD^+ $NADP^+$, CoA, B_6-PO_4 etc.

Q12. Which enzymes do not require coenzymes for their activity?

The extracellular enzymes.

Q13. What is holoenzyme?

Apoenzyme + coenzyme = Holoenzyme.
(Protein part) (Non-protein part)

Q14. What is Isozymes or isoenzymes?

The enzymes that occur in a number of different forms having the same catalytic activity and differ each other chemically, immunologically, and electrophoretically are called isozymes or isoenzymes.

Q15. Give some examples of isozymes.

Oxidases, transaminases, phosphatases, aldolases.

Q16. Name the isoenzymes of lactate dehydrogenase.

I_1 .... HHHH
I_2 .... HHHM
I_3 .... HHMM
I_4 .... HMMM
I_5 .... MMMM

Q17. What is Lysozyme?

(a) Lysozyme is an enzyme present in tears, nasal mucous, sputum, milk, and egg-white.

(b) Its molecular weight is about 15,000.

(c) It destroys the cell walls of many airborne grampositive bacteria in tears and nasal mucus.

Q18. What are ribozymes?

(a) The RNAs which satisfy fully the definition of enzymes are termed **ribozymes.**

(b) Ribozymes catalyze the hydrolysis of phosphodiester bonds in RNA molecules. These reactions are facilitated by free –OH groups.

(c) These play key roles in the intron splicing events essential for the conversion of pre-mRNA to mature mRNAs.

Q19. Name the factors influencing enzyme action.

(a) Contact between enzyme and substrate, (b) Concentration of enzyme and substrate. (c) Temperature. (d) P^H (e) Oxidation. (f) Radiation. (g) Coenzymes and activators. (h) Inhibiting agents. (i) Anti-enzymes.

Q20. What is Michaelis constant?

The substrate concentration that produces half maximal velocity is termed K_m value or Michaelis constant.

Q21. Mention the Michaelis-Menten equation.

$$v = \frac{V_{max}\,[S]}{K_m + [S]}$$

Q22. When does the maximum activity of the enzymes occur?

The enzymes show maximum activity at the optimum temperature in the region of 40°C in case of animal enzymes and 60°C in case of plant enzyme urease, optimum P^H in the region of 5 to 9 except a few enzymes like pepsin.

Q23. What do you mean by competitive and non-competitive inhibition?

In case of competitive inhibition, the inhibitor competes with the substrate for the active group, eg, malonic acid (the inhibitor) competes with succinic acid (the substrate) for which succinate dehydrogenase is inhibited and succinate cannot be converted to fumaric acid.

Some enzymes are inhibited when an inhibitor becomes attached not to the active goup but to some other group of the enzymes. Under such conditions, the activity of the active group is affected, so that union with the substrate occurs less readily or not at all. This type of inhibition is known as non-competitive inhibition.

Q24. What are anti-enzymes?

If certain enzymes are repeatedly injected into an animal, there is the production in the animal's serum of a substance which prevents the normal action of the enzyme injected. This substance is called an anti-enzyme, eg, anti-pepsin, anti-renin.

Q25. What is enzyme induction?

The concentration of a given metabolite within the cell affects the activity of enzyme system which are involved in the synthesis or metabolism of metabolite itself. This phenomenon is known as enzyme induction.

Q26. What is meant by constitutive enzymes?

Enzymes whose concentration in a cell is independent of an added inducer are termed constitutive enzymes.

Q27. What do you mean by feedback inhibition?

$$A \xrightarrow{EnZ_1} B \xrightarrow{EnZ_2} C \xrightarrow{EnZ_3} D$$

In the above biosynthesis reaction for A to D catalyzed by enzymes EnZ_1, EnZ_2, EnZ_3, a high concentration of D will inhibit conversion of A to B. D specially binds to and inhibits EnZ_1 and thus acts as feedback inhibitor. This phenomenon is known as feedback inhibition.

Q28. Why are the enzymes of hot spring not destroyed?

Because the optimum temperature for the enzymes of hot spring is 100°C.

Q29. What do you mean by Allosteric and isosteric site?

Allosteric site : The site of an enzyme which can bind other substances.

Isosteric site : The site of an enzyme which can bind the substrate or other molecules structurally similar to it.

Q30. State the diagnositic importance of serum enzymes.

Serum enzymes	Concentration increased in
Lipase	Acute pancreatitis and pancreatic carcinoma.
Amylase	Intestinal obstruction, acute pancreatitis, and diabetes mellitus.
Aldolase	Muscular dystrophy, cancer, diabetes mellitus, leukemia, hepatitis, glycogen storage disease.
Alkaline phosphatase	Obstructive jaundice, metastatic carcinoma, rickets.
Acid Phosphatase	Prostatic carcinoma with metastases.
Cholinesterase	Nephrotic syndrome.
LDH	Leukemia, acute hepatitis.
SGOT	Myocardial infarction.
SGPT	Acute hepatic disease.
IDH	Cerebral tumors, meningitis.
CPK	Myocardial infarction.
Ceruloplasmin (ferroxidase activity)	Cirrhosis, bacterial infarction, pregnancy.

Q31. Does an enzyme gain its activity when kept at optimum temperature 0°C when kept for sometime?

Yes, the enzyme gains its activity.

Q32. How is the activity of many enzymes regulated in humans?

The activity of many enzymes is regulated by covalent modification most commonly by ATP-dependent phosphorylation, with subsequent hydrolytic removal as inorganic orthophosphate.

Q33. What are the clinical importance of recombinant DNA technology?

(a) It offers to understand the molecular basis of a number of diseases, like sickle cell disease, thalassemia, cystic fibrosis, familial hypercholesterolemia.

(b) Using this technology, human proteins can be produced in abundance for therapy (eg, insulin, growth hormone).

(c) Using this, proteins for vaccines and for diagnostic test (eg, for AIDS) can be obtained.

CHAPTER 9

BIOLOGIC OXIDATION

Q1. What is oxidation and reduction?

Oxidation means removal of electrons and hydrogen atoms and addition of O_2.

Reduction means gain of electrons and addition of hydrogen atoms.

Q2. What is redox potential?

In oxidation and reduction reactions, the free energy exchange is proportionate to the tendency of reactants to donate or accept electrons. This is expressed as redox potential.

Q3. State the clinical importance of biologic oxidation.

(a) The life of higher animals is fully dependent on the supply of oxygen which is used in respiration. The respiration is a process by which cells derive energy in the form of ATP liberating water.

(b) Molecular oxygen is incorporated into oxygenases. Many drugs, pollutants, and chemical carcinogens (xenobiotics) are metabolized by enzymes of this class known as **cytochrome P-450 system.**

(c) Administration of oxygen is most essential in case of circulatory failure.

Q4. Define respiratory chain.

The sequence of enzymes and carriers responsible for the transport of reducing equivalents from substrate to molecular oxygen forming ATP and water is known as the respiratory chain.

Q5. Classify oxido reductases.

(a) Oxidases, (b) Aerobic dehydrogenases (c) Anaerobic dehydrogenases (d) Hydroperoxidases (e) Oxygenases.

Q6. What are cytochromes?

(a) The cytochromes are classified as dehydrogenases except cytochrome oxidase.

(b) In the respiratory chain, they are involved as carriers of electrons.

(c) They are iron-containing hemoproteins in which the iron atom oscillates between Fe^{+++} to Fe^{++} during oxidation and reduction.

(d) Cytochrome C is only soluble among the cytochromes b, c_1, c, a and a_3.

(e) Cytochromes are also found in other locations, eg, the endoplasmic reticulum (cytochrome P-450), plant cells, bacteria, and yeast.

Q7. What is the important role of hydroperoxidases?

Hydroperoxidases protect the body against harmful peroxides. Otherwise, accumulation of peroxides can lead to generation of free radicals which in turn can disrupt membranes and can cause cancer and atherosclerosis.

Q8. What is xanthine oxidase?

(a) It is widely distributed in milk, small intestine, kidney, and liver.

(b) It contains molybdenum and plays an important role in the conversion of purine bases to uric acid.

(c) It is also significant in the liver and kidney of birds which excrete uric acid as the main nitrogenous end product.

(d) It is also involved in protein and amino acid catabolism.

Q9. What is cytochrome oxidase?

(a) It is a hemoprotein widely distributed in many tissues.

(b) It is the terminal component of the carriers of the respiratory chain of mitochondria.

(c) This enzyme is poisoned by carbon monoxide (CO), cyanide, and hydrogen sulphide.

(d) More recent studies show that cytochrome a and cytochrome a_3 are combined with a single protein, and the complex is known as cytochrome aa_3.

(e) It contains two molecules of heme, each having one Fe atom that oscillates between Fe^{+++} and Fe^{++} during ,oxidation and reduction. Two atoms of copper (Cu) are also present, each is associated with a heme unit.

Q10. What is catalase?

(a) Catalase is a hemoprotein containing 4 heme groups.

(b) It uses one molecule of H_2O_2 as a substrate electron donor and another molecule of H_2O_2 as oxidant or electron acceptor.

$$2H_2O_2 \xrightarrow{\text{Catalase}} 2H_2O + O_2$$

(c) It is found in blood, bone marrow, mucous membranes, kidney, and liver.

(d) It destroys hydrogen peroxide formed by the action of oxidases.

(e) **Microbodies** or **Peroxisomes,** found in liver, are rich in oxidases and catalases.

Q11. Why are the microsomal cytochrome P-450 monooxygenase systems so important?

(a) The monooxygenases are found in the microsomes of the liver together with cytochrome P-450 and cytochrome b_5.

(b) Both NADH and NADPH donate reducing equivalents for the reduction of these cytochromes, which in turn are oxidized by substrates in a series of enzymatic reactions collectively known as the hydroxylase cycle.

$$\text{DRUG–H} + O_2 + 2Fe^{++} + 2H^+ \ \text{(P-450)}$$

$$\downarrow \text{Hydroxylase}$$

$$\text{DRUG–OH} + H_2O + 2Fe^{+++} + 2H^+ \ \text{(P-450)}$$

(c) The drugs metabolized by this system are benzpyrene, aniline, morphine, and benzphetamine.

(d) Many drugs such as phenobarbital have the ability to induce the formation of microsomal enzymes and of cytochrome P- 450.

Q12. What is superoxide?

(a) Oxygen is potentially a toxic substance, the toxicity of which has been attributed to the formation of H_2O_2. The toxicity of oxygen is due to its conversion to superoxide.

(b) Superoxide is formed when reduced flavins, present in xanthine oxidase, are reoxidized by molecular oxygen.

$$\text{EnZ – Flavin} - H_2 + O_2$$

$$\downarrow$$

$$\text{EnZ – Flavin} - H + O_2^- + H^+$$

(c) Superoxide can reduce oxidized cytochrome C or be removed by the presence of the specific enzyme superoxide dismutase.

$$O_2^- + \text{Cyt C } (Fe^{+++}) \rightarrow O_2 + \text{Cyt C } (Fe^{++})$$

$$O_2^- + O_2^- + 2H^+ \xrightarrow{\text{Superoxide dismutase}} H_2O_2 + O_2$$

In this reaction, superoxide acts as both oxidant and reductant.

(d) It has been proposed that O_2^- bound to cytochrome P-450 is an intermediate in the activation of oxygen in hydroxylation reactions.

Q13. What is the role of superoxide dismutase?

(a) Superoxide dismutase can protect aerobic organisms against the deleterious effects of superoxide.

(b) The enzyme is present in different compartments of the cell. The cytosolic enzyme is composed of 2 similar subunits, each one containing one equivalent of Cu^{++} and Zn^{++}, whereas the mitochondrial enzyme contains Mn^{++}.

(c) Animals exposed to an atmosphere of 100% oxygen cause an increase of the enzyme, particularly in the lungs, prolonged exposure leads to lung damage and death. Antioxidants, eg, vitamin E, reduce the toxicity of oxygen.

Q14. What do you mean by oxidative phosphorylation?

The system in mitochondria that couples respiration to the generation of ATP from ADP and Pi is termed oxidative phosphorylation.

Q15. Justify the statement that mitochondrion has been termed the "powerhouse" of the cell.

(a) Within the mitochondria, most of the useful energy derived from oxidation within the tissues is captured in the form of high energy intermediate (ATP).

(b) Mitochondria contain the enzyme systems responsible for producing the reducing equivalents in the first place i.e. the enzymes of β-oxidation and the citric acid cycle. The citric acid cycle is the common metabolic pathway for the oxidation of all the major foodstuffs.

(c) All the useful energy formed from oxidation of fat, protein, and carbohydrate are available within the mitochondria.

(d) Mitochondria contain a series of catalysts known as the respiratory chain which are concerned with the transport of hydrogen and electrons and formation of water by their final reaction with oxygen.

Q16. What is coenzyme Q (CoQ)?

(a) Coenzyme Q exists in mitochondria in the oxidized quinone form under aerobic condtions and in the reduced quinol form under anaerobic conditions.

(b) Q is a constituent of the mitochondrial lipids; the other lipids are phospholipids that form the part of the mitochondrial membrane.

(c) The structure of Q is very similar to vitamin K and vitamin E.

(d) It is also similar to plastoquinone found in chloroplasts.

(e) It is a mobile component of the respiratory chain and that it collects reducing equivalents from the more fixed flavoprotein complexes and passes them on to the cytochromes.

Q17. Indicate the role of iron-sulphur protein (FeS; nonheme iron)?

(a) It is associated with the flavoproteins and with cytochrome b.

(b) The sulphur and iron are thought to take part in the oxidoreduction mechanism between flavin and Q which involves only a single e^- change, the iron atom undergoing oxidoreduction between Fe^{++} and Fe^{+++}.

Q18. Why is ATP called the energy "currency" of the cell?

ADP is a molecule that captures, in the form of high-energy phosphate, a significant proportion of the free energy released by catabolic processes. The resulting ATP passes on this free energy to drive those processes requiring energy. That's why ATP has been called the energy "currency" of the cell.

Q19. Exhibit the respiratory chain with its different sites of phosphorylation in the mitochondria.

See Chart 1 below.

Q20. Why does oxidation of succinate via ETC produce 2 ATP?

Because Site I is bypassed by the flavoprotein linked succinate dehydrogenase for which 2 ATPs only formed at Site II and Site III.

Q21. Name the inhibitors of respiratory chain at different sites.

Site I	:	Inhibited by amobarbital, piericidin, rotenone
Site II	:	Inhibited by dimercaprol and antimycin A
Site III	:	Inhibited by dimercaprol and antimycin A

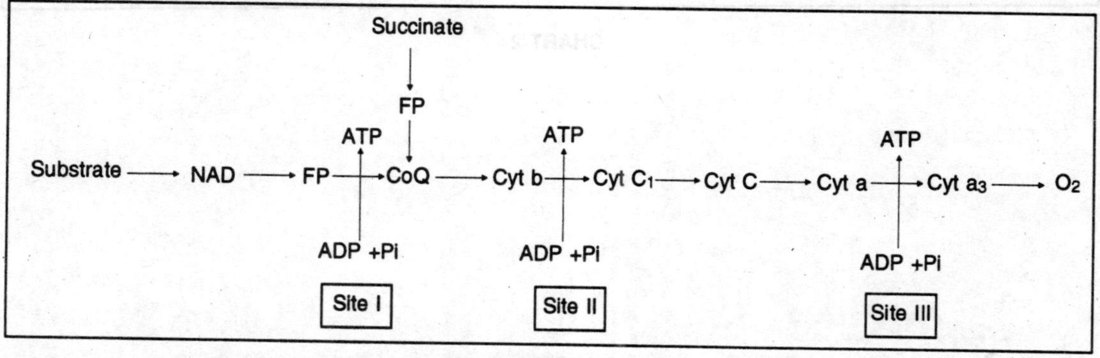

CHART 1

Q22. Name the antibiotic which competely blocks oxidation-reduction and phosphorylation in intact mitochondria.

Oligomycin.

Q23. In the presence of which uncoupler oxidation proceeds without phosphorylation?

Dinitrophenol.

Q24. Name the substance which inhibits oxidative phosphorylation.

Atractyloside.

Q25. What do you mean by oxidative phosphorylation?

The oxidation of hydrogen atoms of foodstuffs carried by different carriers of the respiratory chain takes place by the molecular oxygen at the end of the chain and the energy in the form of ATP is liberated by ADP and Pi at three sites of the chain in the mitochondria.

Q26. Mention the action of the uncouplers.

(a) Its action is to dissociate oxidation in the respiratory chain from phosphorylation.

(b) This causes respiration becoming uncontrolled, since the concentration of ADP or Pi no longer limits the rate of respiration.

Q27. Name some uncouplers.

Dinitrophenol, dinitrocresol, pentachlorophenol, and CCCP (m-chlorocarbonyl cyanide phenylhydrazone). The latter is 100 times as active as dinitrophenol.

Q28. What are the antibiotics responsible for K^+ penetration through the mitrochondrial membrane?

Valinomycin and Nigericin.

Q29. What is the principle of chemiosmotic theory?

This theory states that the components are oxidized in the respiratory chain generating hydrogen ions that are ejected to the outside of a coupling membrane in the mitochondria.

The electrochemical potential difference resulting from the asymmetric distribution of the hydrogen ion is used to drive the mechanism for the formation of ATP.

Q30. What do you mean by MELAS?

MELAS (mitochondrial myopathy, encephalopathy, lactacidosis, and stroke) is an inherited condition due to NADH : ubiquinone oxidoreductase (complex I) or cytochrome oxidase deficiency.

Q31. Mention the different complexes in the respiratory chain.

See Chart 2 below.

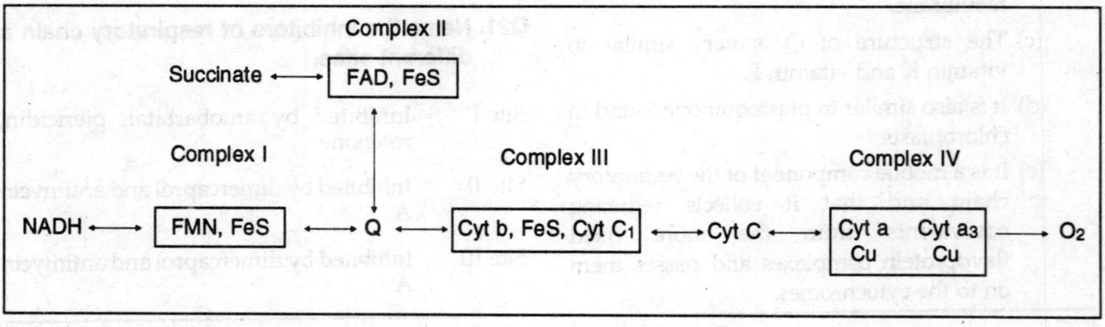

CHART 2

CHAPTER 10

VITAMINS

Q1. What are vitamins?

In addition to carbohydrates, fats, proteins, and minerals in the diets other factors present in natural foods are also required although in small quantities to sustain life. These "accessory food factors" are called vitamins.

Q2. Classify vitamins.

(a) **Fat-soluble vitamins** : Vitamins A, D, E and K.

(b) **Water-soluble vitamins** : Vitamins C and B-complex.

B-Complex : Thiamine (B_1), Riboflavin or lactoflavin (B_2), Niacin (P-P-factor), Pyridoxine (B_6), Pantothenic acid, Lipoic acid, Biotin (H), Folic acid group, inositol, P-amino benzoic acid (PABA), cyanocobalamine (B_{12}) or antipernicious anaemia factor.

Q3. State the clinical significance of fat-soluble vitamins.

(a) Conditions such as steatorrhea and disorders of the biliary system affecting the digestion and absorption of the fat-soluble vitamins can lead to deficiency. These deficiencies or dietary insufficiency cause syndromes incapable of carrying out their physiologic functions.

(b) Toxicity may result from excessive intake of vitamins A and D.

(c) Vitamin A, provitamin A (β-carotene) and vitamin E can prevent cancer.

Q4. What is meant by the term retinoids?

It denotes both the natural forms and the synthetic analogs of retinol.

Q5. What are carotenoids?

β-carotene-like compounds are known as carotenoids.

Q6. How does vitamin A toxicity occur?

Vitamin A toxicity occurs when the capacity of RBP (Aporetinol binding protein) has been exceeded and the cells are exposed to unbound retinol. This occurs with excessive use of vitamin A supplements and has been noted in Arctic explorers consuming polar bear liver, polar bears being at the end of the vitamin A food chain.

Q7. Why is it said that retinol acts as a steroid hormone?

When retinol is taken up into CRBP, it is transported around the cell and binds to nuclear proteins, where it is involved in the control of the expression of certain genes. In this respect, vitamin A behaves similarly to steroid hormones.

Q8. Mention the role of retinoic acid.

(a) Retinoic acid is involved in promoting growth and differentiation of tissues.

(b) Retinoyl phosphate acts as a carrier of oligosaccharides across the lipid bilayer of the cell. It indicates that retinoic acid is involved in glycoprotein synthesis.

Q9. Name the substances that have anticancer activity.

Retinoids and carotenoids.

Q10. Enumerate two lipid-soluble vitamins that have anticancer activity.

Vitamin A and vitamin E because of their antioxidant properties.

Q11. Why is vitamin D said to be steroid prohormone?

Because vitamin D gives rise to a hormone known as calcitriol.

Q12. Show the difference between ergosterol and 7-dehydrocholesterol.

Ergosterol	7-dehydrocholesterol
(a) Occurs in plants	(a) Occurs in animals.
(b) Side chain unsaturated and contain an extra methyl group	(b) Side chain saturated and contains no extra methyl group.
(c) By ultraviolet irradiation from sunlight it is converted to **ergocalciferol (vitamin D_2)** in plant.	(c) By ultraviolet irradiation from sunlight it is converted to **cholecalciferol (Vitamin D_3)** in animals in exposed skin.
(d) Gives rise to D_2 calcitriol	(d) Gives rise to D_3 calcitriol.

Q13. How is the major form of vitamin D formed?

25-hydroxy D_3, the major form of vitamin D is formed in the liver by the hydroxylation on 25 position of vitamin D_3 by vitamin D_3-25 hydroxylase, an enzyme of the endoplasmic reticulum.

Q14. How is the steroid hormone calcitriol formed?

In the renal tubules, bone, and placenta, the 25-hydroxy D_3 is hydroxylated in position 1 by 25-hydroxy $D_3$1-hydroxylase, a mitochondrial enzyme producing 1α, 25-dihydroxy D_3 **(Calcitriol)**.

Q15. How many molecules of vitamin A are formed from one molecule of β-carotene?

Two molecules.

Q16. In what situations vitamin E deficiency is observed?

In chronic steatorrhea, abetalipoproteinemia, cholestatic liver disease, cystic fibrosis.

Q17. How can you prove that vitamin E is a natural antioxidant?

Vitamin E acts as defense against peroxidation of polyunsaturated fatty acids contained in cellular and subcellular memebrane phospholipids.

Q18. Show that vitamin E and selenium act synergistically.

(a) Selenium is a component of glutathione peroxidase which acts as defense against hydroperoxides before they can damage membranes and other cell components. Both vitamin E and selenium act against lipid peroxides.

(b) Selenium is required for normal pancreatic function which is necessary for the digestion and absorption of lipids including vitamin E.

(c) Vitamin E also reduces selenium requirements by preventing loss of selenium from the body.

Q19. What does insufficiency of vitamin E cause?

Insufficiency of vitamin E may cause anemia due to decreased production of hemoglobin and a shortened erythrocyte life span.

Q20. How can the requirement of vitamin E be increased?

By the greater intake of polyunsaturated fat.

Q21. How is vitamin E deficiencies caused?

Intake of mineral oils, exposure to oxygen, or inefficient lipid absorption due to diseases.

Q22. How is vitamin E destroyed?

By commercial cooking and food processing including deep freezing.

Q23. Mention the reactive oxygen species and their antioxidants.

	Reactive species	Antioxidant
O_2	Single oxygen	Vitamin A, β-carotene, Vitamin E.
O_2^-	Superoxide free radical	Superoxide dismutase, Vitamin E, β-carotene
OH	Hydroxyl free radical	—
RO	Alkoxyl free radical	—
ROO	Peroxyl free radical	Vitamin E, Vitamin C.
H_2O_2	Hydrogen peroxide	Catalase, glutathione, peroxidase.
LOOH	Lipid peroxides	Glutathione peroxidase.

Q24. What are the names of the vitamins belonging to the K group?

(a) Vitamins belonging to the K group are polyisoprenoid-substituted naphthoquinones.

(b) Menadione (K_3), the parent compound of the K series, is not found in nature. But if administered, it is alkylated in vivo to one of the menaquinones (K_2).

(c) Phylloquinone (K_1) is the major form of vitamin K found in plants.

(d) Menaquinone-7 is one of the series of polyprenoid unsaturated forms of vitamin k found in animal tissues and synthesized by bacteria in the intestine.

Q25. How is menadione absorbed?

Menadione being water-soluble is absorbed even in the absence of bile salts.

Q26. How does vitamin K effect the blood clotting factors?

Vitamin K is involved in the maintenance of normal levels of blood clotting factors II, VII, IX, and X, all of which are synthesized in the liver initially as inactive precursor proteins.

Q27. What is the role of vitamin K as cofactor?

Carboxylation of glutamate is catalyzed by vitamin K-dependent carboxylase.

The reaction is inhibited by dicumarol types of anticoagulant such as warfarin.

Q28. Why newborn infants have more deficiency of vitamin K rather than adults?

Because the placenta does not pass the vitamin K to the fetus efficiently and the gut is sterile immediately after birth.

Q29. What are the causes of the vitamin K deficiency?

(a) Fat malabsorption which is associated with pancreatic dysfunction, biliary disease, atrophy of the intestinal mucosa, or any cause of steatorrhea.

(b) Sterilization of the large intestine by antibiotics can cause the deficiency if the dietary intake is limited.

Q30. What is Wernicke's encephalopathy?

(a) This is associated with thiamin deficiency.

(b) It is frequently found in chronic alcoholics consuming little other food.

(c) Certain raw fishes contain thiaminase that destroys thiamin, but this is not cirtical in human nutrition.

Q31. Where is riboflavin synthesized?

Riboflavin is synthesized by plants and microorganisms but not by mammals.

Q32. Name the factors that affect the conversion of riboflavin to its cofactor forms.

(a) Hormones (eg, thyroid hormone and ACTH)

(b) Drugs (eg, chlorpromazine, a competitive inhibitor).

(c) Nutritional factors.

Q33. Why does riboflavin deficiency occur in newborn infants?

Because of the light sensitivity to riboflavin causing hyperbilirubinemia.

Q34. What is Niacin?

It is vitamin B_3 and the generic name for nicotinic acid, and nicotinamide. Nicotinic acid is a monocarboxylic acid derivative of pyridine.

Q35. What is vitamin B_6?

Vitamin B_6 consists of three closely related pyridine derivatives : Pyridoxine, pyridoxal, and pyridoxamine.

Q36. Enumerate the role of pyridoxal phosphate in glycogenolysis.

Pyridoxal phosphate is an integral part of the mechanisms of action of phosphorylase which is responsible for the breakdown of glycogen. Muscle phosphorylase contains 70–80% of total body vitamin B_6.

Q37. Mention the circumstances associated with B_6 deficiency.

(a) B_6 deficiency is recognized in nursing infants whose mothers are depleted by the vitamin owing to long-term use of oral contraceptives.

(b) Alcoholics may be deficient owing to metabolism of ethanol to acetaldehyde which stimulates hydrolysis of the phosphate of the coenzyme.

(c) Isoniazid—the antituberculosis drug—can cause vitamin B_6 deficiency by forming a hydrazone with pyridoxal.

Q38. How does consumption of raw eggs cause biotin deficiency?

Egg white contains a heat-labile protein, avidin, which combines very tightly with biotin preventing its absorption resulting in biotin deficiency.

Q39. What is intrinsic factor?

(a) It is a specific glycoportein, secreted by parietal cells of the gastric mucosa.

(b) It helps in the absorption of vitamin B_{12}.

Q40. What is folacin?

Folacin is the generic name of folic acid and related substances having the biochemical activity of folic acid.

Q41. How is folic acid composed of?

Pteridine, a base + PABA + glutamic acid.

Q42. What is the active form of folic acid?

Tetrahydrofolate (H_4 folate).

Q43. Name the inhibitor of folate reductase.

Methotrexate.

Q44. Why is vitamin C not synthesized in humans, guinea-pigs, and birds?

Due to the absence of the enzyme L-gulonolactone oxidase.

Q45. Is ascorbic acid a water-soluble antioxidant?

Vitamin C is a water-soluble antioxidant and may inhibit the formation of nitrosamines during digestion.

Q46. Why is vitamin C a reducing substance?

Since it is a hexauronic acid.

Q47. Why is vitamin B_{12} called lactoflavin?

It was first obtained from milk; hence the earlier name was lactoflavin.

Q48. When does the demand for vitamin B-complex increase?

During pregnancy and lactation, hemorrhage, injury, burn, fever, administration of antibiotics, and in under-nourished subjects.

Q49. What are the sources of vitamins?

See Chart 1 below.

Q50. Mention the daily requirements of vitamins.

See Chart 2 below.

A	D	E	K
All yellow vegetables (sweet potatoes, carrots etc), green leafy vegetables contain β-carotenes which are converted to vit. A. Cod liver oil, milk, butter, eggs.	7-dehydrocholesterol present in the skin is converted to vitamin D_3 by sunlight. Oily fish, egg yolk, butter, and liver.	Fish liver oil, wheat germ, sunflower seed, safflower seed oils, corn, and soyabean oil.	Plant and animal tissues used as food production of the vitamin by microflora of the intestine.

B_1	B_2	Niacin (B_3)	Pyridoxine (B_6)	Pantothenic acid (B_5)	Folic acid	Biotin	B_{12}	Vitamin C
Unrefined cereal grains, meat. All plant and animal tissues used as food.	Yeast, liver, kidney germinating wheat, milk.	Yeast, milk, liver kidney, fish tomato, potato etc.	Yeast, liver, egg yolk, meat, bananas etc.	Cereals, legumes, yeast, liver etc.	Yeast, liver, leafy vegetable, tomatoes etc.	Yeast, liver egg yolk, kidney, tomatoes etc.	Liver, kidney, milk, eggs, meat, fish etc.	Amla, citrous fruits, tomatoes, bananas, peppers etc.

CHART 1

A	D	E	K	B_1	B_2	Niacin (B_3)	Pantothenic acid (B_5)
Adults : 5,000 I.U. During pregnancy & lactation : 6,000-8,000 I.U. Children : 2,000-3,000 I.U. 1 I.U. = 0.3 μg of retinol.	Infants & children : 400 I.U. Adults : 200 I.U. During pregnancy & lactation : 400 I.U. 1 I.U. = 0.025 μg. of cholecalciferol.	15-30 mg.	1-2 mg.	Adults : 1.2 1.5 mg. Children : 0.7 - 1.2 mg. During pregnancy & lactation : 1.3 - 1.5 mg.	Adults : 1.5-1.8 mg. Children : 0.8-1.2 mg. During pregnancy & lactation : 1.5-1.8 mg	Adults : 16-20 mg. Children : 9-16 mg. During pregnancy & lactation : 16-20 mg	Adults : 5-10 mg. Children : 4-5 mg. During pregnancy & lactation : 10-15 mg

Pyridoxine (B_6)	Folic acid	Biotin	B_{12}	Vitamin C
Adults : 1.6-2.0 mg. Children : 0.6-1.2 mg. During pregnancy & lactation : 2-2.5 mg	Adults : 400 μg. Children : 100-300 μg. During pregnancy & lactation : 600-800 μg.	Adults : 50-60 μg. Children : 20-40 μg.	Adults : 3.0 μg. Children : 1-2 μg. During pregnancy & lactation : 4.0 μg	Adults : 45 mg. Children : 40 mg. During pregnancy & lactation : 60-80 mg

CHART 2

Q51. Name the vitamins which are synthesized in the body.

K, B_1, Niacin, Pyridoxine, Biotin, PABA, B_{12}.

Q52. Which vitamins are stable?

A, D, Niacin, B_1, B_{12} (destroyed by temperature above 100°C).

B_{12} (slight loss in ordinary cooking).

C (partly destroyed by oxidation and completely destroyed by processing).

Q53. Why is it called vitamin K?

Since it is a coagulation factor it is termed as vitamin K.

Q54. State the coenzyme activities of vitamins.

(a) **Thiamin in the form of TPP :**

(i) In oxidative decarboxylation of α-keto acids (eg, pyruvate, α-ketoglutarate, and the α-keto analogs of leucine, isoleucine, valine).

(ii) In transketolase reactions (eg, in the pentose phosphate pathway).

(b) **Riboflavin in the form of FAD and FMN :**

FAD and FMN coenzymes act as prosthetic groups of oxidoreductase enzymes. These enzymes are known as flavoproteins. Many flavoprotein enzymes contain one or more metals like molybdenum and iron as essential cofactors and are known as metalloflavoproteins.

(i) FAD or FMN sets as a prosthetic group of α-amino acid oxidase in the oxidative deamination of α-amino acid.

(ii) FAD acts as a prosthetic group of succinate dehydrogenase in TCA cycle, acyl-CoA dehydrogenase in β-oxidation of fatty acids, dihydrolipoyl dehydrogenase in the oxidative decarboxylation of pyruvate and α-ketoglutarate, xanthine oxidase in purine degradation, aldehyde dehydrogenase in the degradation of aldehydes, mitochondrial glycerol-3-phosphate dehydrogenase in transporting reducing equivalents from the cytosol into mitochondria.

(c) **Niacin (B_3) in the form of NAD^+ and $NADP^+$.**

(i) NAD^+ acts as a prosthetic group of lactate dehydrogenase in the cytosol, malate dehydrogenase within the mitochondria.

(ii) Generally, NAD-linked dehydrogenases catalyze oxidoreduction reactions in oxidative pathway (eg, the citric acid cycle).

(iii) NADP-linked dehydrogenases or reductases are found in pathways concerned with reductive syntheses (eg, the pentose phosphate pathway).

(iv) These coenzymes are the key components of metabolic pathways affecting carbohydrate, lipid, and amino acid metabolism.

(d) **Pantothenic acid (B_5) in the form of CoA :**

(i) This coenzyme is involved in the reactions of TCA cycle, fatty acid oxidation and synthesis, acetylation reactions (eg, of drugs), cholesterol synthesis.

(ii) The Acyl carrier protein (ACP) is involved in reactions concerned with fatty acid synthesis.

(iii) The structure of the free CoA is CoA.SH in which SH group is the reactive group of the coenzyme.

(e) **Pyridoxine (B_6) in the form of (B_6-PO_4) :**

(i) This coenzyme acts as a prosthetic group of transaminases in transamination reactions, dcarboxylases in decarboxylation reactions, desulphydrases in desulphydration reactions.

(ii) As a prosthetic group of kynureninase in tryptophan catabolism.

(iii) It is involved in the conversion of cystathionine to cysteine.

(f) **Biotin in the form of Biotin :**

(i) Biotin is involved in the fixation of CO_2 (eg, in the conversion of pyruvate to oxaloacetate, in fatty acid synthesis, and in citrullin formation).

(ii) It influences succinate dehydrogenase, decarboxylase, and deaminases.

(g) **Lipoic acid in the form of lipoic acid :**

It acts as a prosthetic group of dihydrolipoyl transacetylase in the conversion of pyruvate to acetyl-CoA, dihydrolipoyl transsuccinylase in the oxidation of α- ketoglutarate to succinyl-CoA.

(h) **Folic acid in the form of tetrahydrofolate (H_4 folate) :**

This coenzyme is involved in one carbon moiety, in the synthesis of purines, thymine, serine, and methionine.

(i) **B_{12} in the form of methylcobalamin and deoxyadenosylcobalamin :**

This coenzyme is involved in

(i) Transmethylation reaction.

(ii) The biosynthesis of nucleic acid.

(iii) The conversion of RNA to DNA.

(iv) The conversion of methyl malonyl-CoA to succinyl-CoA.

(j) **Vitamin C in the form of ascorbic acid with Cu^+ and Fe^{2+} :**

(i) This coenzyme is required for hydroxylation of proline in collagen synthesis.

(ii) This is involved in the oxidation of P-hydroxyphenylpyruvate to homogentisate in the degradation of tyrosine requiring the cofactor copper in the reduced state. The subsequent step is catalyzed by homogentisate dioxygenase requiring the coenzyme ascorbic acid and the cofactor Fe^{2+}.

(iii) It is required at the dopamine β-hydroxylase step in the synthesis of epinephrine from tyrosine.

(iv) It is required at the 7α-hydroxylase step in bile acid formation.

(v) The absorption of iron is enhanced by the presence of ascorbic acid.

(vi) It is also required in steroidogenesis.

(vii) It acts as general water soluble antioxidant and may inhibit the formation of nitrosamines during digestion.

(k) Vitamin K acts as cofactor of the carboxylase that forms γ-carboxyglutamate in precursor proteins.

Q55. State the deficiency manifestations of vitamins.

Vitamin A : Night blindness, xerophthalmia, keratinization, toad skin, decreased resistance to bacterial infection.

Vitamin D : (i) Rickets in children.

Symptoms : Toneless muscles, deformities of the legs, pigeon chest, frontal bossing of the skull, failure to sit and stand up.

(ii) Osteomalacia in adults.

Symptoms : Spontaneous fractures, softening of bones, pain in the ribs.

Vitamin E : Anemia, neurologic disorder, abortion in women, muscular dystrophy.

Vitamin K : Hemorrhagic disease of the new born.

Vitamin B_1 : Beriberi.

Symptoms : Loss of appetite, weakness of the legs, edema of the legs and face, palpitation and cardiovascular, neurologic and muscular degeneration.

Wernicke's encephalopathy is found in chronic alcoholics.

Vitamin B_2 : Characteristic lesions of the lips and tongue, fissures at the angles of the mouth, dermatities of the face.

Niacin (B_3) : Pellagra.

Symptoms : Dermatities of face, neck, knees, and breast, digestive disorder, weight loss, depression, and dementia (headache, forgetfulness).

Pantothenic acid (B_5) : Burning foot syndrome, inadequate growth, fatty liver, failure in gaining weight, gastrointestinal disorder.

Pyridoxine (B_6) : Anemia, retarded growth, skin lesions, failure in gaining weight.

Biotin : Depression, hallucination, muscle pain, and dermatitis.

Folic acid : Megaloblastic anemia (the neuclei of the neutrophil contain more than the normal number of lobes), retardation of growth, weakness, infertility, inadequate lactation in females.

Vitamin B_{12} : Pernicious anemia, glossitis (inflammation of the tongue), diseases of central and peripheral nervous system.

Vitamin C : Scurvy

Symptoms : Bleeding of the gum, loosening of teeth, swelling at the end of the long bones, anemia, edema, delayed wound healing, subcutaneous and internal hemorrhage.

Q56. Name the sulphur containing vitamins.

Thiamin (B_1), Biotin, Lipoic acid.

Q57. Name one vitamin which has got reducing property.

Vitamin C.

Q58. What does happen in the excess intake of vitamins?

Vitamin A : Drowsiness, severe headache, and vomiting.

Vitamin D : Nausea, vomiting, diarrhea.

Vitamin E : Nausea, Vomiting.

Vitamin K : Hyperbilirubinemia.

Excess water soluble vitamins are excreted out.

CHAPTER 11

HEMOGLOBIN

Q1. What is hemoglobin?

(a) It is a conjugated protein consisting of heme and the protein globin.

(b) Its molecular weight is 64,450.

(c) It contains 4 gram atoms of iron per mole in the ferrous (Fe^{2+}) state.

Q2. State the clinical significance of hemoglobin.

(a) It functions in oxygen binding, oxygen transport, electron transport, and photosynthesis.

(b) It provides information regarding genetic diseases such as sickle cell disease and the thalassemias.

(c) Carbon monoxide disrupts the physiologic functions of hemoglobin.

(d) Oxidation of Fe^{2+} to Fe^{3+} form destroys biologic activities.

Q3. What does myoglobin do in severe exercise?

Myoglobin of red muscle tissue stores oxygen. During severe exercise this stored oxygen is released by muscle mitochondria for the use of oxygen-dependent synthesis of ATP.

Q4. What is myoglobin?

(a) It is a single polypeptide chain of molecular weight 17,000 consisting of 153 amino acids.

(b) It is a compact spherical molecule measuring 4.5 × 3.5 × 2.5 nm.

(c) Seventy five per cent of the residues are present in 8 righthanded alpha helices from 7 to 20 residues in length.

Q5. What are the two major transport functions of hemoglobin?

(a) Transport of O_2 from the respiratory organ to the peripheral tissues.

(b) Transport of CO_2 and protons from peripheral tissues to the respiratory organ for subsequent excretion.

Q6. Show that hemoglobins are tetrameric proteins.

(a) Hemoglobins are made up of two each of two different polypeptides or monomer units (termed α, β, γ, δ, s etc.). HbA means α and β polypeptides of hemoglobin A.

(b) The tetrameric structures of common hemoglobins are HbA (normal adult hemoglobin) = $\alpha_2\beta_2$, HbF (fetal hemoglobin) = $\alpha_2\gamma_2$, HbS (Sickle cell hemoglobin) = $\alpha_2\delta_2$, HbA_2 (a minor adult hemoglobin) = $\alpha_2\delta_2$.

Q7. How does oxygenation of hemoglobin take place?

(a) Hemoglobin binds 4 oxygen molecules per tetramer.

(b) The binding of O_2 depends on whether other O_2 molecules are present on the same tetramer. If O_2 is already present, binding of subsequent O_2 molecules occurs more readily.

Q8. What is meant by P_{50}?

(a) It is the partial pressure of oxygen that half-saturates a hemoglobin.

(b) P_{50} can vary widely depending on the organism.

(c) Human fetal hemoglobin (HbF) gives a good example. For HbA, P_{50} = 26 mm. Hg; for HbF, P_{50} = 20 mm Hg. This difference allows HbF to extract oxygen from the HbA of placental blood.

Q9. Explain *Bohr effect.*

(a) The CO_2 generated in peripheral tissues combines with water to form carbonic acid

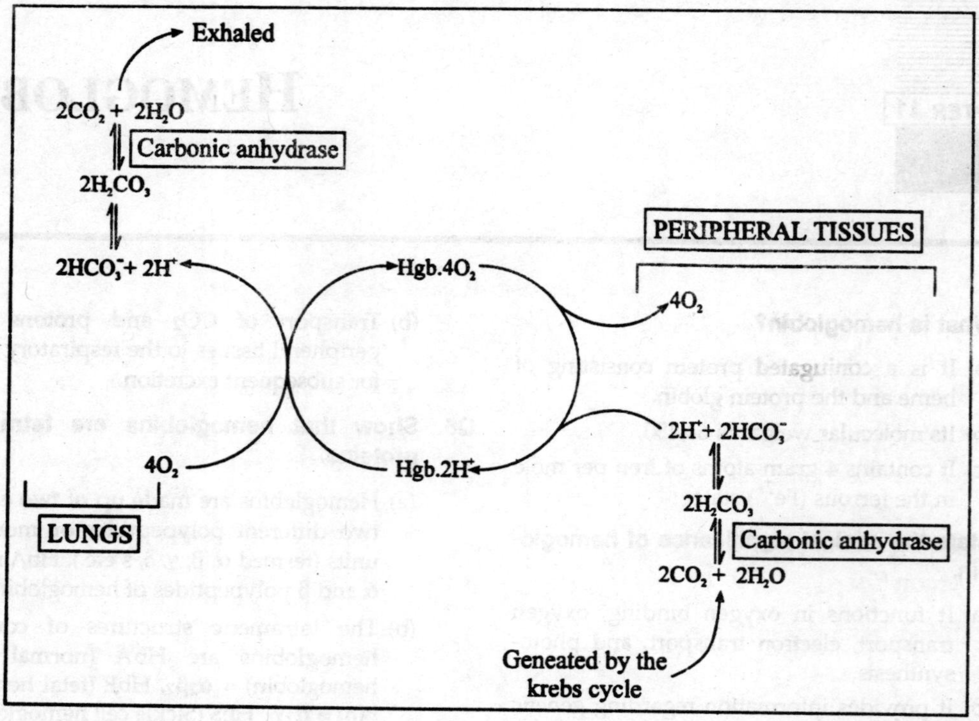

CHART 1

(H_2CO_3) which dissociates into protons and bicarbonate ions.

(b) The deoxygenated hemoglobin acts as a buffer by binding protons and delivering them to the lungs.

(c) In the lungs, the binding of O_2 by hemoglobin releases protons from the hemoglobin.

(d) The protons combine with bicarbonate ion generating carbonic acid which with the help of carbonic anhydrase becomes carbon dioxide.

(e) The carbondioxide is exhaled from the lungs.

See Chart 1 above.

Q10. What is hemoglobinopathy?

The condition in which biologic function is altered due to a mutation in hemoglobin is known as hemoglobinopathy.

Q11. What is meant by polycythemia?

Mutations that favour the R form exhibit increased oxygen affinity. Therefore, they fail to deliver adequate oxygen to peripheral tissues. The resulting tissue hypoxia leads to polycythemia (increased concentration of erythrocytes).

Q12. How is thalassemia caused?

In thalassemia, the synthesis of either the α or β chains of hemoglobin is reduced. This causes anemia which may be severe.

Q13. What is myoglobinuria?

After massive crush injury, myoglobin being released from ruptured muscle fibers apperas in the urine colouring it dark red.

Q14. Mention the normal level of hemoglobin in blood.

In males : 14.5 – 16.5 gms / 100 ml.
In females : 12.0 – 14.5 gms / 100 ml.

Q15. What is porphyrin?

A structure formed by four pyrrole rings by methylene bridges.

Q16. What is chlorophyll?

It is a magnesium containing porphyrin.

Q17. Define porphyria.

It is the condition in which excessive urinary excretion of coproporphyrin and uroporphyrin takes place.

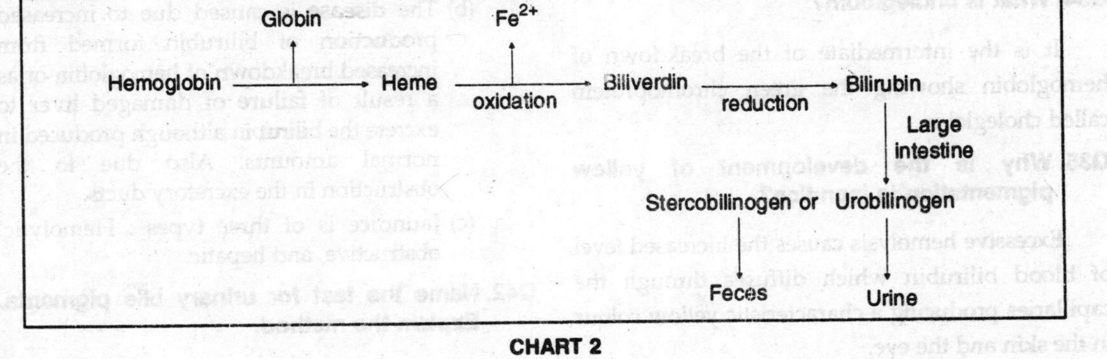

CHART 2

Q18. Explain porphyrinurias.

The conditions like cirrhosis, acute pancreatitis, and hemolytic anemias in which there is increased excretion of coproporphyrin.

Q19. What do you mean by hematin and hemarin?

Oxidized heme is said to be hematin.

Hematin chloride is said to be hemarin.

Q20. Why is hemoglobin an effective buffer?

Due to its high content of histidine.

Q21. What is methemoglobin?

Methemoglobin is formed when blood is treated with $KMnO_4$, chlorates, and nitrates.

Q22. What is meant by carboxyhemoglobin?

(a) Bubbling carbon monoxide (CO) through a solution of oxyhemoglobin results in the formation of carboxyhemoglobin (Hb.CO).

(b) It is more stable than oxyhemoglobin and produces a characteristic red solution usually described as cherry red.

Q23. What is sulphemoglobin?

(a) It is formed by passing H_2S through a solution of hemoglobin.

(b) It is a stable compound.

(c) It forms a dirty green solution.

Q24. What is cyanhemoglobin?

It is formed from methimoglobin and HCN.

Q25. What do you mean by parahematin or cathemoglobin?

It is formed when methemoglobin is denatured directly by treating with chloroform.

Q26. What is hemochromogens?

Besides globin, heme can combine with albumin, histidine, and ammonia etc. Producing a group of compounds known as hemochromogens.

Q27. What is meant by sickle cell hemoglobin?

The abnormal hemoglobin, HbS, has valine at position 6 in the β-chain in place of glutamic acid.

Q28. Mention the basic units of heme synthesis.

Succinyl-CoA and glycine.

Q29. What is erythrocruorins?

(a) They are iron porphyrinoproteins occurring in the blood and tissue fluids of some invertebrates.

(b) Their function is corresponding to hemoglobin.

Q30. What are catalases?

(a) They are iron porphyrine enzymes.

(b) Their molecular weight is about 225,000.

(c) They contain 4 gram atoms of iron per mol.

(d) Their activity is minimal in plants.

Q31. State the breakdown of hemoglobin.

See Chart 2 above.

Q32. How much hemoglobin is broken down each day?

(a) About 8 grams of hemoglobin are broken down each day in the normal adult producing 27 mg of iron.

(b) The protoporphyrin yields about 300 mg of bilirubin.

Q33. What is meconium?

The presence of biliverdin and porphyrin which causes the dark green colour of the feces of the new-born infant.

Q34. What is choleglobin?

It is the intermediate of the breakdown of hemoglobin showing the green chromoprotein called choleglobin.

Q35. Why is the development of yellow pigmentation in jaundice?

Excessive hemolysis causes the increased level of blood bilirubin which diffuses through the capillaries producing a characteristic yellow colour in the skin and the eye.

Q36. How is hemolysis caused?

(a) Caused by destruction of RBC when kept in much hypotonic solution.
(b) Caused by hemolysins present in snake venom and certain bacteria.
(c) Caused by drugs like quinine, phenacetic, and lipid solvents like ether, chloroform etc.

Q37. Explain hemophilia.

(a) It is a hemorrhagic disease of males although transmitted by females on hereditary basis.
(b) The deficiency of antihemophilic globulin cause the disease.
(c) Blood clotting time is prolonged in hemophilia.

Q38. What is meant by hemoglobinuria?

Hemoglobin being set free from RBC due to hemolysis is passed to the urine leading to this condition.

Q39. What is the normal level of serum bilirubin?

0.1 to 0.5 mg / 100 ml.

Q40. State porphyrias.

(a) It is the condition in which the excretion of both coproporphyrin and uroporphyrin is increased.
(b) Porphyrias are of two types : hereditary and familial (congenital), and acquired.

Q41. Explain jaundice.

(a) It is the disease in which there is increased level of serum bilirubin which is diffused into the tissues producing yellow pigmentation.
(b) The disease is caused due to increased production of bilirubin formed from increased breakdown of hemoglobin or as a result of failure of damaged liver to excrete the bilirubin although produced in normal amounts. Also due to the obstruction in the excretory ducts.
(c) Jaundice is of three types : Hemolytic, obstructive, and hepatic.

Q42. Name the test for urinary bile pigments. Explain the method.

Fouchet's test.

Method :

(a) Acidify 10 ml. urine with a few drops of dilute acetic acid.
(b) Add 5 ml. 10% $BaCl_2$ solution.
(c) Add 2 drops of saturated solution of $MgSO_4$.
(d) Mix and allow to stand for a few minutes.
(e) Filter and unfold the filter paper.
(f) Add one drop of Fouchet's reagent to the precipitate.
(g) The development of green colour indicates the presence of bile pigments.

Q43. What is oxygen capacity of blood?

It is the number of ml. of oxygen in 100 ml. of blood when it is fully saturated with air.

Q44. What is oxygen combining power of blood?

It is the number of ml. of oxygen combined with hemoglobin of 100 ml. of fully saturated blood.

Q45. What is meant by oxygen dissociation curve?

This curve shows relationship between partial pressure of oxygen (PO_2) and percentage saturation of hemoglobin with oxygen.

Q46. Why is oxygen dissociation curve S-shaped?

It happens so due to the combination of one molecule of hemoglobin with 1 to 4 molecules of oxygen under physiologic conditions.

Q47. What does the lower part of S-shaped curve indicate?

It indicates a fall in partial pressure of oxygen and casues a much greater dissociation of oxygen from oxyhemoglobin.

Q48. What is the percentage of oxygen saturation?

The amount of oxygen combined with hemoglobin in 100 ml. blood is expressed as percentage of the combining power of that blood.

Q49. What does happen when CO_2 is lost from blood?

Bicarbonate level decreases and chloride ions increases.

Q50. What is Gilbert's disease?

(a) It is a heterozygous group of diseases associated with unconjugated hyperbilirubinemia.

(b) Defect in the hepatic clearance of bilirubin.

(c) The bilirubin UDP-glucuronyltransferase activities in the liver are reduced.

Q51. What is Crigler-Najjar syndrome, Type I?

(a) It is a rare disorder of humans owing to primary metabolic defect in the conjugation of bilirubin due to the absence of UDP-Glucuronyltransferase in hepatic tissues.

(b) Serum bilirubin exceeds 20 mg/dl on untreatment.

Q52. What is Crigler-Najjar syndrome, Type II?

(a) Due to the milder defect in the bilirubin conjugated system.

(b) Serum bilirubin level does not exceed 20 mg/dl. All of the bilirubin is of unconjugated type.

(c) The patient with this syndrome responds to large doses of phenobarbital.

Q53. What is Lucey-Driscoll syndrome?

(a) A transient familial neonatal nonhemolytic unconjugated hyperbilirubinemia and jaundice.

(b) Healthy women can give birth to infants with jaundice.

(c) The only defect is that there is a progestational steriod derivative present in mother's blood which can inhibit the glucuronyl transferase activity.

Q54. What do you know about Dubin-Johnson syndrome?

(a) This is an autosomal recessive disorder characterized by conjugated hyperbilirubinemia and jaundice in childhood and during adult life.

(b) The defect signifies that conjugation of bilirubin takes place but there is irregularity in hepatic secretion of conjugated bilirubin in bile.

Q55. What is neonatal physiological jaundice?

(a) There is unconjugated hyperbilirubinemia and jaundice in neonates.

(b) There is accelerated hemolysis due to immature hepatic system for uptake, conjugation, and secretion of bilirubin.

(c) Glucuronyl transferase activity is also reduced.

Q56. What do you mean by bilirubin encephalopathy (Kernicterus)?

(a) It is a serious condition developing in "unconjugated hyperbilirubinemias".

(b) It is specifically seen in "hemolytic disease of the new born".

Q57. Name the drug which increases risk of Kernicterus in jaundice neonates.

Administration of sulphonamides to pregnant women in terminal stage and neonates increases the risk of "Kernicterus", since sulphonamides are bound to albumin. Hence the binding capacity of unconjugated bilirubin is developed.

Q58. Name the drugs that increase the activity of glucuronyl transferase with reasons.

(a) Aminoquinolines, chlorcyclizine, benzpyrene, and phenobarbitones increase the activity of hepatic glucuronyl transferase of normal adults and neonates.

(b) These drugs also increase the synthesis of this enzyme.

(c) Phenobarbitone is the most commonly clinically used drug.

Q59. What is the main function of BPG (2, 3-Bisphosphoglycerate).

It stabilizes the T or deoxygenated from of hemoglobin.

Q60. State the function of deoxyhemoglobin S.

The sticky patch of it can bind to the complementary patch on another deoxygenated Hbs molecule. This binding causes polymerization of deoxyhemoglobin S forming long fibrous precipitates. These extend throughout the erythrocyte and mechanically distort it causing lysis.

Q61. Mention the importance of glycosylated hemoglobin (HbA_{1C}).

(a) Normally 5 per cent glycosylated hemoglobin is proportionate to blood glucose concentration. The measurement of it thus provides useful information for the management of diabetes mellitus.

(b) The raised HbA_{1C} can guide the physician to select suitable treatment (eg, increased insulin dosage).

Q62. Point out the body secretions where bilirubin is found.

(a) Found in body fluids like CSF, fluid, cyst.

(b) Absent from true secretions like tears, saliva, pancreatic juice.

Q63. Where is unilateral jaundice found?

(a) In Patients with hemiplegia.

(b) In presence of unilateral edema.

Q64. Give examples of hemolytic jaundice.

(a) Thalassemia,

(b) Mismatched blood transfusion,

(c) Snake bite,

(d) Malaria,

(e) Rh incompatibility,

(f) Glucose-6-phosphate dehydrogenase deficiency.

Q65. Give examples of hepato-cellular jaundice.

(a) Viral hepatitis (type A, B, C commonly),

(b) Drugs—Rifampicin,

(c) Poisons—$CuSO_4$,

(d) Pregnancy—Acute fatty liver of pregnancy.

(e) Alcoholic hepatitis

(f) Weil's disease.

Q66. Give examples of obstructive jaundice

(A) Intrahepatic (medical)

(i) Viral hepatitis

(ii) Chronic active hepatitis.

(iii) Cirrhosis of liver

(iv) Drugs.

(B) Extrahepatic (Surgical)

(i) Gall stone impaction in common bile duct.

(ii) Carcinoma of head of pancreas.

(iii) Carcinoma of gall bladder.

(iv) Stricture of common bile duct.

Q67. Mention the conditions where latent jaundice is found.

Mitral stenosis, myocardial infarction, cirrhosis of liver, Pernicious anemia, congestive cardiac failure.

Q68. What do you mean by latent jaundice?

It is clinically nonevident, but only detected by serum analysis when serum bilirubin level is in between 1 to 2.5 mg/dl.

Q69. Show the difference between conjugated and unconjugated bilirubin.

	Unconjugated	Conjugated
(a) Water solubility	0	+
(b) Affinity for lipids	+	0
(c) Renal excretion	0	+
(d) Vanden Bergh reaction	Indirect	Direct

Q70. What is Van-den-Bergh test?

This test helps in differentiation of hyperbilirubinemia due to accumulation of conjugated bilirubin from unconjugated bilirubin.

Q71. How will you interpret the result of Van-den-Bergh test?

(a) Normal serum — Negative.

(b) Obstructive jaundice — Direct positive.

(c) Hemolytic jaundice — Indirect positive.

(d) Hepatocellular jaundice — Biphasic positive.

CHAPTER 12

WATER & MINERAL METABOLISM

Q1. Show the distribution of water in the body.

(a) The average body water is 60 to 70 per cent of the total body weight. Females contain a little less than males.

(b) Intracellular fluid contains 50 per cent of the total body weight.

(c) Extracellular fluid contains 20 per cent of the total body weight. This 20 per cent is further divided into the following compartments :

(i) Plasma (the extracellular fluid of the blood) : 4.5 per cent.

(ii) Interstitial and lymph fluids : 8.0 per cent.

(iii) Dense connective tissue, cartilage, and bone : 6.0 per cent.

(iv) Transcellular fluid (found in pancreas, liver, skin) : 1.5 per cent.

Q2. Mention the factors influencing the distribution of body water.

(a) Organic compounds of small molecular size (glucose, urea, amino acids etc.)

(b) Organic substances of large molecular size (proteins).

(c) Inorganic electrolytes.

Q3. Show the daily water intake and output.

Water intake in ml.		Water output in ml.	
Drinks	1,350	Lungs	500
Solid food	900	Skin	700
Oxidation of food	450	Urine	1,400
		Feces	100
	2,700		2,700

Q4. State the internal secretion of water and its fate.

(a) The continuous secretion of water in the alimentary canal in the form of digestive juices is reabsorbed along with the water of food and drink.

(b) The amount of internal secreted water is 7 to 10 litres per day. All of this amount is reabsorbed except about 100 ml. which is excreted in feces.

Q5. State the physiological functions of water.

(a) **Solvent power :** Water is the best solvent for various substances to be dissolved. Even the water-insoluble substances are made water-soluble by hydrotropic action.

(b) **Dielectric constant :** Oppositely charged particles can coexist in water. Therefore, it is a good ionized medium.

(c) **Catalytic action :** A large number of chemical reactions in the body are accelerated by water due to its ionizing power.

(d) **Lubricating action :** : Water acts as a lubricant in the body to prevent friction of joints.

(e) **Latent heat of evaporation :** Water has the highest latent heat of evaporation for which maximum cooling occurs on evaporation.

(f) **Specific heat :** More heat is required to raise the temperature of 1 gram of water through 1°C. So, the heat produced in the body by chemical reactions is utilized in water for which body temperature cannot rise.

Q6. What is dehydration?

The condition in which the body is in negative water balance due to the more or less of water than the intake is known as dehydration.

Q7. How to correct dehydration?

(a) Ordinary sodium chloride solution may be given parenterally.

(b) A mixture of $\frac{2}{3}$ isotonic saline solution and $\frac{1}{3}$ sodium lactate solution should be administered intravenously in case of prolonged diarrhea.

(c) Dehydration due to diabetes mellitus, uremia, extensive burns cannot be corrected by means of above two processes.

Q8. How is water intoxication caused?

Water intoxication is caused by the excess of water retention in the body due to renal failure, hypersecretion of ADH, and excessive administration of fluids parenterally.

Q9. Mention the symptoms of water intoxication.

Headache, nausea, and muscular weakness.

Q10. Classify mineral elements.

(a) **Principal mineral elements (macronutrients) :** Calcium, phosphorus, magnesium, sodium, potassium, sulphur, and chlorine.

(b) **Trace elements.**

(i) Essential trace elements (micronutrients) : Iron, copper, zinc, cobalt, manganese, molybdenum, chromium, selenium, fluorine and iodine.

(ii) Possible essential trace elements : Nickel, tin, vanadium, and silicon.

(iii) Nonessential trace elements : Aluminium, cadmium, germanium, lead, arsenic, boron, and mercury.

Q11. Mention the factors for the absorption of calcium.

(a) Vitamin D promotes absorption of calcium.

(b) High protein diet helps to increase the absorption of calcium.

(c) Phytic acid present in cereal grains decrease the absorption of calcium.

(d) Excess fibres in the diet interfere absorption of calcium.

(e) Fatty acids and oxalic acid decrease absorption of calcium.

(f) Calcium and phosphorous absorption is the most convenient if the ratio of these two becomes 1 : 1 or 2 : 1 or 1 : 2.

Q12. What are the fractions of calcium normally in the plasma per 100 ml?

Ionized (diffusible) calcium	2 mg.
Protein bound (Nondiffusible)	5 mg.
Complex form (as citrate and phosphate)	2 mg.

Q13. State the factors influencing blood calcium level.

(a) Vitamin D : It maintains the normal blood calcium level by helping the absorption of calcium.

(b) Plasma proteins : It maintains the normal level of blood calcium.

(c) Plasma phosphate : If the plasma phosphate concentation decreases, serum calcium level increases.

Q14. How is calcium excreted daily?

In urine	: 200 mg.
In feces	: 70 – 90 per cent of total excreted calcium.
In sweat	: 15 mg.

Q15. State the daily losses of sodium.

Urine	5 – 35 mg.
Stool	10 – 125 mg.
Skin	25 – 25 mg.
Total :	40 – 185 mg.

Q16. How is iron distributed?

See Chart 1, next page.

Q17. State the factors effecting iron absorption.

(a) High phosphate diet decreases iron absorption.

(b) Phytic acid present in cereals and oxalates interfere iron absorption.

(c) Copper deficiency causes decrease in iron absorption.

(d) Alcohol ingestion favours iron absorption.

Q18. State the iron absorption and transport mechanism.

See Chart 2, next page.

Q19. Mention the daily excretion of iron.

Adult male	0.5 – 1.5 mg. in the urine.
Women during mensturation	1.0 – 3.0 mg. in the urine.
Adult (Both male and female)	0.5 – 1.0 mg. in the sweat.
Adult (Both male and female)	0.3 – 0.75 mg. in the feces.

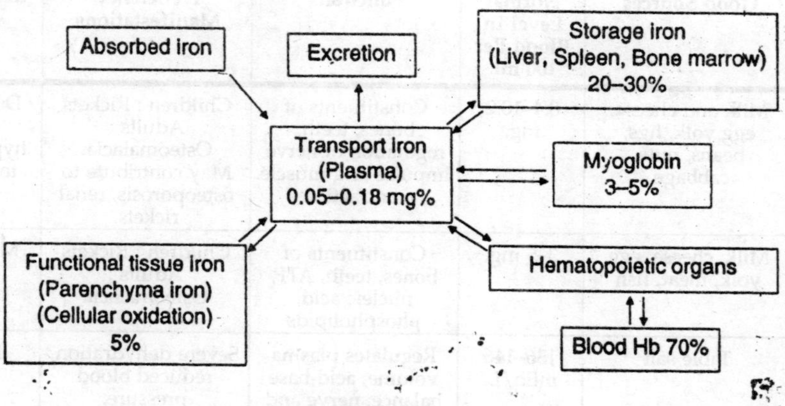

CHART 1

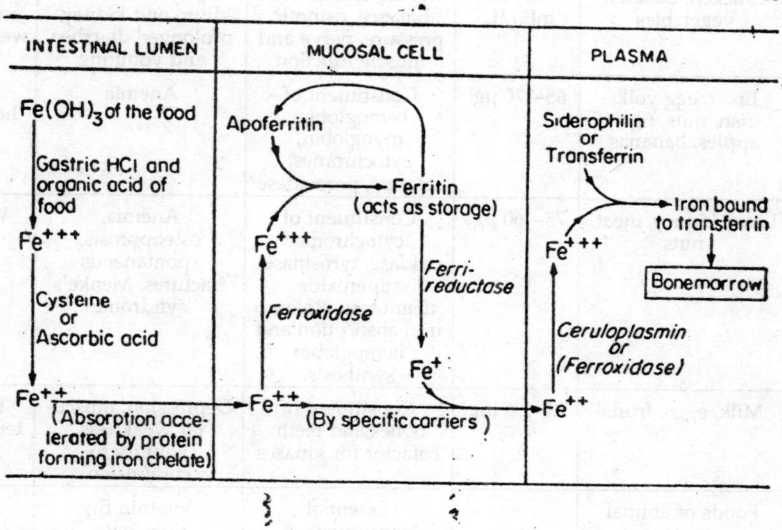

CHART 2

Q20. What is ceruloplasmin?

(a) It is copper binding protein.

(b) It has got ferroxidase activity.

Q21. What is Wilson's desease?

(a) It is a rare hereditary disorder of copper metabolism. In this diseae, absorption of copper is very high.

(b) In this disease, ceruloplasmin formation is very less and excessive deposition of copper in the liver causes hepatic cirrhosis.

(c) The symptom of this disease is defects in renal tubular reabsorption and glomerular filtration.

(d) The disease can be treated by the administration of penicillamine.

Q22. What are the goitrogenic substances in foods?

Cabbage, cauliflower, and radish contain substances 1 to 5, vinyl-2-thio-oxazolidone which react with iodine of the food and make it unavailable to the body. These substances are known are "Goitrogenic" substances.

Q23. How is fluorosis prevented?

By removing fluorides from the water by treatment with activated carbon.

Q24. Show the summary of major characteristics of some minerals.

See next page.

Minerals	Good Sources	Normal Level In Blood Per 100 ml.	Functions	Deficiency Manifestations	Toxicity Disease or Symptoms
Calcium	Milk and cheese, egg yolk, figs, beans, nuts, cabbage.	8.4–10.2 mg.	Constituents of bones, teeth; regulation of nerve impulse and muscle contraction	Children : Rickets Adults : Osteomalacia. May contribute to osteoporosis, renal rickets.	Due to hypervitaminosis D or hypercalcemia due to hyperparathyroidism.
Phosphorus	Milk, cheese, egg yolk, meat, fish	3–5 mg.	Constituents of bones, teeth, ATP, nucleic acid, phospholipids	Children : Rickets Adults : Osteomalacia	May lead to bone loss.
Sodium	Table salt	136–146 mEq/L.	Regulates plasma volume, acid-base balance, nerve and muscle function.	Severe dehydration, reduced blood pressure, circulatory failure	Hypertension.
Potassium	Chicken, banana, vegetables	3.5–4.5 mEq/L.	Regulates acid-base balance, osmotic pressure, nerve and muscle function	Injury to myocardium and kidney, prolonged diarrhea and vomiting	Cardiac arrest, mental confusion, weakness, paralysis.
Iron	Liver, egg yolk, fish, nuts, figs, apples, bananas	65–170 µg.	Constituent of hemoglobin, myoglobin, cytochromes, catalase, peroxidase	Anemia	Siderosis, hemochromatosis.
Copper	Liver, kidney, meat, nuts	75–160 µg.	Constituent of cytochrome oxidase, tyrosinase, superoxide dismutase. Role in iron absorption and hemoglobin synthesis.	Anemia, osteoporosis, spontaneous fractures. Menke's syndrome	Wilson's disease
Magnesium	Milk, eggs, fruits	1.7–2.6 mg.	Constituent of bones and teeth; cofactor for kinases	Depression, muscular weakness, liability to convulsion	Depressed deep tendon reflexes and respiration
Cobalt	Foods of animal origin	—	Essential constituent of vitamin B_{12}	Vitamin B_{12} deficiency	—
Zinc	Meat, eggs, milk, liver	50–150 µg.	Essential constituents of carbonic anhydrase, lactate dehydrogenase, alkaline phosphatase	Hypogonadism, growth failure, impaired wound healing, anemia	Gastrointestinal irritation, vomiting
Selenium	Meat, plants, but varies with soil content	—	Constituent of glutathione, peroxidase	Protein energy malnutrition, parenteral nutrition	Hair loss, cirrhosis of liver, anemia dermatitis, irritability
Iodine	Sea foods and vegetables, marine vegetation	4–10 mg.	Constituent of thyroxine, triiodothyronine	Children : Cretinism Adults : Goiter, hypothyroidism, myxedema.	Thyrotoxicosis, goiter
Fluorine	Drinking water	—	Involved in the development of bones and teeth	Dental caries	Dental fluorosis
Chromium	Meat, liver, nuts, cheese	0.009–0.055 parts per million	Constituent of glucose tolerance factors, potentiates insulin	Impaired growth malnutrition	Growth depression

Q25. Mention the daily requirements of minerals.

	Calcium (mg)	Phosphorus (mg)	Sodium (mg)	Potassium (mg)	Iron (mg)	Copper (mg)	Magnesium (mg)
Infants	360—540	240—400	250—750	500—1300	10—15	0.6—0.8	100—150
Children	800—1200	800—1200	900—2700	1500—3000	12—18	1.5—2.5	150—200
Adults	800	800	5000—15000	1800—5700	14—20	1.6—3.0	200—300
During pregnancy and lactation	1200	1200	5000—15000	1800—5700	15—24	2.0—4.0	400—600

	Cobalt (μg)	Zinc (mg)	Selenium (μg)	Iodine (μg)	Fluorine (mg)	Chromium (mg)
Infants	0.02—0.04	5	10—15	40—50	0.1—1.0	0.01—0.06
Children	0.045—0.09	10	20—30	70—120	0.5—2.5	0.02—0.2
Adults	0.045—0.09	15	60—70	140—160	1.5—4.0	0.05—0.2
During pregnancy and lactation	0.06—0.12	16—20	65—75	175—200	2.0—6.0	0.15—0.3

Q26. What is Addison's disease?

It is characterized by a high elevation of serum potassium accompanied by a high intracellular potassium of patients with renal failure, advanced dehydration or shock.

Q27. What is Hemochromatosis?

(a) Some individuals have got much capacity for the absorption of iron from the intestine (about 20 to 45 per cent iron absorbed). In such cases a very large excess (as much as 40 to 80 grams) of iron accumulates in the tissue after many years.

(b) The hemosiderosis may be accompanied by a bronze pigmentation of the skin, hemochromatosis.

(c) There may be liver damage with signs of cirrhosis, diabetes, and a pancreatic fibrosis. This condition is referred to as bronze diabetes.

Q28. What is erythrocuprein?

A copper containing protein found in the RBC.

Q29. What is Menke's disease?

(a) It is an x-linked inherited disorder of intestinal copper absorption.

(b) Transport of copper through intestinal mucosal cell membrane is defective.

(c) If not treated at birth, the disease produces mental retardation and unsatable temperature, abnormal bone formation, and susceptibility to infection.

(d) The other name of this disease is steel hair syndrome.

Q30. State Acrodermatitis enteropathica.

(a) It is a rare inherited disorder characterized by the inability to absorb zinc.

(b) The disease is also manifested by dermatologic, ophthalmologic, gastrointestinal and neuropsychiatric features alongwith growth retardation and hypogonadism.

Q31. What do you know about zinc-finger motif?

(a) The protein TF III A is the positive regulator of 5S RNA transcription. Each molecule contains 9 zinc ions.

(b) There are two types of zinc-finger motif : Cys-Cys-Zinc finger and Cys-His Zinc finger.

Q32. Mention the role of chromium in carbohydrate metabolism?

Chromium is the true potentiator of insulin and hence it is called as **glucose tolerance factor** (GTF).

Q33. What is the occupational hazard with chromium?

Hexavalent chromium is toxic and chronic occupational exposure to chromate is associated with increased risk of lung cancer.

Q34. Mention the molybdenum requiring enzymes.

(a) Aldehyde oxidase.

(b) Sulfite oxidase.

(c) Xanthine Oxidase.

(d) NADH-nitrate reductase.

Q35. State the sources of selenium.

(a) It is available in plants growing in selenium-containing soil.

(b) Its concentration in plants depends on the concetration of the soil in which the plants are grown.

Q36. Enumerate the biological forms of selenium occuring in the body.

These are selenium associated analogues of sulphur containing amino acids :

(a) Seleno - methionine,

(b) Seleno - Cysteine,

(c) Seleno - Cystine.

Q37. State the importance of selenium containing enzyme.

(a) The selenium-containing enzyme is **glutathione peroxidase** which is present in mitochondria and cytosol of the cell.

(b) This enzyme acts as an intracellular antioxidant. It reduces hydroperoxides.

$$ROOH + 2G\text{–}SH \xrightarrow[\text{(Selenium-containing)}]{\text{Glutathione peroxidase}} R.OH + H_2O + G\text{–}S\text{–}S\text{–}G$$

Q38. Mention the daily requirement of selenium.

Adults : 60 to 100 μg.

Children and infants : 10 to 30 μg.

During pregnancy and lactation : 65 to 120 μg.

Q39. State the features of selenium toxicity in humans.

(a) The features of selenium toxicity are chronic dermatitis, loss of hair, and brittle nails.

(b) Likely cause of selenium toxicity is occupational exposure to electronics, glass, and paint industries.

Q40. How can selenium toxicity be treated?

By the administration of halogenated aromatic hydrocarbons which disconnects selenium from its association with sulphur-containing amino acids.

Q41. State the role of selenium in cancer.

(a) Selenium is a cancer protective agent. High selenium content in foods and high selenium levels in blood have less occurrence of cancer.

(b) Selenium brings about changes in carcinogen metabolites.

(c) It protects from carcinogen-induced oxidant damage.

Q42. What is keshan disease?

(a) The disease is caused by selenium deficiency in humans.

(b) The manifestation of the disease is cardiomyopathy, acute or chronic cardiac enlargement, arrythmias, and ECG changes.

(c) It affects mainly children and younger women.

(d) This disease is prevalent in keshan country of north east china where the the soil is deficient in selenium.

Q43. What do you know about Kaschinbeck disease?

This disease in found in different parts of the eastern Asia and is manifested by degenerative osteoarthritis affecting children between 5 and 13 years of age.

CHAPTER 13

ENERGY REQUIREMENTS

Q1. How are foodstuffs used within the body?

Foodstuffs, by catabolism, provide energy and raw material for building up the tissues.

Q2. How is the energy utilized in the body?

(a) For the production of heat.

(b) For muscular work.

(c) For incorporating new material into new tissue in growth or maintenance.

Q3. How is the heat value of food measured?

In kilocalories.

One kilocalorie is the amount of heat required to raise the temperature of 1,000 gms of water from 15°C to 16°C. It is 1,000 times the small calorie.

Q4. How can the caloric value of a food be found?

By burning a known weight of the food in a bomb calorimeter and measuring the heat given off.

Q5. Show the caloric values of the foodstuffs.

Foods	Physiological values. (when food burnt within the body) (K. cal/gm).	Physical values (As determined by Bomb calorimeter) (K. cal/gm)
Carbohydrte	4	4.1
Fat	9	9.4
Protein	4	5.6

Q6. How is work measured?

In terms of ergs. (1 Kcal = 427 kgm = 4.19×10^{10} ergs).

Q7. What is Basal Metabolic Rate (BMR)?

The total heat production of a person lying comfortably on bed at complete mental and physical rest at least 12 hours after the previous meal. This is known as BMR.

It is usually about 40 Kcal /sq. meter/hour.

Q8. What is Du Bois Surface Area formula?

$$A = H^{0.725} \times W^{0.425} \times 71.84$$

Q9. Mention the conditions for the measurement of BMR.

(a) Post-absorptive state. (b) Awakeness of patient. (c) Mental and physical relaxation. (d) Environmental temperature must be between 20°C and 25°C. (e) Recumbent position during the test.

Q10. Name the conditions in which BMR is high or low.

High : Smaller individuals, at the age of 5, male, cold climate, good nutrition, thyroxine and epinephrine high, infectious diseases.

Low : Larger individuals, new born and old age, female, hot climates, starvation, and malnutrition.

Q11. State the significance of BMR.

(a) It is essential to know the BMR to have an idea of the total amount of food or calories required to maintain body weight or caloric balance.

(b) It is the principal guide for diagnosis and treatment of thyroid disorders.

(c) The BMR is above normal in fever, diabetes insipidus, and leukemia.

(d) In starvation and under nutrition, the BMR is below normal.

Q12. What do you mean by respiratory quotient (R.Q.)?

It is the ratio of the volume of CO_2 formed in the body to the volume of O_2 utilized by the body.

Q13. Show the significance of RQ.

In case of carbohydrate :

$$C_6H_{12}O_6 + 6O_2 = 6CO_2 + 6H_2O.$$

$$\therefore \; R.Q. = \frac{6CO_2}{6O_2} = 1$$

In case of fat :

$$2C_{57}H_{110}O_6 + 163O_2 = 114CO_2 + 110H_2O.$$

$$\therefore \text{ R.Q.} = \frac{114CO_2}{163O_2} = 0.70$$

In case of Protein : 0.80

In case of mixed diet : 0.85

Q14. What is "Protein Sparing Action"?

Carbohydrate and fat "spare" protein and thus make it available for anabolic purposes.

Q15. What do you mean by Specific Dynamic Action (SDA)?

It is the extra heat production over and above the caloric value of a given amount of food when used by the body.

SDA for protein : 30%, fat : 30%, carbohydrate : 5%, mixed diet : 10%.

Q16. How is heat lost?

(a) 80% of the total heat loss occurs through the cutaneous circulation and sweat glands.

(b) About 15% is lost by evaporation of water and CO_2 in respiration.

(c) About 5% in warming the food and inspired air.

Q17. How can man live in very high external temperature?

(a) If the air is dry and sweat is evaporated from the surface of the body, a large amount of heat can be abstracted.

(b) If the air is moist as well as hot, evaporation of sweat is slower and may not be sufficient to keep the body temperature down. In this case death ensues from heat stagnation.

Q18. What are the effects of an increase in the body temperature?

(a) Quickening of pulse and respiration, headaches, and weakness.

(b) Consciousness is lost if the temperature continues to rise and death occurs if temperature above 109°F (43°C) is persisted.

Q19. Does the metabolic rate vary with the body temperature?

(a) The rate of most chemical reaction is increased by a rise of temperature—a rise of 10°C increases the rate about 2 to 3 times.

(b) In fever in man, the metabolic rate is increased.

Q20. What is "Stoker's cramp"?

(a) Cramping pain occurs in the muscles of individuals working in hot environment when large amount of sodium containing fluid are lost in the sweat.

(b) The fluid loss is replaced by drinking plain water only.

Q21. What does cause shivering?

(a) It can occur when the body temperature is low by a painful or a cold stimulus applied to the skin.

(b) It may occur when the body temperature rises.

CHAPTER 14

NUTRITION

Q1. State the clinical importance of nutrition.

(a) Some degree of nutritional deficiency are observed in the poor or the elderly and in groups having nutritional requirements, eg, growing children, pregnant or lactating women, ill and convalescing patients, alcoholics, or patients on intravenous feeding or persons completely on vegetarian diets.

(b) Deficiencies are more prevalent in more deprived populations, eg, deficiency of protein (kwashiorkor), of vitamin A (xerophthalmia), of mineral iron (anemia), and of energy (starvation).

(c) Malabsorption causes deficiency of nutrients leading to pathologic conditions, eg, vitamin B_{12} malabsorption causes anemia.

(d) Excess intake of particular nutrients being associated with certain diseases is well recognized, eg, with atherosclerosis and coronary heart disease, diabetes, cancer of the breast and colon, cerebrovascular disease and strokes, and cirrhosis of the liver.

Q2. Enumerate the essential nutritional requirements of humans.

(a) Essential amino acids (Methionine, valine, threonine, phenylalanine, tryptophan, leucine, isoleucine, lysine, histidine.)

(b) Essential fatty acids (linoleic acid, linolenic acid, and arachidonic acid).

(c) Vitamins (water-soluble and fat-soluble vitamins).

(d) Minerals [Macrominerals, and microminerals (trace elements)].

(e) Fibers (required for optimal health).

(f) Water (The most critical component of the diet).

(g) Energy(Utilization of carbohydrates, fats, and proteins in variable proportions.)

Q3. Mention the energy intake for men and women.

Category	Age (years)	Weight [kg(lb)]	Energy needs (Mean) (kcal)
Men	23-50	70(154)	2,900
Women	23-50	55(120)	2,200
Pregnant			+300
Lactating			+500

Q4. State the factors involving expenditure of energy.

(a) The BMR (the energy expenditure) is higher in males than females, in young children, and in people with fever and hyperthyroidism, but lower in hypothyroidism and in starvation.

(b) The SDA (thermogenic effect) of food is equivalent to about 5-10% of total energy expenditure and is involved in the energy expenditure due to digestion and to any stimulation of metabolism caused by the entry of new substrate.

(c) Physical activity is the maximum affecting energy expenditure showing maximum athletic activity.

(d) In case of low environmental temperature, increased energy expenditure is caused due to shivering. The expenditure of extra energy is required in cooling in case of temperature above blood heat.

Q5. Enumerate the requirement and intake of protein and aminoacids in humans.

	Requirements (mg/kg. body weight/day)			Intake g/day.
	Infant (4-6 months)	Child (10-12 years)	Adult	Adult (70 kg) Allowance
Protein :	1,100	1,000	800	56
Animal	....	....	....	....
Vegetable	....	....	....	....
Essential amino acids :	(3-4 months)			
Histidine	28	12	10	0.70
Isoleucine	70	28	10	0.70
Leucine	161	42	14	0.98
Lysine	103	44	12	0.84
Methionine (and cysteine)	58	22	13	0.91
Phenylalanine (and tyrosine)	125	22	14	0.98
Threonine	87	28	07	0.49
Tryptophan	17	03.3	03.5	0.25
Valine	93	25	10	0.70
Data from recommended dietary allowances, 10th ed. National Academy of Sciences, 1989.				

Q6. What is the average caloric requirement of an adult male?

About 3,000 Kilocalories per day.

Q7. What are the total daily requirements of different grades of work?

Category	Kilocalories
Sedentary (Tailors, shoemakers)	2,500
Light work (Carpenters, painters)	3,000
Moderate work (Farmers, soldiers)	3,500
Heavy work (Navies, stonemasons)	4,000
Very heavy work (Lumberman)	8,000
Long bicycle races	10,000

Q8. Mention the average kilocaloric requirements of children and adolescents.

Age (years)	Kilocaloric requirements for	
	Boys	Girls
1	1,000	1,000
2-6	1,500	1,500
7-10	2,000	2,000
11-14	2,750	2,750
15-19	3,500	2,500

Q9. State the importance of fiber.

(a) Dietary fiber cannot be digested by an animal's own enzyme.

(b) In herbivores, fiber is a major source of energy after it has been digested by rumen microorganisms to acetate, propionate, butyrate which are absorbed.

(c) In humans, a high fiber-diet helps water retention during passage of food along the gut producing larger, softer feces.

(d) A high fiber diet in humans causes the reduction of cancer of the colon, cardiovascular disease, and diabetes mellitus.

(e) The more soluble fibers found in fruit lower blood cholesterol level by binding the dietary cholesterol.

Q10. What do you mean by a suitable diet?

(a) It must be capable of digestion, absorption, and assimilation.

(b) It must be palatable.

(c) It must contain sufficient protein to provide for growth, waste, and repair.

(d) It must supply sufficient energy.

(e) It must have sufficient amount of fat, carbohydrate, water, and salts in suitable proportions.

(f) It must contain ample vitamins.

Q11. Mention the minimum requirement of protein, fat and carbohydrate.

Proteins	100 gms.
Fats	100 gms.
Carbohydrates	400 gms.
∴ The requirement is 1 : 1: 4.	

Q12. What is meant by first class and second class protein?

The proteins that contain a higher proportion of essential amino acids and capable of supporting life and growth are said to be first class protein, eg, egg, milk, and meat proteins.

The proteins that contain a lower proportion of essential amino acids and cannot support life and growth are said to be second class proteins, eg, vegetable proteins.

Q13. What is PEM?

Protein-energy malnutrition (PEM) is a range of disorders of starvation and malnutrition that involve deficiencies of other nutrients such as vitamins and minerals in addition to protein. Two extreme forms are marasmas and kwashiorkor.

In marasmas, there is the deficiency of both energy and protein. In kwashiorkor, there is a deficiency in both the quality and the quantity of protein.

Q14. Why is the intake of carbohydrate so high?

(a) It is the cheapest form of food.

(b) It is palatable, readily digested, and absorbed.

(c) It ensures complete oxidation of fat.

(d) Natural forms of carbohydrate are sources of minerals and vitamins.

Q15. What do you mean by protective substances?

The protective substances in a diet are those which are essential for specific purposes other than energy production, i.e. first class proteins, mineral elements, and vitamins; lack of which may cause deficiency disease.

Q16. What are the advantages of mixed diet?

(a) It supplies everything essential to the body.

(b) Less chance of iron deficiency and less wastage of energy on specific dynamic action of food.

Q17. What is the principal function of linoleic acid?

Arachidonic acid, formed from linoleic acid, serves as a precursor of leukotrienes, prostaglandins, and thromboxanes which function as local hormones.

Q18. Mention the association between fat consumption.

(a) High fat consumption is associated with cancer of the breast and colon.

(b) Cholesterol, abundant in egg yolk, is responsible for atherosclerosis.

Q19. State the vitamin-responsive syndrome.

Vitamin	Disease	Biochemical defect
Biotin	Propionic acidemia	Propionyl-CoA carboxylase
Vitamin B_{12}	Methylmalonic aciduria	Formation of cobamide coenzyme
Folic acid	Folate malabsorption	Folic acid transport
Niacin	Hartnup disease	Tryptophan transport
Pyridoxine (vitamin B_6)	Infantile convulsions, cystathionuria, Homocystinuria	Glutamic acid decarboxylase, cystathioninase, cystathionine synthase
Thiamin	Hyperalaninemia, thiamin-responsive lactic acidosis	Pyruvate dehydrogenase

Q20. Why is a fairly large fat intake desirable?

(a) Vitamins A, D, E and K are associated with fat.

(b) It is a satisfying meal since it is slowly absorbed during five or six hours after ingestion.

(c) It helps in reducing the bulk of a diet due to its high caloric value.

Q21. What are the essential nutritive principles of human diet?

(a) Energy value.

(b) Quality and Quantity of primary foods, minerals, and vitamins.

(c) Variation in the diet.

(d) Digestibility.

(e) Cooking

(f) Psychological factor.

(g) Cost.

Q22. What do you mean by biological value of protein?

It is an expression of the presence and amounts of the various essential amino acids.

Q23. Mention the normal salt daily requirements.

Sodium chloride	: 5-15 gms.
Calcium	: 1 gm. but 2 gms for pregnant and lactation.
Phosphorus	: 1-2 gms. but 2-4 gms for pregnant and lactation.
Iron	: 5-15 mg. but 15-20 mg. for pregnant and lactation.
Iodine	: 100-150 μg but 175-200 μg for pregnant and lactation.

Q24. What is the quantity of body water?

35-40 litres.

Q25. State the effects of starvation.

(a) There is ever-increasing weakness.

(b) Pulse rate, blood pressure and RBC remain nearly normal until death is near.

(c) Output of feces is greatly reduced.
(d) Steady loss of weight affecting mainly the fatty tissues, muscles and glands.
(e) The heart and brain are practically unaffected.

Q26. What is the effect of giving protein in starvation?

The output of nitrogen is increased as well as the output of energy.

Q27. Mention the cholesterol content of some important human foods.

Foods	Cholesterol (mg per 100 gms edible portion)
Cow's milk	11
Fish	70
Chicken	60
Egg yolk	1,500
Butter	250
Liver	300
Brain	2,000
Vegetable oils	Nil

Q28. What is the percentage of linoleic acid in different edible oil?

Safflower oil	80 per cent
Sunflower oil	70 per cent.
Peanut oil	30 per cent.
Coconut oil	5 per cent.

Q29. Show the nutritional value of some common foodstuffs.

See Chart 1 below.

Q30. Mention the composition of cocoa and chocolate.

See Chart 2 below.

Q31. Show the difference between the composition of human milk and cow's milk.

See Chart 3 & Chart 4 below.

Foodstuffs	Proteins (gram %)	Fat (gram %)	Carbohydrates (gram %)	Fe (mg %)	Ca (mg %)	NaCl (mg %)	Vit. B_1 (mg %)	Vit. B_2 (mg %)	Niacin (mg %)	Vit. C (mg %)	Vit. A I.U.
Cow's milk	3.5	3.9	4.9	0	118	123	46	200	80	0	180
Buffalo's milk	4.2	8.8	5.3	0	21.2	0	0	0	0	0	162
Hen's egg	12.8	11.5	5.7	25	54	141	trace	353	0	0	882
Mutton (Goat)	14.8	10.9	0	1.8	123	141	49	219	5.3	0	28
Butter	0.6	81	0.4	0	20	1764	0	0	0	0	2716
Rice	7.5	1.7	77.0	2.8	39	88	198	53	1.8	0	0
Atta	9.3	2.6	49.0	7.4	56	116	451	120	4.9	0	60
Musoor dal	26.1	0.7	60.1	21	134	88	444	0	1.4	0	250
Beans	1.1	0	2.5	0.7	46	0	71	46	0.4	11	120
Onion	1.1	0	11.6	0.7	169	0	35	11	0.4	11	0
Apple	0.3	0.4	14.9	1.4	6	0	116	28	0.4	4	trace
Banana	1.2	0.2	23.0	0.4	8	282	78	32	0.4	4	trace

CHART 1

	Protein (gm %)	Fat (gm %)	Carbohydrate (gm %)	Calories per 100 gms.
Cocoa	18.1	26.8	40.3	489
Chocolate	4.8	31.1	59.9	555

CHART 2

	Solids (%)	Water (%)	Proteins (gm %)	Fats (gm %)	Carbohydrate (gm %)	Calcium (mg %)	Phosphorus (mg %)	Magnesium (mg%)	Sodium (mg %)	Potassium (mg %)	Chloride (mg%)
Human milk	12.5	87.5	1 to 2.5	1.0 to 8.0	4.5 to 8.0	40	30	5	15	60	40
Cow's milk	13.5	87.0	2.0 to 6.0	1.5 to 6.5	2.0 to 6.0	120	90	20	50	140	110

CHART 3

	Vit. A (μg %)	Vit. D (I.U.)	Vit. C (mg %)	Vit. B_1 (μg %)	Vit. B_2 (μg %)	Niacin (μg %)	Vit. B_6 (μg %)	Pantothenicacid (μg %)	Calories (per 100 gms)
Human milk	50	5.0	4.5	15	45	180	10	200	67
Cow's milk	35	2.5	2.0	45	200	80	50	350	69

CHART 4

Q32. What is colostrum?

The secretion of the lactating mammary glands during the first few days of lactation (first 4 to 5 days) after parturition is called colostrum.

Q33. What are the effects of exercise?

(a) Exercise helps to strengthen pelvic muscles and increase the efficiency of the chest and abdominal muscles during labour.

(b) Exercise prevents the physiologic changes that result from ageing. Regular exercise of the joints avoids stiffness and atrophy of the muscles.

(c) It has got the main role in weight reduction.

(d) It also helps in eliminating obesity, high blood pressure which are the causes of heart diseases.

Q34. Mention the effects of alcohol.

(a) Alcohol damages the liver, stomach, mouth, and throat.

(b) It increases blood pressure and invites psychiatric disorder.

(c) The vital organs of the body are damaged by excessive drinking.

(d) Serious accidents may occur at work and on the journey.

Q35. State the effect of cooking.

(a) Inedible portions are removed.

(b) Harmful bacteria and organisms are destroyed.

(c) Collagen fibres of meat are converted to gelatin for easy digestion.

(d) It increases the taste and flavour of the food by which salivary secretion containing salivary amylase is increased.

(e) It increases the water content and digestibility of vegetables.

(f) Fats are little changed in the process of cooking.

(g) Vitamin B_1 and C are specially destroyed on cooking.

Q36. Show the chemical composition of tea and coffee.

Tea : Theophylline (1, 3-dimethylxanthine).

Coffee : Caffeine (1, 3, 7-trimethylxanthine).

Neither tea nor coffee can be regarded as food unless taken with milk and sugar.

Tea after repeated boiling forms tannic acid which prevents appetite.

Q37. What do you mean by a balanced diet?

A balanced diet is composed of all the necessary substances required for the growth and maintenance. These substances are protein, fat, carbohydrate, vitamins, minerals, and water.

Q38. What is the effect of giving carbohydate and fat in starvation?

Carbohydrate reduces nitrogen excretion and fat increases energy yield.

CHAPTER 15

BLOOD, LYMPH AND C.S.F.

Q1. Show the clinical aspects of blood.

(a) Anemias result in the reduction of the number of RBCs and their content of Hb.

(b) Some of the blood group systems, present on the membranes of RBCs and other blood cells, are of extreme importance in relation to blood transfusion and tissue transplantation.

(c) Neutrophils play a central role in acute inflammation and other WBCs, such as lymphocytes, play important roles in chronic inflammation.

(d) Leukemias can affect precursor cells of any of the major classes of WBCs. Combination chemotherapy, using combination of various chemotherapeutic agents, has been remarkably effective in the treatment of certain types of leukemias.

Q2. State the structure of RBC.

(a) It is essentially composed of a membrane surrounding a solution of Hb.

(b) There are no intracellular organelles, such as mitochondria, lysosomes, or golgi apparatus. Cells are nonnucleated.

(c) It is not metabolically inert, ATP, synthesized from glycolysis, is important in processes that help the RBC to maintain its biconcave shape and also in the regulation of the transport of ions and of water in and out of the cell.

(d) The biconcave shape increases the surface-to-volume ratio of the RBC, thus facilitating gas exchange.

(e) It contains cystoskeletal components that play an important role in determining its shape.

Q3. Mention the characteristic of RBC.

The normal level of RBC	
in adult males	: 4.5 – 6.2 millions/μl.
in adult females	: 4.0 – 5.5 millions/μl.
The total number of RBCs in the circulation	: 2.5×10^{13}.
The life span of normal RBC	: 120 days
The normal level of Hb	
for males	: 14.5 – 16.5 grams/100 ml.
for females	: 12.0 – 14.5 grams/100 ml.
The hematocrit values	
for men	: 42 – 52 %
for women	: 37 – 47 %

Q4. State the causes of some important disorders affecting RBCs.

Disorder	Major Cause
(a) Iron deficiency anemia	Inadequate intake or excessive loss of iron.
(b) Sickle cell anemia	Sequence for codon 6 of the β chain changed from GAG to GTG in the sickle cell gene resulting in substitution of valine for glutamic acid.
(c) α-thalassemias	Mutations in the α-globin genes, mainly unequal crossing-over and large deletions and less commonly nonsense and frameshift mutations.
(d) β-thalassemias	A very wide variety of mutations in the β-globin gene including deletions, nonsense, and frameshift mutations.
(e) Glucose-6-phosphate dehydrogenase (G6PD) deficiency	A variety of mutations in the gene for G6PD, mostly single point mutations.
(f) Pyruvate kinase (PK) deficiency	A variety of mutations in the gene for the R (red cell) isozyme of PK.
(g) Megaloblastic anemias :	
(i) Deficiency of Vitamin B_{12}	Decreased absorption of B_{12} often due to a deficiency of intrinsic factor.

Disorder	Major Cause
(ii) Deficiency of Folic acid	Decreased intake, defective absorption, or increased demand for folate.
(h) Methemoglobinemia	Intake of excess oxidants (various chemicals and drugs). Genetic deficiency in the NADH-dependent methemoglobin reductase system. Inheritance of HbM.

Q5. What is erythropoietin?

(a) Human erythropoietin (EPO) is a glycoprotein of 166 amino acids having molecular weight 34,000.

(b) It is the major regulator of human erythropoiesis.

(c) It is synthesized mainly by the kidney and is released in response to hypoxia into the blood stream in which it travels to the bone-marrow.

(d) The major use of recombinant erythropoietin (rEpO) is in the treatment of a small number of anemic states, such as that due to renal failure.

Q6. What is the composition of blood?

(a) Proteins (Albumin, globulin, fibrinogen, glycoproteins, lipoproteins, enzymes, hormones).

(b) Lipids.

(c) Organic and inorganic substances.

Q7. State the important aspects of the metabolism of the red blood cell.

(a) Glucose is the energy source of RBC whose membrane contains high affinity glucose transporters.

(b) Glycolysis is the site of production of ATP. There is no production of ATP by oxidative phosphorylation since there are no mitochondria in RBCs.

(c) The RBC has a variety of transporters that maintain ionic and water balance.

(d) The pentose phosphate pathway is operative in the RBC. Hemolytic anemia is caused due to the deficiency of the activity of glucose-6-phosphate dehydrogenase.

(e) RBC can synthesize GSH and the reduced glutathione is important in the metabolism of RBC.

(f) The RBC contains certain enzymes of nucleotide metabolism. Deficiencies of these enzymes cause hemolytic anemia.

Q8. Mention some properties of the glucose transporter of the membrane of the RBC.

(a) It exhibits specificity for glucose.

(b) Five distinct glucose transporter have been detected in mammalian tissues, of which the red cell transporter is one.

(c) It is not dependent upon insulin.

(d) It transports glucose when inserted into artificial liposomes.

(e) It contains 12 transmembrane helical segments.

Q9. What is the effect of reticulocytes in protein synthesis?

(a) The mature red blood cells cannot synthesize protein.

(b) Reticulocytes are active in protein synthesis.

(c) The reticulocytes, if they enter the circulation, they loose their intracellular organelles (ribosomes, mitochondria etc.) within about 24 hours becoming young RBCs and loosing their ability to synthesize protein.

Q10. How does the deficiency of glucose-6-phosphate dehydrogenase cause hemolytic anemia?

Mutations in the gene for G6PD

↓

Decreased activity of G6PD

↓

Decreased level of NADPH

↓

Decreased regeneration of GSH from GSSG by glutathione reductase (which uses NADPH)

↓

Oxidation due to decreased levels of GSH and increased levels of intracellular oxidants and of membrane proteins

↓

Hemolysis

Q11. Mention the biochemical information regarding the membrane of the human red blood cell.

(a) The membrane is a bilayer composed of 50% lipid and 50% protein.

(b) The phospholipids and cholesterol are present on the membrane. The phospholipids are phosphatidyl choline (PC), phosphatidyl ethanolamine (PE), phosphatidyl serine (PS) and sphingomyelin.

(c) Many of the proteins are glycoproteins.

Q12. Show the nature of methemoglobinemia.

(a) The inherited type is due to deficient activity of methemoglobin reductase.

(b) Ingestion of certain drugs or chemicals can cause acquired methemoglobinemia.

Q13. What is hereditary elliptocytosis?

(a) It is a genetic disorder that is similar to hereditary spherocytosis except that affected RBCs assume an eliptic, dislike shape.

(b) It is also due to abnormalities in spectrin.

Q14. What is blood group?

It is the defined system of red blood cell antigens (blood group substances) controlled by a genetic locus having a variable number of alleles (eg, A, B and O in the ABO system).

Q15. What is H substance?

(a) It is the precursor of both the A and B substances and it is the blood group substance found in persons of type O.

(b) H substance is formed by the action of a fucosyltransferase.

Q16. Show that neutrophils are key players in the body's defence against bacteria.

(a) Neutrophils are mobile phagocytic cells that play a key role in acute inflammation.

(b) When bacteria enter tissues, a number of phenomena result that one collectively known as the acute inflammatory response. They include

(i) increase of vascular permeability.

(ii) entry of activated neutrophils into the tissues.

(iii) activation of platelets.

(iv) spontaneous subsidence of the invading microorganisms have been dealt with successfully.

Q17. What is the effect of mutations in the genes for components of the NADPH-Oxidase system?

Mutations in genes for the protein components (ie., cytochrome b_{558} and 2 cytosolic polypeptides) of the NADPH-oxidase system.

↓

Diminished production of superoxide ion and other active derivatives of oxygen.

↓

Diminished killing of certain bacteria

↓

Recurrent infections and formation of tissue granulomas in order to wall of surviving bacteria

Q18. State the function of myeloperoxidase.

(a) The enzyme myeloperoxidase, present abundantly in neutrophil granules, can act on H_2O_2 to produce hypohalous acids :

$$H_2O_2 + X^- + H^+ \xrightarrow{\text{Myeloperoxidase}} HOX + H_2O$$

[X^- = Cl^-, Br^-, I^- or SCN^-

HOCl = Hypochlorous acid]

(b) HOCl, the active ingredient of household liquid bleach, is a powerful oxidant and is highly microbicidal.

Q19. Enumerate the actions of Proteinases of neutrophils.

(a) The proteinases of neutrophils can hydrolyze elastin, various types of collagens and other proteins. If such enzymatic action is allowed to proceed unopposed, can cause serious damage to tissues.

(b) Most of these proteinases are lysosomal enzymes and exist mainly as inactive precursors in normal neutrophils.

(c) The action of proteinases are checked by antiproteinases.

Q20. What is the impact of recombinant DNA (rDNA) technology on hematology?

(a) The bases of thalassemias and of many disorders of coagulation have been greatly clarified using cloning and sequencing.

(b) The study of oncogenes has given understanding of the leukemias.

(c) Deficiency of adenosine deaminase, which affects lymphocytes particularly, is the first disease to be treated by gene therapy.

Q21. What is the P^H of blood?

7.4.

Q22. What is plasma and serum?

Plasma : Fibrinogen is separated from uncoagulated blood by centrifugation.

Serum : The clear liquid after the centrifugation of clotted blood.

Q23. Mention the importance of viscosity of blood.

Determines blood pressure.

Q24. Where are antibodies formed?

Liver, splen, bone marrow, and lymph nodes.

Q25. Show the main clotting mechanism of blood.

Prothrombine + thromboplastin + Ca^{++} → Thrombin.

Thrombin + Fibrinogen → Fibrin (clot).

Q26. Name the anti-coagulants of blood.

(a) Oxalates of potassium or ammonium.

(b) Fluoride and citrates of sodium.

(c) Heparin.

Q27. Indicate the varieties of globulins.

α_1, α_2, β and γ.

Q28. What is CSF?

(a) It is a clear, colourless fluid having low viscosity, and does not coagulate.

(b) It is formed by the choroid plexus and finally passes to the subarachnoid space.

(c) In meningitis, the glucose content of it is decreased.

(d) Its specific gravity is 1.006 to 1.007.

Q29. Mention the normal levels of different constituents of CSF.

Appearance	: Clear and colourless
Specific gravity	: 1.003 – 1.008
Cells	: < 5/mm^3, all lymphocytes
pH	: 7.34 – 7.43
Pressure	: 50 – 180 mm H_2O
Glucose	: 40 – 70 mg/dL or 60% of serum glucose
Chloride	: 118 – 132 mEq/L
Protein, Lumber	: 15 – 45 mg/dL.
Albumin	: 58%
α_1-globulins	: 9%
α_2-globulins	: 8%
β-globulins	: 10%
γ-globulins	: 10%
Protein, Cisternal	: 15 – 25 mg/dL
Protein, Ventricular	: 5 – 15 mg/dL

Q30. What is synovial fluid?

It is a clear, light yellow, viscous fluid which serves to lubricate the articular surfaces of joints.

Q31. Mention the compostion of synovial fluid.

Cells	: 200 cells/mm^3
Polymorphonuclear cells	: 25%
Fibrin clot	: None
Crystals	: None
Glucose	: Same as serum
Hyaluronic acid	: 2.45 – 3.97 g/L
pH	: 7.32 – 7.64
Protein	: 2.5 g/dL
Albumin	: 63%
α_1-globulins	: 7%
α_2-globulins	: 7%
β-globulins	: 9%
γ-globulins	: 14%
Uric acid	: Same as serum

Q32. State the clinical aspects of CSF.

(a) Glucose completely disappears in CSF in coccal meningitis due to meningococci, staphylococci, pneumococci etc.

(b) Glucose content varies from 10 to 40 mg per 100 ml. in tuberculous meningitis.

(c) In viral meningitis, glucose concentration is often normal but occasionally as low as 20 mg/100 ml.

(d) Small increases of glucose are found in encephalitis, poliomyelitis, and cerebral abscess.

(e) Protein level of CSF is markedly increased in acute meningitis, polyneuritis, tumor, multiple myelomatosis.

(f) Decreased level of CSF chloride is found in meningitis. The reduction is more marked in tubercular meningitis.

(g) Increased level of CSF chloride is found in hyperchloraemic acidosis in chronic renal failure.

(h) Copper level in CSF is increased in Wilson's disease.

Q33. What is semen?

(a) Semen is a suspension of spermatozoa in a seminal plasma which is a mixture of secretions from the prostate, seminal vesicles, urethral glands, Cowper's glands.

(b) It is viscous and clots readily due to the presence of fibrinogen but the clot is rapidly broken down by fibrinolysin of prostatic fluid.

(c) It contains sodium, potassium, proteins, lipids, cholesterol, glucose (trace), fructose, ascorbic acid etc.

(d) One ml. semen contains 40-100 million spermatozoa.

(e) The seminal vesicles are also the main source of the prostaglandins found in the semen.

CHAPTER 16

CANCER, ONCOGENES, AND GROWTH FACTORS

Q1. How does radiant energy cause cancer?

(a) Ultraviolet rays, X-rays, and γ-rays cause single and double-strand breaks or cross-linking of strands. Briefly speaking, damage to DNA is the basic mechanism of carcinogenicity with radiant energy.

(b) Besides the direct effect of DNA, X-rays and γ-rays cause the formation of free radicals in tissues. The resultant OH, superoxide, and other radicals can interact with DNA and other macromolecules leading to molecular damage showing the carcinogenic effects of radiant energy.

Q2. Show that many chemicals are carcinogenic.

(a) The organic carcinogens (MAB, dimethylnitrosamine, diethylstibesterol, afflatoxin B_1) are found to interact directly with target molecules (direct carcinogens) but others require prior metabolism to become carcinogenic (procarcinogens).

Procarcinogen → Proximate carcinogen → Ultimate carcinogen.

The enzymes monooxygenases and transferases responsible for metabolic activation of procarcinogens are mainly species of cytochrome P-450, located in the endoplasmic reticulum.

(b) The chemical carcinogens bind covalently to cellular macromolecules, including DNA, RNA, and proteins.

(c) These organic and inorganic carcinogens are found to interact with the purine, pyrimidine, or phosphodiester groups of DNA. The most common site of attack is guanine.

Q3. Is DNA the critical macromolecule in carcinogenesis?

(a) Cancer cells beget cancer cells; i.e., the changes responsible for cancer are transmitted from mother to daughter cells.

(b) Both irradiation and chemical carcinogens damage DNA.

(c) Many tumor cells exhibit abnormal chromosomes.

(d) Purified DNA (oncogenes) from cancer cells can transform normal cells into cancer cells.

Q4. Are some DNA and RNA viruses carcinogenic?

(a) Polyomavirus and SV40 viruses are involved in oncogenesis. Under certain circumstances, infection caused by these viruses can result in malignant transformation.

(b) Epstein-Barr virus is associated with Burkitt's lymphoma and nasopharyngeal carcinoma in humans.

(c) Hepatitis B virus is associated with some causes of liver cancer in humans.

Q5. Establish that oncogenes play a crucial role in carcinogenesis.

(a) Many normal cells contain protein-tyrosine kinase activity. The amount of phosphotyrosine is low in most normal cells but is usually elevated in cells transformed by an oncogenic virus containing a protein-tyrosine kinase.

(b) Viruses incorporated cellular genes into their genomes during their passages through cells. The genes present in normal cells are said to be proto-oncogenes, and their products are believed to play important roles in normal differentiation and other cellular processes.

Q6. Mention the different mechanisms by which proto-oncogenes are activated to oncogenes.

(a) Promoter insertion.

(b) Enhancer insertion.
(c) Chromosomal translocations.
(d) Gene amplification.
(e) Single-point mutation.

Q7. Show some polypeptide growth factors, their sources and functions.

Growth factor	Source	Function
Erythropoietin	Kidney, urine	Regulates development of early erythropoietic cells.
Epidermal growth factor (EGF)	Mouse salivary gland	Stimulates growth of many epidermal and epithelial cells.
Nerve growth factor (NGF)	Mouse salivary gland	Tropic effect on sympathetic and certain sensory neurons.
Platelet-derived growth factor (PDGF)	Platelets	Stimulates growth of mesenchymal and glial cells.
Insulin like growth factors 1 and 11	Serum	Stimulate sulfate incorporation into cartilage and exert insulin like effects on many cells.
Transforming growth factor (TGF-β)	Kidney, platelets	Exerts both stimulatory and inhibitory effects on certain cells.

Q8. What do you mean by endocrine, paracrine, and autocrine?

Endocrine : Like hormones, the growth factors are synthesized elsewhere in the body and pass in the circulation to their target cells.

Paracrine : The cells that synthesize the growth factor are not themselves affected due to the lack of suitable receptors. This mode of action is called paracrine.

Autocrine : Growth factors can affect the cells that synthezise them. This mode of action is called autocrine.

Q9. What is P^{53}?

(a) It is a tumor suppressor gene of mol. weight 53,000.

(b) It is located on short arm of chromosome 17.

(c) It binds various viral proteins forming inactive oligomeric complexes.

(d) It acts as transcriptional regulator, presumably regulating genes that participate in cell division.

Q10. State the major features of P^{53} mutations found in human tumors.

(a) Mutations in the P^{53} gene are the most common genetic alterations in human cancer and are frequent in colon, breast, and lung cancers.

(b) This gene contains about 100 different mutations.

(c) The mutational spectrum differs among different cancers.

(d) Transversions are frequent in lung cancer and liver cancer.

Q11. Enumerate the biochemical changes often found in fast-growing tumor cells.

(a) Increased activity of ribonucleotide reductase.

(b) Increased synthesis of RNA and DNA.

(c) Decreased catabolism of pyrimidines.

(d) Increased rates of aerobic and anaerobic glycolysis.

(e) Synthesis of fetal proteins (eg, carcinoembryonic antigen).

(f) Loss of differentiated biochemical functions (eg, diminished synthesis of specialized proteins).

(g) Inappropriate synthesis of certain growth factors and hormones.

Q12. What do you mean by resistance?

The drugs are effective initially to cancer chemotherapy but after several months these drugs become ineffective by the mechanisms developed by the tumor cells producing acquired resistance.

Q13. Mention some drugs used in cancer chemotherapy.

Class of compound	Example	Site of action	Treatment use
Sex homones	Estrogens	Block effects of androgens in prostatic tumor	Cancer of the prostate
Corticos-teroids	Predni-sone	Inhibits proliferation of lymphocytes	Myeloma
Folate antago-nists	Metho-texate	Inhibits dihydrofolate reductase	Choriocar-cinoma
Pyrimi-dine antago-nists	Fluorou-racil	Inhibits thymidilate synthetase	Colorectal cancer

Class of compound	Example	Site of action	Treatment use
Purine antagonists	Mercaptopurine	Inhibits purine synthesis	Acute myelocytic leukemia
Alkylating agents	Melphalan	Alkylates DNA and other molecules	Myeloma.
Other agents	Cisplatin	Causes strand breakage in DNA	Carcinoma of the lung.
Antitumor antibiotics	Doxorubicin	Intercalates in DNA and stabilizes DNA-topoisomerase II complex	Hodgkin's disease.

Q14. What is multidrug resistance (MDR)?

Resistance is not only to the drug but also to other structurally unrelated anticancer agents; eg, if resistance to methotrexate develops, the tumor cells may also be found to be resistant to antitumor antibiotics such as doxorubicin and to plant compounds such as vincristine. This is known as multidrug resistance (MDR).

Q15. What is meant by P-glycoprotein?

(a) Phosphorylated glycoprotein of mol.wt. 1,70,000 present in the plasma membrane contains 1,280 amino acids.

(b) It acts as energy-dependent efflux pump expelling a variety of drugs and thus mediating MDR.

(c) It is the member of multigene family.

(d) It is also present in some normal cells of kidney, gut, etc.

(e) It can also be induced by other stimuli such as heat shock.

(f) It can be inhibited by chemosensitizers such as verapamil.

Q16. What do you mean by metastasis?

It is the spread of cancer cells from a primary site of origin to other tissues where they grow as secondary tumors. It is a complex phenomenon to analyze in humans. Many changes are found at the surfaces of malignant cells, although not all are directly related to the problem of metastasis. Metastasis occurs due to the changes in the oligosaccharide chains of cells glycoproteins.

Q17. State the changes that have been detected at the surfaces of malignant cells.

(a) Appearance of new antigens.

(b) Loss of certain antigens.

(c) Alterations of surface charge.

(d) Alterations of permeablity.

(e) Alterations in transport properties.

(f) Diminished adhesions.

(g) Increased agglutinability by many lectins.

(h) Changes of glycolipid constituents.

(i) Alterations of the oligosaccharide chains of glycoproteins.

Q18. What are tumor markers?

Many cancers are associated with the abnormal production of enzymes, proteins, and hormones which can be measured in plasma or serum. These molecules are known as tumor markers.

Q19. Mention the clinically useful tumor markers.

Marker	Associated cancer
Carcinoembryonic antigen (CEA)	Colon, lung, breast, pancreas.
Alfa-fetoprotein (AFP)	Liver, germ cell.
Human chorionic gonadotropin (hCG)	Tropoblast, germ cell.
Calcitonin (CT)	Thyroid (medullary carcinoma).
Prostatic acid phosphatase (PAP)	Prostate

Q20. Show the applications of tumor markers.

Process	Conditions
Detection	Screening in asymptomatic persons.
Diagnosis	Differentiating malignant from benign conditions.
Monitoring	Predicting effect of therapy and detecting recurrent cancer.
Classification	Choosing therapy and predicting tumor behaviour.
Staging	Defining extent of disease.
Localization	Nuclear scanning of injected radioactive antibodies.
Therapy	Cytotoxic agents directed to marker containing cells.

Q21. How will you explain cancer?

(a) Cancer is a cellular tumor that can metastasize and can attack the surrounding and distant tissues.

(b) Cancer cells are characterized by diminished control of growth, the capacity to invade local tissues, and capacity to spread to the distant parts of the body by metastasis.

Q22. Mention the morphological changes shown by cancer cells.

(a) Cells are round shaped and bigger than normal cells.

(b) Cells show hyperchromatism, altered nuclear and cytoplasmic ratio, extreme mitosis.

(c) Cells form multilayers.

Q23. Name the agents that cause cancer.

(a) **Physical** : Radiant energy.

(b) **Chemicals** : Different chemical compounds. Some act directly and others act as procarcinogens.

(c) **Biological** : Oncogenic viruses.

Q24. Enumerate some cancers that are found commonly in children.

(a) Retinoblastoma.

(b) Neuroblastoma.

(c) Wilms' tumor

(d) Certain tumors like lymphomas and leukemias.

(e) Sarcomas of bones and skeletal muscles.

Q25. What do you mean by metabolic activation?

It is the process by which a procarcinogen is converted to **ultimate carcinogen** by enzyme-catalyzed reactions in the body.

Q26. What do you know about HTLV-1 virus?

(a) This virus associates with human leukemias and lymphomas.

(b) It contains a segment called **tat**. The proteins coded by "tat" gene is responsible for transformation. They affect the transcription of certain growth factors and receptors such as IL-2 and IL-2R.

Q27. Explain the chromosomal translocation found in Burkitt's Lymphoma.

(a) Burkitt's Lymphoma is a fast growing cancer of human B Lymphocytes.

(b) The reciprocal translocation in this lymphoma takes place between chromosomes 8 and 14 in 90 per cent of cases and 8 and 2 or 8 and 22 in 10 per cent of cases.

Q28. What is philadelphia chromosoma?

The philadelphia chromosome involves chromosomes 9 and 22 in chronic granulocytic leukemia.

Q29. Mention some cancers produced by loss of antioncogenes.

(a) Retinoblastoma.

(b) Wilm's tumors of kidney.

(c) Oat cell carcinoma of lungs.

(d) One type of breast cancer.

Q30. What is Rb gene?

(a) It is an antioncogene located on chromosome 13 q 14.

(b) Normally, it throws an inhibitory effect.

Q31. What do you know about interferons?

(a) Interferons are cellular glycoproteins formed immediately after the virus infects the cells.

(b) Its molecular weight is about 20,000 to 25,000.

(c) Inactivated viruses, fungal extracts, and synthetic polynucleotides etc. are the inducers for interferons namely IFN-α, IFN-β, and IFN-γ. IFN-α is produced by utilizing recombinant DNA technology and is used therapeutically.

Q32. Explain the mechanism of action of interferons.

(a) Interferons induce an enzyme called oligo-2′5′ adenylate synthetase which activates endonuclease that degrades the viral mRNA.

(b) It inactivates IF-2 required for protein synthesis.

Q33. Mention the functions of interferons.

(a) It has antiviral action which inhibits multiplication of viruses.

(b) It has also anticancerous action.

Q34. Enumerate the therapeutic uses of interferons.

(a) Interferons are used in multiple myeloma, low grade non-Hodgkin's lymphoma, and hairy-cell leukemia (HCL).

(b) They are used in myelo proliferative disorders (MPD) and in regressing hemangiomas of infants and yound adults.

(c) They are also used in solid tumors like Kaposi's sarcoma, renal cell carcinoma, breast cancer.

(d) They are used in connective tissue disorders and chronic neurological diseases.

Q35. Mention some natural anticancer substances.

(a) β-carotene and vitamin A.

(b) Vitamin E.

(c) Lycopene (a carotenoid present in ripe tomatoes).

(d) Vitamin C

(e) Selenium and zinc.

Q36. What are telomeres?

(a) Telomeres are composed of DNA and proteins that are located at the ends of chromosomes of lower and higher eukaryotes. They exhibit stability on the ends of chromosomes.

(b) In humans, telomeres consists of 1000 or more arrays of TTAGGG repeats at the terminals of 3′ ending strands.

(c) Telomeres are found to shorten during sequence of primary cultures of cells.

(d) These are proved to be useful targets for anticancer therapy.

Q37. What do you know about telomerase?

(a) Telomerase is the enzyme which is active in normal or malignant cells with a high proliferation rate.

(b) The measurement of its acitivity may help in the staging of certain tumors.

(c) Telomerase activity is found to be lacking in cultured primary cells.

(d) It is the useful target for anticancer therapy.

Q38. Mention some new and developing cancer therapies.

(a) Cell cycle inhibitors.

(b) Metastasis inhibitors.

(c) Protein kinase inhibitors.

(d) Ribozymes.

(e) Vaccines.

(f) Telomerase inhibitors.

(g) Glycosylation inhibitors.

CHAPTER 17

DETOXICATION

Q1. Define detoxication.

It is the process by which toxic substances are converted into soluble non-toxic forms by oxidation, reduction, hydrolysis, and conjugation for removal from the body by excretory routes.

Q2. Mention the sites for detoxication in man.

Mainly liver, kidney and other tissues also participate to some extent.

Q3. Name some substances that are destroyed in the body by oxidation.

(a) Meprobamate, a tranquilizer used in psychiatric disorders, by oxidation is converted into hydroxymeprobamate.

(b) Chloral, used as hypnotic, is oxidized to form trichloroacetic acid which is excreted as its salts.

Q4. Name some drugs which undergo hydrolysis in the liver of man.

Digitalis → Sugar + Aglycone.

Atropine → Tropic acid + tropine.

Procaine → P-aminobenzoic acid + Diethylaminoethanol.

Q5. What are the characteristics of conjugation?

(a) Conjugation mainly occurs in liver and to some extent in kidneys.

(b) Different conjugating agents are available in the body and some of these are synthesized in the body; eg, glucoronic acid formed by uronic acid pathway.

(c) Conjugation forms less toxic and more soluble compounds which are easily excreted.

(d) Conjugation takes place independently or it is followed by oxidation, reduction, or hydrolysis of a substance.

Q6. Give some examples of conjugation.

Glycine + Benzoic acid → Hippuric acid.

Glycine + Salicylic acid → Salicyluric acid.

Sulphuric acid + Indoxyl → Indican.

Bilirubin + Glucuronic acid → Bilirubin glucuronides.

Q7. How is selenium poisoning detoxified?

Selenium replaces sulphur in cysteine and methionine in the body and interferes the availability of S-containing amino acids. This can be detoxified by administering P-bromobenzene. Selenium containing species now forms mercapturic acid type of compound with P-bromobenzene and thus excreted in urine.

Q8. How does conjugation take place by acetylation?

(a) Acetylation is done by active acetate (acetyl-CoA) by the help of the enzyme acetyl transferase present in the cytosol.

(b) In humans, sulphadrugs are conjugated by acetylation and 50 per cent of sulphadrugs are excreted as acetylated derivatives.

Examples :

$$\text{Sulphanilamide + Acetyl-CoA} \xrightarrow{\boxed{\text{Acetyl transferase}}} \text{Acetylated sulphanilamide} \downarrow \text{Excreted in urine.}$$

$$\text{PABA + Acetyl-CoA} \xrightarrow{\boxed{\text{Acetyl transferase}}} \text{Acetylated PABA} \downarrow \text{Excreted in urine}$$

Q9. How is antitubercular drug isoniazid detoxicated?

The antitubercular drug isoniazid is detoxicated by acetylation.

Q10. How are mercapturic acid formed in the humans?

Mercapturic acids are formed by the conugation of cysteine with aromatic compounds like bromobenzene naphthelene with the help of acetic acid.

Examples :

Cysteine + Bromobenzene + acetic acid ⟶ Bromophenyl mercapturic acid

Cysteine + naphthalene + acetic acid ⟶ Napthyl mercapturic acid

Q11. State the role of glutamine in conjugation reaction.

In humans and in primates, glutamine conjugates phenyl acetic acid formed in phenylketonuric patients to produce phenyl acetyl glutamine which is excreted in urine. The mousy odour of the urine of phenylketonuric patients is due to such production.

Q12. Explain the role of BAL (British anti-Lewesite) in detoxication.

(a) BAL is chemically 2, 3-mercaptopropanol.

(b) Heavy metals like As, Cd, Hg are found to bind the –SH groups of enzymes and the enzymes become inactivated.

(c) BAL has a greater affinity for these heavy metals and pulls away the metal ions from their enzyme combination forming a complex which is readily excreted in urine.

Q13. How are cyanides detoxicated in the body?

(a) Highly toxic cyanides are available in the body in less quantities from some fruits, proteins, and tobacco smokes.

(b) These are conjugated by thiosulphates or colloidal sulphur in the liver being catalyzed by the enzyme rhodanase

$$HCN + S \xrightarrow{\text{Rhodanase}} HCNS + Na_2SO_3$$

(Rhodanase is also called thiosulphate cyanide sulphur transferase).

(c) These are also detoxicated by cysteine.

(d) Traces of these are also detoxified by the formation of cyano-cobalamine.

Q14. State the importance of cytochrome P_{448} system.

(a) Cytochrome P_{448} species is also termed as aromatic hydrocarbon hydroxylase (AHH).

(b) The higher concentration of this enzyme is found in cells and tissues of cigarette smokers than nonsmokers. This enzyme is involved in the conversion of "inactive PAHS" (procarcinogens) present in the lungs of cigarette smokers to "active carcinogens" by hydroxylation reactions.

(c) Elevated activity of this enzyme is observed in the placentae of pregnant women who are cigarette smokers. Therefore, the foetus becomes exposed to harmful metabolites.

CHAPTER 18

METABOLISM OF XENOBIOTICS

Q1. What is xenobiotic?

(a) A xenobiotic (xenos means strange) is a compound that is foreign to the body.

(b) The principal classes are drugs, chemical carcinogens, and various compounds that are available in our environments. More than 200,000 man-made environmental chemicals exist.

(c) Most of the compounds are metabolized in the human body in the liver. Sometimes a xenobiotic may be excreted unchanged.

Q2. How are the xenobiotics metabolized?

(a) The major reaction in metabolism is hydroxylation catalyzed by a class of enzymes named as monooxygenases or cytochrome P-450 species. These enzymes can also catalyze deamination, desulfuration, peroxygenation, dehalogenation, epoxidation, and reduction reactions.

(b) The hydroxylated products are converted to various metabolites by specific enzymes by conjugation with glucuronic acid, sulfate, acetate glutathione, or certain amino acids or by methylation.

The original xenobiotics are referred to as prodrugs or procarcinogens.

Q3. What are the characteristics of cytochrome P-450?

(a) They are hemoproteins.

(b) They are present in high concentrations in the membranes of the endoplasmic reticulum of liver. They are also found in other tissues.

(c) In the adrenal, they are found in mitochondria as well as in the E.R.; the various hydroxylases present in that organ play an important role in cholesterol and steroid biosynthesis.

(d) The mitochondrial cytochrome P-450 system uses an NADPH-linked flavoprotein.

(e) There are at least six closely related species of cytochrome P-450 present in liver E.R. that act on a wide variety of drugs, carcinogens, and other xenobiotics.

(f) The lipid, preferably lecithin, is the component of the cytochrome P-450 system.

(g) Cytochrome P-450 system can metabolize the blood anticoagulant dicumarol.

(h) Recent studies have shown that individual species of cytochrome P-450 frequently exist in polymorphic forms, some of which exhibit low catalytic activity.

Q4. State the role of glutathione in xenobiotic metabolism.

(a) Glutathione is commonly abbreviated to GSH and the SH group is the business part of the molecule. Certain toxic electrophilic xenobiotics are conjugated to the nucleophilic GSH by the enzyme glutathione S-transferase present in liver cytosol.

$R + GSH \rightarrow R\text{–}S\text{–}G$ [R = an electrophilic xenobiotics]

GSH is, therefore, an important defense mechanism against certain toxic compounds, such as some drugs and carcinogens.

(b) It participates in the decomposition of potentially toxic hydrogen peroxide in the reaction catalyzed by glutathione peroxidase.

(c) It is an important intracellular reductant helping to maintain essential SH groups of enzymes in their reduced state.

(d) It acts as a carrier in the transport of certain amino acids across membranes in the kidney.

Amino acid + GSH $\rightarrow$ γ-glutamyl amino acid + cysteinylglycine.

Q5. Show acetylation and methylation reactions of xenobiotics.

(a) X + Acetyl-CoA → Acetyl-X + CoA.
The drug isoniazid, used in the treatment of tuberculosis, is subjected to acetylation. The enzyme acetyltransferase, present in liver, influences the rate of clearance of this drug.

(b) The xenobiotics are also subjected to methylation by methyltransferases, using S-adenosylmethionine as methyl donor.

Q6. Mention the various factors that affect the activities of the enzymes metabolizing xenobiotics.

(a) Species may cause the variation in the activities of these enzymes.

(b) Genetic factors are involved in causing the significant difference in enzyme activities among individuals.

(c) Age and sex are other factors in the variation of the activities of these enzymes.

(d) Metabolites of certain xenobiotics can inhibit the activities of xenobiotic metabolizing enzymes.

Q7. What is the role of epoxide hydrolase?

(a) This enzyme can exert a protective effect against certain carcinogens.

(b) Epoxides, formed from some procarcinogen substrates, are highly reactive. Epoxide hydrolase, present in the membranes of E.R., converts epoxide into much less reactive dihydrodiols.

CHAPTER 19

HORMONES

Q1. What are hormones?

Substances, which are secreted by the ductless or endocrine glands and excite some other organs to specific activity being discharged into the blood, are called hormones.

Q2. State the clinical aspect of hormones.

(a) Diseases of the endocrine system occur due to the excessive or deficient production of hormones. This become the example of the application of basic principles to clinical medicine.

(b) The endocrine disease syndrome can be better understood by knowing the general aspects of hormone action and understanding the physiologic and biochemical effects of the individual hormones.

Q3. What are receptors?

(a) Receptors are proteins.

(b) Most of the receptors are present in very small amounts.

(c) The insulin receptors is a heterotetramer ($\alpha_2\beta_2$) linked by multiple disulfide bonds.

(d) The receptors for insulinlike growth factor 1 (IGF-I), epidermal growth factor (EGF), and low density lipoprotein (LDL) are generally similar to the insulin receptor.

(e) Receptors of the steroid/thyroid type constitute a large superfamily. Many related members of this family have no known ligand, hence are called orphan receptors. These receptors have several functional domains.

(f) The glucocorticoid receptor has several functional domains :

(i) A hormone binding region in the carboxy-terminal portion.

(ii) An adjacent DNA-binding region.

(iii) A specific region is located in the amino-terminal half of the molecule.

(iv) One or more regions that activate or repress gene transcription.

(v) Two regions responsible for transloca-tion of the receptor from the cytoplasm to the nucleus.

Q4. Show that cAMP is the second messenger for many hormones.

(a) The intracellular level of cAMP is increa-sed or decreased by various hormones and this effect varies from tissue to tissue.

(b) Epinephrine causes large increases of cAMP in muscle and relatively small changes in liver.

(c) Epinephrine, ACTH, TSH, glucagon, MSH, and ADH stimulate adenylate cyclase and increase cAMP.

Q5. Mention the general features of hormone classes.

	Group I	Group II
Types	Steroids, iodothyroinines, calcitriol	Polypeptides, proteins, glycoproteins, catecholamines.
Solubility	Lipophilic	Hydrophilic
Transport proteins	Yes	No.
Plasma half-life	Long (hours in days)	Short (minutes)
Receptor	intracellular	Plasma membrane
Mediator	Receptor-hormone complex	cAMP, Ca^{++}, metabolites of complex phosphoinosi-tides, others.

Q6. What is calmodulin?

(a) It is the calcium-dependent regulatory protein of molecular weight 17,000.

(b) It has 4 Ca^{++} binding sites and most of the molecule assumes an alphahelical structure.

(c) It is one of numerous subunits of complex proteins and is involved in regulating various kinases.

(d) Ca^{++}/Calmodulin regulates the activity of many structural elements in cells.

Q7. Show that calcium is a mediator of hormone action.

(a) The effect of many hormones

(i) is blunted by Ca^{++}-free media.

(ii) can be mimicked by agents that increase cytosolic Ca^{++}.

(iii) influences cellular calcium flux.

(b) A good number of metabolic enzymes are regulated by Ca^{++}, phosphorylation, or both.

Q8. Explain that the intracellular messenger is unknown for some hormones.

The hormones which have no intracellular messenger cluster into two groups. One group consists of insulin, the insulin-like growth factors (IGF-I and IGF-II) and a variety of other growth factors. The other major group consists of proteins from the growth gene family. The large group of hormones employ different mechanisms of intracellular signalling.

Q9. How is hormone secretion regulated?

(a) Cholinergic sympathetic fibres stimulate catecholamine secretion from adrenal medulla.

(b) Either the stimulatory or inhibitory hormones from a controlling gland can control the endocrine releases, eg, corticotropin can control the corticosteroids and thyrotropin can control the thyroid hormones.

(c) When the concentration of a target gland hormone is increased, it inhibits the secretion of tropic hormone stimulating the gland by feedback mechanism.

(d) Certain cyclic rhythm are associated with the secretion of hormones over a period of time. In the early hours of deep sleep the concentration of growth hormone and prolactin is increased.

Q10. State the clinical significance of growth hormone.

(a) The level of growth hormone is raised in malnutrition with kwashiorkor due to abnormal glucose concentration at the control sites.

(b) Dwarfism is caused due to the hyposecretion of growth hormone and gigantism in children and acromegaly in adults are due to hypersecretion of GH.

Q11. What do you mean by a tropic hormone?

(a) A tropin or a tropic hormone is the hormone which influences the activities of other endocrine glands which are involved in stress and reproduction.

(b) The tropic hormones undergo feedback inhibition at the pituitary or hypothalamic level by hormone product of the final target gland.

Q12. Show the clinical significance of Cushing's disease.

The disease is caused by the overproduction of corticotropin because of tumor or hyperplasia of β-cells of the anterior pituitary. This leads to hypersecretion of glucocorticoids. The symptoms developed are :

(a) Hyperglycemia and glycosuria.

(b) Muscle wasting and atrophy of skin.

(c) High level of free fatty acids and high uninary NPN.

(d) Abnormal retention of fat forming the moon-shaped face appearance.

(e) Retention of Na^+ and water and development of hypertension.

Q13. What is the nature of MSH of middle lobe of pituitary?

(a) There are two types of MSH–α-MSH and β-MSH. The β-MSh consists of 13 amino acids. α-MSH is present in larger quantities.

(b) MSH causes the darkening of the skin and skin pigmentation by deposition of melanin by melanocytes.

Q14. Mention the clinical significance of MSH.

(a) MSH secretion is inhibited by cortisone and hydrocortisone, epinephrine and norepinephrine.

(b) Excess MSH production increases the synthesis of melanin causing brown pigmentation of skin. This happens so in **Addison's disease.**

(c) Pigmentation does not occur in the lack of MSH due to the deficiency of the function of hypophysis.

Q15. What is the nature of oxytocin and vasopressin?

(a) Both are small peptides containing 9 (nine) amino acids. Oxytocin differs from vasopressin in respect of 3rd and 8th amino acid residues.

(b) They are stored in the pituitary in association with two proteins neurophysin 1 and 11 with molecular weights of 19,000 and 21,000, respectively.

Q16. State the clinical importance of vasopressin.

(a) The failure in secretion of vasopressin causes diabetes insipidus which is characterized by increased volume of urine output up to 20-30 litres per day with low specific gravity and excessive thirst.

(b) Unusual vasopression secretion causes hypertonic urine, low plasma Na^+ level, and water intoxication having drowsiness, irritability, nausea, vomiting, convulsion, coma.

(c) It stimulates the contraction of smooth muscles and thus causes vasoconstriction by increasing cytosolic Ca^{++} concentration.

(d) It has glycogenolytic effect.

Q17. Mention the clinical significance of oxytocin.

(a) Oxytocin causes contraction of smooth muscle having effects on mammary glands and uterus.

(b) Estrogens increases the secretion of oxytocin while progesterone decreases and inhibits the same.

(c) Oxytocin is secreted during coitus by the female uterus which accelerates the aspiration of semen into the uterus.

Q18. Mention the principal abnormalities of pituitary function.

(a) **In hyperpituitarism :**

(i) Gigantism, (ii) Acromegaly, (iii) Cushing's syndrome.

(b) **In hypopituitarism :**

(i) Dwarfism, (ii) Myxedema.

(c) **In hypoactivity :**

(i) Simmond's disease in adults, (ii) Frolish's syndrome & Lorain-levy type childhood.

Q19. What is panhypopituitarism?

(a) This is the condition in which there is the deficiency of the function of the pituitary. This results from the destruction of the gland due to age or infarct.

(b) This also results from the pressure of a tumor on the pituitary.

(c) There is hypoglycemia with increased sensitivity to insulin.

(d) There is the reduction in the excretion of sex hormone products (17-ketosteroids) in urine.

Q20. What is Lorain-Levy type?

This is the condition in which the following are found :

(a) Cessation of skeletal growth and nonappearance of secondary sexual characters.

(b) Ischemic necrosis and rare changes of simple cysts.

(c) Destructive lesion of pituitary.

Q21. Name the principal hormones of thyroid gland secretion.

Thyroxine (T_4); Triiodo thyronine (T_3); Reverse T_3.

Q22. Name the main carrier proteins.

(a) Thyroxine-binding globulin (TBG).

(b) Thyroxine binding prealbumin (TBPA).

Q23. Mention the symptoms of Frohlich's syndrome.

(a) The children become stunted and idiotic.

(b) The genitals are hypoplastic and there are diffuse deposits of fats.

Q24. State the chemistry of Insulin.

(a) It is a heterodimeric protein consisting of 51 aminoacids.

(b) It requires Zn^{++} for its crystallization.

(c) It is composed of two polypeptide chains A and B.

A chain contains 21 amino acids and B chain contains 30 amino acids.

(d) Both the chains are held together by two S-S linkages. Cys 7 and cys 20 and A chain are joined to cys 7 and cys 19 of B chain, respectively.

(e) A chain carries an intra-chain S-S linkages between cys 6 and cys 11.

(f) In A-chain, N-terminal amino acid is glycine and C- terminal is asparagine.

(g) in B-chain, N-terminal amino acid is phenylalanine and C-terminal is threonine.

(h) Its molecular weight is 5,734.

Q25. Why is not insulin tablet given orally for the treatment of diabetic patients?

Since insulin is protein it is digested by proteolytic enzyme insulinase and hence it cannot perform its action.

Q26. What is the role of δ-cells of pancreas?

(a) δ-cells constitute of 1-8% of the cells.

(b) They are responsible for secretion of somatostatin and a pancreatic polypeptide.

Q27. State the factors affecting secretion of insulin.

(a) Stimulation of right vagus causes secretion of insulin.

(b) Hyperglycemia directly stimulates β-cells of pancreas for synthesis and release of insulin.

(c) Fructose and mannose can also stimulate release of insulin.

(d) Glucagon, secreted from α-cells of pancreas, is also a potent stimulant of insulin secretion.

(e) Fatty acid such as octanoic acid can stimulate β-cells to secrete insulin.

(f) Gastrin, CCK-PZ, and secretin can directly stimulate insulin secretion.

(g) Leucine and arginine can stimulate insulin secretion.

(h) Epinephrine is highly effective inhibitor of insulin secretion.

Q28. Discuss the metabolic role of insulin.

(a) Insulin increases glucose uptake.

(b) It increases glycolysis, conversion of pyruvate to acetyl-CoA, and stimulates glycogenesis.

(c) It decreases gluconeogenesis, glycogenolysis, but increases HMP-shunt.

(d) It increases fatty acid synthesis, synthesis of triacyl glycerod (TG); but decreases lypolysis, ketogenesis.

(e) It increases amino acid uptake and protein synthesis.

Q29. What are insulin-like Growth Factors?

(a) Insulin-like growth factors, IGF-1 and IGF-11, are not of pancreatic origin but are produced by liver and other tissues.

(b) IGFs are involved more in stimulating growth.

(c) They are carried by plasma proteins.

Q30. Mention the clinical nature of glucagon.

(a) It is a polypeptide containing 29 amino acids having 15 different amino acids. Histidine is the N-terminal and threonine is the C-terminal amino acids.

(b) Its molecular weight is 3,485.

Q31. Enumerate the factors controlling glucagon secretion.

(a) Hypoglycemia increases glucagon secretion from α-cells of pancreas. Glucagon secretion is directly inhibited by glucose.

(b) Fatty acids also inhibit glucagon release but exercise stimulates the release of glucagon.

(c) Most amino acids, growth hormone and CCK-PZ, and calcium stimulate the secretion of glucagon.

(d) Emotion and stress also stimulate the secretion of glucagon.

Q32. State the nature of somatostatin.

(a) It is a peptide consisting of 14 amino acids.

(b) It is secreted from hypothalamus, δ-cells of islet of langerhans of pancreas, and also produced by D-cells of antral mucosa of stomach and also duodenal mucosa.

(c) It acts as a regulator of growth hormone secretion and inhibits growth hormone release.

(d) It serves as a neurotransmitter substance in the brain.

(e) It inhibits both insulin and glucagon secretion and serves as an intraislet (paracrine) regulator of secretion of these hormones.

Q33. Explain Addison's disease.

(a) Hypoadrenocorticism produces Addison's disease.

(b) This disease is also caused by tubercolosis and malignancy.

(c) Other rare causes of this disease are amyloidosis, fungus infection, hemosiderosis, and leukemic infiltrations.

(d) The clinical features and biochemical changes are decreased Na^+ reabsorption, decreased E.C.F. volume, development of acidosis, reduced urinary NPN, hyponatremia, and hyperkalemia.

Q34. Why is epinephrine said to be catecholamine?

In the structure of epinephrine catechol ring is attached to amine, hence it is said to be catecholamine.

Q35. State the physiological actions of epinephrine.

(a) It causes vasodilatation of the arterioles of skeletal muscles, dilatation of the pupil.

(b) It is a stimulant of heart action and increases the coronary blood flow.

(c) It stimulates glycogenolysis, increases breakdown of triglyceride and hepatic gluconeogenesis, and inhibits insulin release.

(d) It also increases blood lactic acid level by promoting muscle glycolysis.

Q36. What is the nature of parathormone?

(a) Parathrome is a protein consisting of 83 amino acids with a molecular weight of about 9,500 secreted by the parathyroid gland.

(b) It increases the plasma calcium level and the tubular reabsorption of calcium by the kidneys and also decreases the reabsorption of phosphate by the kidneys.

Q37. What is the nature of calcitonin?

(a) Calcitonin is a protein consisting of 32 amino acids with a molecular weight of about 3,000 secreted by the parathyroid gland.

(b) The secretion of calcitonin is stimulated by glucagon and cAMP. Its release is directly proportional to the plasma Ca^{++} concentration.

(c) Gastrin and CCK-PZ are the potent stimulators of calcitonin secretion.

(d) It decreases plasma phosphorus and calcium level.

(e) Calcitonin level in serum is increased in medullary thyroid carcinoma.

(f) It is administered in the treatment of paget's disease.

Q38. What is the main role of mineralocorticoid hormone?

It helps in the regulation of balance between Na^+ and K^+ in blood and cells by promotion of reabsorption of Na^+ and inhibition of reabsorption of K^+ by distal tubules of kidney.

Q39. State the role of cortisol in the body.

(a) It is responsible for the conversion of amino acids into carbohydrates, helps in maintaining normal synthesis of glucose by the liver, inhibits fatty acid synthesis in the liver, inhibits the synthesis of RNA and DNA in connective tissues.

(b) It inhibits bone formation, increases the peripheral arteriolar response to vasoconstrictor, increases GFR and free water clearance.

Q40. Mention the important action of aldosterone.

It regulates sodium and potassium excretion by the kidney.

Q41. Enumerate the types of sex hormones.

(a) Androgens (male hormones)

(b) Oestrogens (female hormones)

(c) Gestogens (progestational hormones).

Q42. How is testosterone secretion regulated?

It is regulated by leutenizing hormone (LH) of the anterior pituitary. A high blood level of testosterone exerts "feedback" inhibition of LH secretion.

Q43. State the metabolic role of androgens.

(a) Androgens can cause the decrease in urinary urea, increase in body weight.

(b) They promote protein synthesis, increase fructose production.

(c) They stimulate the growth of bones and reduce the excretion of citrates in the urine.

Q44. Name the naturally occurring estrogens in humans.

(a) β-Estradiol.
(b) Estrone.
(c) Estriol.

The most active form of the estorgen is β-estradiol which is in metabolic equilibrium with estrone.

Q45. State the nature of relaxin.

(a) Relaxin is a hormone concerned with the relaxation of pelvic tissues and cavity in conjugation with other factors.
(b) It is produced during pregnancy by corpus luteum and also by placenta and the production is stimulated by progesterone.
(c) It is formed by two peptide chains consisting of 22 and 26 amino acid residues having molecular weight of about 9,000 and is inactivated by proteolytic enzymes.
(d) It causes dilatation and softening of uterine cervix facilitating child-birth.

Q46. Name the hormones secreted by human placenta.

(a) **Peptide hormones :**
  (i) Human chorionic gonadotropin hormone (HCG).
  (ii) Chorionic somatomammotropin (CS) or Placental lactogen (PL).
(b) **Ovarian steroid hormones :**
  (i) Progestins.
  (ii) Estrogens, chiefly estriol.

Q47. State the chemical nature of HCG.

(a) It is a glycoprotein consisting of two chains α and β. α-chain is constituted of 92 amino acids and β-chain is constituted of 145 amino acids.
(b) α-chain carries two asparagine linked oligosaccharides. β-chain contains two asparagine linked oligosaccharides and four serine linked oligosaccharides.

Q48. Mention the actions of HCG.

(a) This hormone causes the enlargement of corpus luteum and stimulates its secretion. It maintains the secretory corpus luteum in first three months of pregnancy.
(b) It also produces testosterone.

Q49. Name the G.I. hormones.

(a) Gastrin family : Gastrin, CCK.
(b) Secretin family : Secretin, GIP, VIP, glucagon including glicentin.
(c) Neurocrine peptides : Neurotensin, substance P, somatostatin, bombesin like peptides (BLI).

Q50. State the physiological functions of CCK.

(a) It causes gall-bladder contraction.
(b) It stimulates pancreatic enzyme secretion.
(c) It stimulates the growth of pancreatic acinar tissue.

Q51. What is MAO?

(a) Monoamine oxidase is an oxido reductase that deaminates monoamines.
(b) It occurs in highest concentration in liver, stomach, kidney, and intestine.
(c) MAO inhibitors are used to treat hypertension and depression.

Q52. What does happen in menopause?

(a) Menopause occurs with the loss of all follicles and ovarian estrogen production.
(b) There is no alternative source of progesterone but substantial amounts of a weak estrogen, estrones are produced. The levels of estrone are not sufficient to suppress pituitary gonadotropin levels and marked increases of LH and FSH are charactersistic of the postmenopausal years. GnRH may also be involved in the onset of menopause.
(c) Postmenopausal women are prone to two problems associated with tissue catabolism. Estrone cannot always prevent the atrophy of the epithelium of the lower urinary tract and vagina. Osteoporosis is a major health problem in older individuals.

Q53. Name the synthetic antagonists to prevent conception and inhibit tumor growth.

(a) The first developed synthetic antagonist was **diethylstilbestrol**. The other modified steroids are 17α-ethinyl estradiol and mestranol which are used in oral contraceptives.
(b) Clomiphene citrate (clomid) can cause multiple follicles and multiple pregnancies can occur.

(c) Nafoxidine and tamoxifen combine with the estrogen receptor to form very stable complexes with chromatin. These antagonists are used in the treatment of estorgen receptor dependent breast cancer.

Q54. Enumerate the hormones related to female reproductive system causing pathophysiology.

(a) Polycystic ovary syndrome is associated with the overproduction of androgens that causes obesity, irregular menses, and impaired fertility.
(b) The rare leydig cell and arrhenoblastoma tumors produce testosterone, granulosa-theca cell tumors produce estrogens, and intraovarian adrenal rests produce cortisol.

Q55. State the nature of pancreatic polypeptide.

(a) It is secreted by the pancreatic F cells which is recently discovered.
(b) It consists of 36 amino acids with a molecular weight of 4,200.
(c) Its secretion in humans is increased by a protein meal, fasting, exercise, and acute hypoglycemia and is decreased by somatostatin and intravenous glucose.
(d) It has effects on hepatic glycogen levels and gastrointestinal secretion.

Q56. What are the characteristics of insulin-like growth factors?

(a) The other names IGF-1 and IGF-11 are somatomedin C and multiplication stimulating activity (MSA), respectively.
(b) IGF-1 consists of 70 amino acids and IGF-11 consists of 67 amino acids.
(c) The sources of IGF-1 are liver and other tissues and of IGF-11 are diverse tissues.
(d) The plasma levels of both are ng/ml range.
(e) The major physiologic role of IGF-1 is skeletal and cartilage growth and that of IGF-11 is embryonic development.
(f) They have plasma binding protein.

Q57. Name the substances that affect glucagon secretion.

Amino acids, fatty acids, ketones, neurotransmitter, and gastrointestinal tract hormone.

Q58. What is Grave's disease?

(a) This disease is the result of hyperthyroidism and is caused by the production of thyroid-stimulating I_gG.
(b) The disease is manifested by rapid heart rate, widened pulse pressure, nervousness, inability to sleep, weight loss, weakness, excessive sweating, sensitivity to heat, and moist skin.
(c) The disease is treated by blocking hormone production with an antithyroid drug.

Q59. State the role of 1, 25 $(OH)_2 - D_3$.

(a) It is the only hormone that stimulates intestinal absorption of calcium and phosphate.
(b) It enhances the actions of parathyroid hormones on renal calcium reabsorption.
(c) It acts at the cellular level just like steroid hormones.

Q60. How is mammary gland developed?

(a) Glandular tissue is grown during puberty with the deposition of adipose tissue.
(b) Extensive development occurs during pregnancy when glandular tissue is exposed to high concentration of estradiol and progesterone.

Q61. State the causes of lactation.

(a) The rate of synthesis of milk is increased by prolactin, glucocorticoids, and insulin by raising the amounts of the specific mRNA.
(b) Lactation starts when the level of progesterone is decreased abruptly after delivery.
(c) Suckling enhances lactation by the release of oxytocin from the posterior pituitory. Oxytocin stimulates contraction of the myoepithelial cells that surround the alveolar ducts, thus expelling milk from the gland.

Q62. How does menopause occur?

(a) Menopause occurs with the loss of all follicles and ovarian estrogen production.
(b) No progesterone is produced and the levels of LH and FSH are highly increased.
(c) GnRH is also involved at the beginning of menopause.
(d) Postmenopausal women face problems associated with tissue catabolism. Osteoporosis is a major health problem in older individuals.

Q63. What is Turner's syndrome?

It is a genetic disorder in which there is female internal and external genitalia, several development abnormalities, and delayed puberty.

Q64. What is meant by polycystic ovary syndrome?

In this syndrome, there is obesity, irregular menses, and impaired fertility.

Q65. What is cretinism?

(a) This occurs in hypothyroidism in children.

(b) This is manifested by slow growth, dwarfism, mental retardation, dry skin, puffy lips, scanty hair, pot bellies, and vacant look.

Q66. What do you know about myxoedema?

(a) This is caused by hypothyroidism in adults.

(b) This diease is having the characteristics of dry coarse skin, puffiness of face, falling of hairs of eyebrows, anemia, obesity, low BMR and hypercholesterolemia.

(c) The patients show sensitivity to cold and lowered body temperature.

Q67. Mention the principal anabolic hormones.

(a) Insulin,

(b) Growth hormone,

(c) Testosterone.

These three hormones stimulate protein synthesis and bring about a positive nitrogen balance by retaining nitrogen.

Q68. What do you mean by encephalins?

Encephalins are the pentapeptides in the brain. They are :

(a) Methionine—encephalin, and

(b) Leucine—encephalin.

Q69. What is thyroperoxidase?

(a) It is an enzyme which requires H_2O_2 for its activity.

(b) It is a tetramer having molecular weight 90,000.

Q70. Which form of thyroid hormone is metabolically most active?

"Free" T_4 and T_3 are metabolically active hormones. "Free" T_3 is more active than T_4.

Q71. Mention the action of antithyroid drugs.

(a) They inhibit conversion of iodide to "active iodine".

(b) They inhibit iodination and coupling.

(c) They inhibit hormonal release.

Q72. What are the antithyroid drugs that inhibit the conversion of T_4 to T_3 in target tissues?

Propyl thiouracil, Propanolol.

Q73. What is meant by Wolff-Chaikoff's effect?

Serum iodide concentration when exceeds 30 mg/dl, it produces antithyroid effects and even produces goitre. This is known as **Wolff-Chaikoff's effect.**

Q74. What is Von-Recklinghausen's disease?

(a) It is the effect of hyperparathyroidism on bone.

(b) It signifies increased osteoclastic activity resulting in resorption of bones forming cyst like structures.

Q75. What do you mean by Trousseau's sign?

(a) It is found in tetany.

(b) Application of pressure over arm with blood pressure cuff results in muscular spasm causing accoucheur hand (carpopedal spasm).

Q76. What do you know about C–cells?

(a) C–cells are derived from "neural crest" and are found in thyroids, parathyroids, and thymus.

(b) Calcitomin is mainly produced by "C–cells of thyroids". Hence, it is also called as **thyrocalcitonin.**

Q77. What is entero-glucagon?

(a) It is formed by gastric or duodenal "A" cells and pancreatic α-cells.

(b) It is less active in stimulating adenyl cyclase and hence can not perform similar functions as that pancreatic glucagon.

Q78. State Conn's syndrome.

(a) Conn's syndrome is due to increased secretion of aldosterone and due to tumors of adrenal cortex.

(b) It is manifested by hypernatremia, hypokalamia, and alkalosis.

CHAPTER 20

GLYCOPROTEINS

Q1. What are glycoproteins?

Glycoproteins are proteins that have oligosaccharides (glycan) chains covalently attached to their polypeptide backbones. They are conjugated proteins. Almost all the plasma proteins of human except albumin are glycoproteins.

Q2. State the clinical significance of glycoproteins.

(a) Many proteins of cellular membranes contain significant amounts of carbohydrates.

(b) Some blood group substances are glycoproteins and others are glycosphingolipids.

(c) Certain hormones are also glycoproteins.

(d) Changes in the structure of glycoproteins and other glycoconjugates on the surfaces of cancer cells are responsible for the phenomenon of metastasis.

Q3. Discuss the functions of glycoproteins.

(a) Glycoproteins are constituents of cell walls, collagen, elastin, fibrins, bone matrix.

(b) They are also present in mucins and mucous secretions.

(c) They are involved in transporting vitamins, lipids, minerals and trace elements.

(d) They constitute immunoglobulins and interferon.

(e) They are the constituents of certain hormones like chorionic gonadotropin, thyrotropin (TSH) and certain enzymes like proteases, nucleases, glycosidases, hydrolases etc.

(f) They are also involved in cell attachment/recognition sites such as cell-cell, virus-cell, bacterium-cell, hormone receptors.

Q4. Mention the range of carbohydrate present in glycoprotein.

1% to over 85% by weight.

Q5. What are the functions of the oligosaccharide chains of glycoproteins?

(a) Oligosaccharides protect from inside and outside the cell against proteolysis.

(b) They are involved in biologic activity.

(c) They affect embryonic development and differentiation.

(d) They may affect sites of metastasis selected by cancer cells.

(e) They affect insertion into membranes, intracellular migration, sorting, and secretion.

(f) They affect proteolytic processing of precursor proteins to smaller products.

Q6. How many monosaccharides are found in the oligosaccharide chains?

Seven monosaccharides.

Q7. Where and how are the nucleotide sugars formed?

Formed in cytosol of the cell

UTP + Glucose-1-phosphate

↓ UDP-Glc Pyrophosphorylase

UDP-Glc + Pyrophosphate

UDP-Glc ⟷ (UDP-Glc epimerase) UDP-Gal

Q8. Where does glycosylation reaction take place?

Within the lumen of the Golgi apparatus.

Q9. State the nature of lectins.

(a) Lectins are proteins of plant origin that bind one or more specific sugars.

(b) Many lectins are used in purification and analysis of glycoproteins.

Q10. In how many classes glycoproteins are divided?

They are divided into four classes on the basis of the nature of the linkage between their polypeptide chains and their oligosaccharide chains.

(a) Those containing a ser-GalNAC linkage.

(b) Proteoglycans containing a ser-xyl linkage.

(c) Collagens containing a hydroxylysine-Gal linkage.

(d) Glycoproteins containing an Asn-GlcNAC linkage.

Classes 1, 2, and 3 are joined to the corresponding amino acids via 0-glycosidic linkages. The fourth class involves a N- glycosidic linkage. Since classes 2 and 3 are relatively uncommon the term 0-linked glycoproteins is used. The members of class 4 are called N-linked glycoproteins (a linkage involving the N of the amide group of asparagine and a sugar residue). Some glycoproteins contain both N- and 0-glycosidic linkage.

Q11. By which substance are mucins rich?

0-linked glycoproteins.

Q12. Name the substances required for the biosynthesis of 0-linked glycoproteins.

Nucleotide sugars such as UDP-GalNAC, UDP-Gal, CMP- NeuAc and the membrane-bound enzyme glycoprotein glycosyltransferases.

Q13. Enumerate the major classes of N-linked glycoproteins.

Complex, hybrid, and high-mannose.

Q14. Mention the different factors in the regulation of oligosaccharide processing.

(a) The activities of various hydrolases and transferases.

(b) Certain enzymes can act only if another enzyme has acted previously.

(c) The activities of various transferases can wax and wane systematically during development.

(d) Different glycotransferases have different intracellular locations and play important role in regulation.

(e) Protein conformation is involved in the extent of processing.

(f) Analysis of the activities of glycoprotein-processing enzymes in various types of cancer cells. Cancer cells synthesize different oligosaccharide chains from those made in control cells.

Q15. Name the inhibitors of glycosylation.

Tunicamycin, deoxynojirimycin and swainsonine are the inhibitors.

(a) **Tunicamycin** : Inhibits the enzyme catalyzing addition of GlcNAC to dolichol-P.

(b) **Deoxynojirimycin** : Inhibitor of glucosidases 1 and 11.

(c) **Swainsonine** : Inhibitor of mannosidase 11.

Q16. How does I-cell disease occur?

(a) I-cell disease is a rare condition in which cultured cells lack almost all of the normal lysosomal enzymes; the lysosomes thus accumulate many different types of ungraded molecules forming inclusion bodies.

(b) Plasma of these patients contain very high activities of lysosomal enzymes.

(c) The cultured cells contain a normal receptor for uptake of lysosomal enzymes but these enzymes lack a recognition marker.

Summary of the causation of this disease.

Mutations in DNA.

↓

Mutant GlcNAC phosphotransferase

↓

Lack of normal transfer of GlcNAC-1-P to certain enzymes destined by lysosomes

↓

These enzymes lack Man-6-P and are secreted from cells rather than targeted to lysosomes

↓

Lysosomes are thus deficient in certain hydrolases, do not function properly, and accumulate partly digested cellular material, menifesting as inclusion bodies.

Q17. Enumerate the most prominent diseases due to genetic defects of glycoprotein.

Mannosidosis, fucosidosis, sialidosis, and aspartylglycosaminuria.

Q18. What are lactins?

(a) Lectins are the sugar binding protein that precipitate glycoconjugates.

(b) Enzymes, toxins, and transport proteins can be classed as lectins if they bind carbohydrate.

(c) Lectins such as concanavaline A (Con A) can be attached covalently to sepharose. The resulting sepharose—Con A maybe used for the purification of glycoproteins.

(d) Smaller amounts of certain lectins are required to cause agglutination of tumor cells than of normal cells.

Q19. State the role of tunicamycin.

(a) Tunicamycin is the agent which can be used experimentally to inhibit various stages of glycoprotein biosynthesis.

(b) If cells are grown in the presence of tunicamycin, no glycosylation of their normally N-linked glycoproteins will occur.

(c) This inhibitor of glycoprotein processing does not affect the biosynthesis of O-linked glycoproteins.

Q20. What is the role of PH–30?

The glycoprotin, PH-30, is important in binding of the PM of the sperm of the oocyte and also in the fusion of the two membranes. These interactions enable the sperm to enter and thus fertilize the Oocyte.

Q21. State the role of selectins.

(a) Selectins are single–chain Ca^{2+}-binding transmembrane proteins that contain a number of domains.

(b) Leukocytes and endothelial cells contain specific lectins, called **selectins**, on their surfaces.

(c) Certain compounds are synthesized that block selectin–ligand interactions and thus may inhibit the inflammatory response.

CHAPTER 21

MUSCLE

Q1. State the clinical significance of muscle.

(a) **Malignant hyperthermia,** a serious complication for some patients undergoing certain types of anesthesia, is due to abnormal accumulation of Ca^{++} in the cytoplasm of skeletal muscle cell causing rigor and excessive heat. The increase in Ca^{++} is due to mutation in the structural gene for the Ca^{++} release channel of the sarcoplasmic reticulum.

(b) **Heart failure** is very common medical condition. Cardiomyopathies are the conditions of this heart failure.

(c) Endothelium-derived relaxing factor—a compound synthesized by endothelial cells that regulates smooth muscle tone—is the gas nitric oxide (NO).

(d) Glyceryl trinitrate, the vasodilator, is used in the treatment of angina pectoris.

Q2. What are the requirements met by chemical-mechanical machine?

(a) A constant supply of chemical energy is a must. ATP and creatine phosphate supply chemical energy in vertebrate muscle.

(b) There must be the regulation of mechanical activity and the force of contraction in muscle.

(c) The machine must be connected to an operator.

(d) There must be the means of returning the machine to its original state.

Q3. Name the striated and nonstriated muscle.

Striated muscle : Skeletal and cardiac muscle.

Nonstriated muscle : Smooth muscle.

Skeletal muscle : Under voluntary nervous control.

Cardiac and smooth muscle : Under involuntary nervous control.

Q4. What are the substances present in sarcoplasm?

Glycogen, ATP, phosphocreatine, and the enzymes of glycolysis.

Q5. Express the nature of sarcomere.

(a) It is repeated along the axis of a fibril at distances of 1,500-2,300 nm.

(b) The myofibril consists of alternating dark and light bands (A bands and I bands).

(c) The central region of A band (the H zone) is less dense than the rest of the band.

(d) The I band is bisected by a very dense and narrow z line.

Q6. What is the nature of thick filaments?

(a) The thick filament is confined to the A band and contains the protein myosin.

(b) These filaments are about 16 nm in diameter.

Q7. State the nature of thin filament.

(a) The thin filament lies in the I band and extends also into the A band but not into the H zone of the A band.

(b) It contains the protein actin, tropomyosin, and troponin.

(c) Each thin filament lies between 3 thick filaments and each thick filament is surrounded by 6 thin filaments.

Q8. What is the mass of fresh muscle fibril?

It is made up of 75% water and more than 20% protein.

Q9. What is the nature of muscle protein actin?

(a) Actin (G-actin) is a globular protein of molecular weight 43,000 and makes up 25% of muscle protein by weight.

(b) G-actin, in the presence of magnesium, polymerizes to form an insoluble double helical filament called F-actin.

(c) The F-actin fiber is 6-7 nm thick and has a pitch of 35.5 nm.

(d) Neither G- nor F-actin exhibits any catalytic activity.

Q10. Express the nature of myosin.

(a) Myosin constitutes 55% of muscle protein by weight and forms the thick filaments.
(b) It is a hexomer with a molecular weight of 460,000 and it has a fibrous portion consisting of two interwined helices.
(c) The hexamer consists of one pair of heavy chains (mol. wt. 200,000) and two pairs of light chains (mol. wt. 15,000 – 27,000).
(d) Skeletal muscle myosin exhibits ATPase activity and binds to F- actin.

Q11. What is heavy meromyosin (HMM)?

(a) It is a soluble protein which has both a fibrous portion and a globular portion.
(b) It exhibits ATPase activity and binds to F-actin.
(c) HMM being digested by papain generates S-1 and S-2. S-2 is fibrous in character and exhibits no ATPase activity and does not bind to F-actin.
(d) S-1 has a molecular weight of 115,000 and exhibits ATPase activity.
(e) Both S-1 and HMM although exhibit ATPase activity, that catalytic activity is accelerated 100-200-fold by the addition of F-actin.

Q12. What is α- actinin?

It is a protein in the Z line to which the ends of the F- actin molecules of the thin filaments attach.

Q13. Explain tropomyosin.

(a) It is a fibrous molecule consisting of α- and β- chains that attach to F- actin.
(b) It is present in all muscular and muscle-like structures.

Q14. State the nature of Troponin.

(a) It consists of three polypeptides.
(b) Troponin T(TpT) binds to tropomyosin and to other two troponin components.
(c) Troponin I (TpI) inhibits the F-actin-Myosin interaction and also binds to other components of troponin.
(d) Troponin C (TpC) is a calcium-binding protein that is structurally and functionally similar to calmodulin which is an important calcium-binding protein widely distributed in nature.

Q15. When does relaxation occur?

(a) When sacoplasmic Ca^{++} falls below 10^{-7} mol/L owing to its entry into the sarcoplasmic reticulum by the Ca^{++} ATPase.
(b) $TPC.4Ca^{++}$ loses its Ca^{++}.
(c) Troponin inhibits further myosin head F-actin interaction.
(d) The myosin head detaches from the F-actin in the presence of ATP.

Q16. What is the role of Ca^{++} in muscle contraction?

Ca^{++} plays a central role in regulation of muscle contraction.

Q17. State the hydrolysis of ATP that drives the cyclic association and dissociation of actin and myosin.

See Chart 1 below.

Q18. What is calsequestrin?

Inside the sarcoplasmic reticulum, Ca^{++} is bound to a specific Ca^{++} binding protein designated calsequestrin.

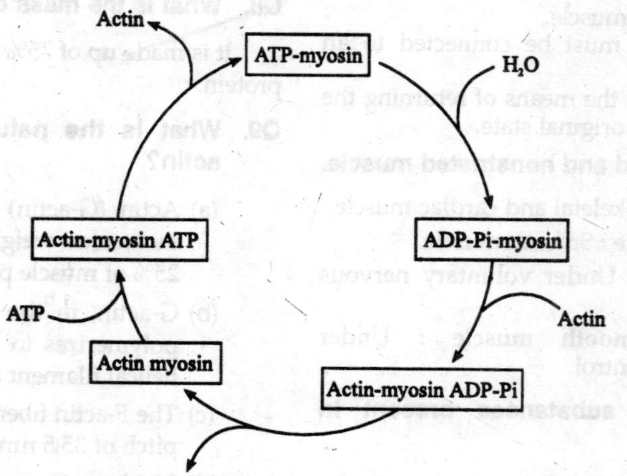

CHART 1

Q19. Show the differences between skeletal, cardiac, and smooth muscle.

Skeletal muscle	Cardiac muscle	Smooth muscle
Striated	Striated	Nonstriated
Small T tubules	Large T tubules	Generally rudimentary T tubules
Plasmalemma lacks many hormone receptors	Plasmalemma contains a variety of receptors	Plasmalemma contains a variety of receptors.
Nerve impulse initiates contraction	Has intrinsic rhythmicity	Contraction initiated by nerve impulses, hormones etc.
Caldesmon not involved	Caldesmon not involved	Caldesmon is important regulatory protein
Very rapid cycling of the cross-bridges	Relatively rapid cycling of the cross-bridges	'Slow cycling of the cross-bridges
Not responsive to NO (endothelial relaxing factor)	Not responsive to NO.	NO plays important role in relaxing vascular smooth muscle.

Q20. State the causation of malignant hyperthermia.

Mutations in the RYR-1 gene

↓

Altered Ca^{++} release channel protein

↓

Mutated channel opens more easily and stays open longer, thus flooding the cytosol with Ca^{++}

↓

High intracellular levels of Ca^{++} stimulate sustained muscle contraction

Q21. What are the chain of events leading to opening of the Ca^{++} release channel?

Depolarization of nerve

↓

Depolarization of skeletal muscle

↓

Depolarization of the transverse tubular membrane

↓

Charge movement of the slow Ca^{++} voltage channel

↓

Opening of the Ca^{++} release channel.

Q22. State the causation of familial hypertrophic cardiomyopathy.

Predominantly missense mutations in the β-cardiac myosin heavy-chain gene on chromosome 14

↓

Mutant polypeptide chains that lead to formation of defective myofibrils

↓

Compensatory hypertrophy of one or both cardiac ventricles

↓

Cardiomeagaly and various cardiac signs and symptoms including sudden death

Q23. Explain the regulation of smooth muscle contraction by Ca^{++}.

See Chart 2 next page.

Q24. State the different sources of ATP in muscle.

See Chart 3 next page.

Q25. Mention the different types of fibers of skeletal muscle with their characteristics.

	Type 1 (Slow twitch)	Type 11 (Fast twitch)
Myosin ATPase	Low	High
Energy utilization	Low	High
Mitochondria	Many	Few
Colour	Red	White
Myoglobin	Yes	No
Contraction rate	Slow	Fast
Duration	Prolonged	Short

Q26. What are the types of muscle fibers and major fuel sources used by a sprinter and by a marathon runner?

Sprinter (100 m)	Marathon Runner
Type 11 (glycolytic) fibers are used predominantly	Type 1 (oxidative) fibers are used predominantly.
Creatine phosphate is the major energy source during the first 4-5 seconds	ATP is the major energy source throughout
Glucose derived from muscle glycogen and metabolized by anaerobic glycolysis is the major fuel source	Blood glucose and free fatty acids are the major fuel sources
Muscle glycogen is rapidly depleted	Muscle glycogen is slowly depleted.

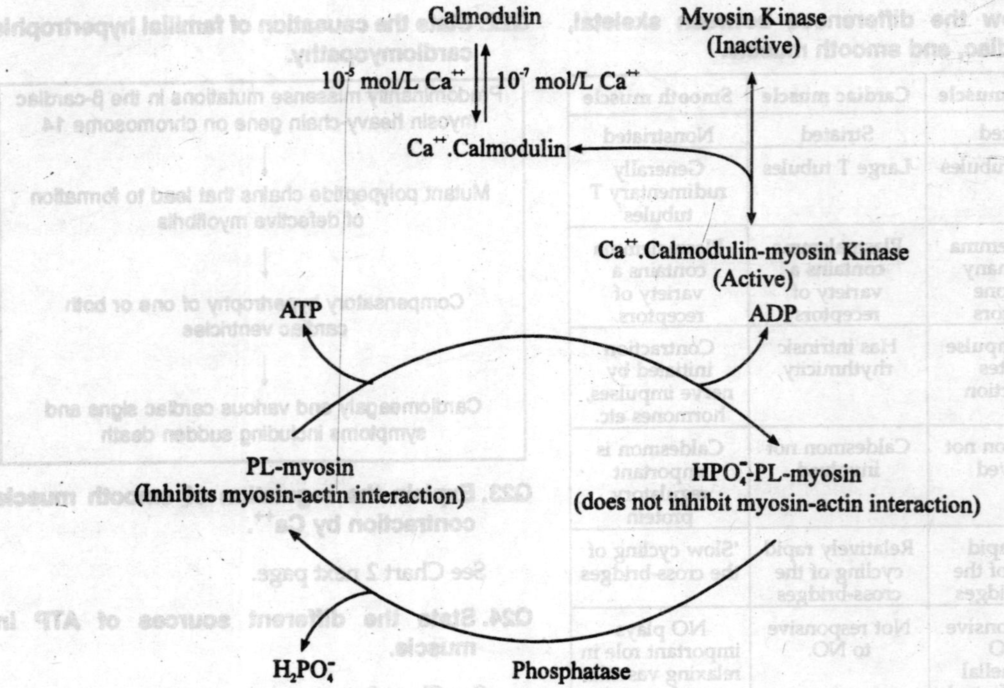

CHART 2

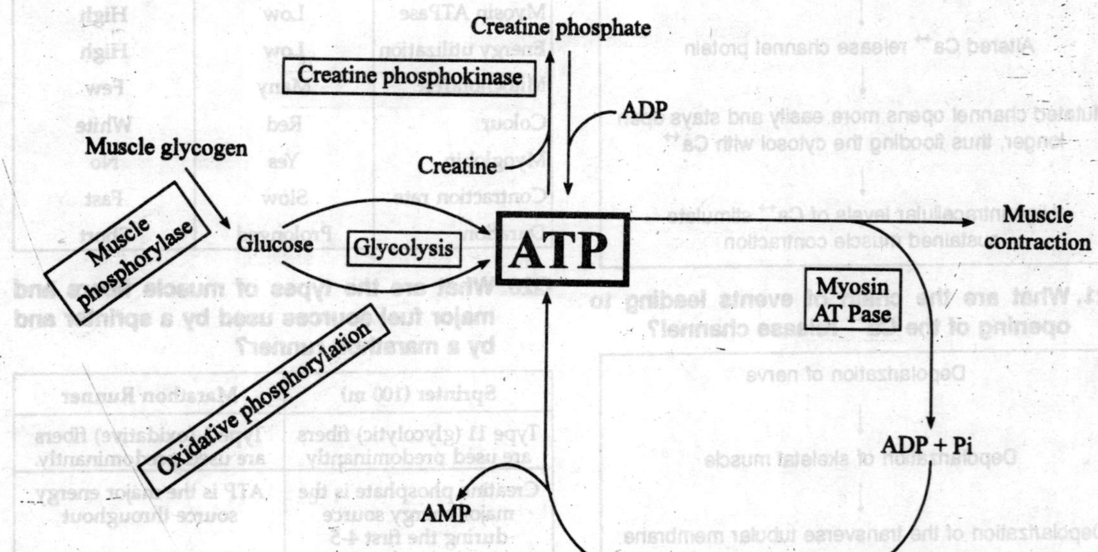

CHART 3

Q27. What is the cause of fatigue of muscles during exercise?

(a) The primary cause is the accumulation of protons but not of lactate in the muscle.

(b) Increase of protons (decreased P^H) can affect the function of muscle in a number of ways :

(i) Lowering the V_{max} of PFK-1.

(ii) Lessening the release of Ca^{++} from the sarcoplasmic reticulum.

(iii) Lessening the activity of the actomyosin ATPase.

(iv) By affecting conformation of some of the muscle proteins.

Q28. Mention the filamentous structures contained in eukaryotic cells.

Three types of filamentous structures :

(a) Actin filaments (7-9.5 nm in diameter, also known as microfilaments.

(b) Microtubules (25 nm)

(c) Intermediate filaments (10 – 12 nm).

Q29. State the nature of actin present in nonmuscle cells.

(a) The G-actin protein of nonmuscle cells has a molecular weight of about 43,000 and contains N-methylhistidyl residues.

(b) In the presence of magnesium and potassium chloride, this actin can polymerize to form the double helical F-actin filaments.

(c) There are two types of actins—β-actin and γ-actin. Both types can coexist in the same cell and polymerize in the same filament.

(d) Actin forms microfilaments of 7 – 9.5 nm.

(e) The bundles of microfilaments are prominent and are referred to as **stress fibers** which disappear as cell motility increases.

Q30. Show the role of skeletal muscle in the reservation of protein in the body.

(a) Skeletal muscle is active in the degradation of certain amino acids and the synthesis of others.

(b) In mammals, muscle is the primary site of catabolism of the branched chain amino acids. Muscle oxidizes leucine to CO_2 and converts isoleucine, valine, aspartate, glutamate into the intermediates of TCA cycle.

(c) Muscle also synthesizes and releases large amounts of alanine and glutamine.

Q31. State the characteristics of microtubules.

(a) Microtubules consist of cytoplasmic tubes 25 nm in diameter and of indefinite length.

(b) These are necessary for the formation of mitotic spindle.

(c) They form the major structural component of cilia and flagella.

(d) They are the major protein component of axon and dendrites.

(e) They consist of dimers of α-tubulin and β-tubulin.

(f) Certain alkaloids can prevent microtubule assembly. These include colchicine and demicoleine, vinblastin, and griseofulvin.

Q32. Name the substances richly present in nervous tissue.

Lecithin, Cephalin, Sphingomyelin, Kerasin etc.

Q33. Mention the composition of heart muscle.

(a) Phospholipids and cholesterol contents are more.

(b) Lower concentration of creatine and carnosine than that of skeletal muscle.

Q34. State the composition of nerve tissue.

Proteins (nucleoprotein, globulin, and neurokeratin), water, phosphatides, cerebrosides, cholesterol, and inorganic substances.

Q35. What is neurokeratin?

It is chemically the inert protein which fulfils the structural function.

Q36. What do you mean by sebum?

It is the secretory product of sebaceous gland containing the mixture of lipids.

Q37. Enumerate the composition of skin.

(a) **Protein :** Keratin.

(b) **Lipids :** Cholesterol, Phospholipids, 7-dehydrocholesterol.

(c) **Carbohydrates :** Glucose, glycogen, mucopolysaccharides.

(d) **Inorganic substances :** Na^+, K^+, Ca^{++}, Mg^{++}, Cl^-.

Q38. Name the pigments of skin.

Melanin, carotenoids.

Q39. Mention the amino acid which is rich in hair.

Keratin, the protein of the hair, is rich in sulphur containing amino acid **cystine**. Hence, the smell of burning of sulphur is available when the hair is burnt.

Q40. Indicate the composition of hair.

Protein (Keratin), lipids, water, ash (10% of it is silica).

Q41. What is the chief constituent of nail?

Keratin.

Q42. What do you mean by acetyl choline?

(a) It is formed by acetyl-CoA and choline.

(b) Stimulation of the nerve endings causes the liberation of it.

Q43. Why do hairs become bright red, grey, and white in colour?

Bright red hair : Due to high content of iron.

Grey hair : Due to the deficiency of melanin formation.

White hair : Due to an increased proportion of calcium carbonate and phosphate.

Q44. State the nature of keratin.

(a) Keratin is of two types—1 and 11, consisting of 10–20 different polypeptides which differ from each other.

(b) The four classes of intermediate filament proteins—namely desmin, vimentin, neurofilament, and glial filament—have a high degree of homology among them.

(c) A keratin filament contains at least 2 different keratin polypeptides, whereas the other four classes of intemediate filaments are homopolymers.

(d) Each of the intermediate filaments consists of 4α-helical domains separated from one another by regions of β-pleated sheets.

CHAPTER 22

EXCRETION

Q1. What are the principal functions of the kidney?

(a) It protects the body by eliminating the waste products of metabolism.

(b) It protects the body from the foreign and nonendogenous substances.

(c) It maintains salt and water balance.

(d) It maintains acid-base balance of the body.

Q2. How is the rate of glomerular filtration influenced?

(a) Variations in BP in glomerular capillary.

(b) Concentration of plasma proteins.

(c) State of blood vessels.

(d) Factors changing intratubular pressure.

Q3. Name the factors influencing the volume of glomerular filtrate.

(a) The volume of blood passing through the glomeruli per minute.

(b) The effective glomerular filtration pressure.

(c) The number of glomeruli functioning at a time.

Q4. Mention the important factors that influence urine formation.

(a) Filtration pressure.

(b) Rate of renal blood flow.

(c) State of nephrons.

(d) Aldosterone and ADH.

(e) Caffeine.

(f) Emotions and exercise.

Q5. Indicate the daily normal range of urine volume.

1,200–1,800 ml. [The average being 1,500 ml. and solids are 60 grams.]

Q6. What is the specific gravity of urine?

1.003 – 1.030.

Q7. How is the volume of filtrate reduced?

(a) In extrarenal conditions, such as dehydration, oligaemic shock, and cardiac failure which diminish the volume of blood passing through the glomeruli or lower the glomerular filtration pressure.

(b) Alteration in the glomeruli which occur in glomerulonephritis.

Q8. What does happen when the volume of glomerular filtrate is reduced?

The kidneys become unable to eliminate waste products which accumulate in blood.

Q9. Indicate the amount of filtrate excreted and absorbed by the glomeruli in 24 hours.

Filtrate : 170 liters.

Excreted in the urine : 1.5 liters.

Filtrate reabsorbed in the tubules : 99%.

Q10. What is the amount of glucose reabsorbed by the tubules per day?

170 grams of glucose.

Q11. Mention the amount of urea filtered and excreted per day.

Filtered : 50 grams.

Excreted : 30 grams.

Q12. Why are certain substances foreign to the body excreted more than the filtrate?

Certain substances foreign to the body, eg, diodrast, para-amino hippuric acid, and phenol red are filtered through the glomeruli as well as secreted by the renal tubules. The total amount is excreted out. Hence, the excreted amount is more than that of the filtrate.

Q13. What do you mean by glomerular filtration rate (GFR)?

Certain substances, eg, inulin, thiosulphate, and mannitol are excreted by the glomeruli. These substances are not reabsorbed or secreted by the tubules. Therefore, the amount of these substances excreted per minute in the urine is the same as that filtered by the glomeruli per minute and this can only show the glomerular filtration rate.

Q14. Mention the conditions in which the volume of urine is increased.

(a) **Physiological conditions :**

(i) Intake of water in the form of drinks, soups or wet foods.

(ii) Cold climate.

(iii) High protein diet.

(iv) Tea, coffee, and cocoa consumption.

(v) Nervousness or excitement.

(b) **Pathological conditions :**

(i) Diabetes mellitus and diabetes insipidus.

(ii) Kidney diseases.

Q15. When does the volume of urine decrease?

(a) Less fluid intake.

(b) Diseases of the heart.

(c) Acute nephritis.

(d) Fever and diarrhea.

Q16. Indicate the colour, odour, and P^H of normal urine.

Colour : Pale yellow.

Odour : Aromatic.

P^H **:** 6.0.

Q17. When does the urine become acidic?

(a) In high protein diet, the urine becomes acidic because of excessive phosphates and sulphates produced in the catabolism of protein.

(b) In acidosis and fever.

(c) In vegetable diet, the urine is alkaline.

Q18. Why does urine become alkaline on standing?

Owing to the conversion of urea to ammonia by bacterial decomposition.

Q19. Why is the colour of normal urine pale yellow?

Owing to the presence of the chief pigment urochrome and small quantities of urobilin and hematoporphyrin are also present.

Q20. Why does the urine become turbid?

Owing to the presence of :

(a) Phosphates in alkaline urine.

(b) Salts of urate in acidic urine.

(c) Precipitation of traces of nucleoproteins.

(d) Fat or pus or bacteria.

Q21. What does excessive urinary urobilinogen indicate?

It indicates excessive hemolysis.

Q22. When does urine contain excess of indican?

(a) In severe constipation.

(b) Gangrene of lung and cholera etc.

Q23. When does phosphate excretion increase?

In bone diseases and wasting diseases of CNS.

Q24. What do you mean by diastatic index?

(a) It is the amount of diastase of 1 ml. urine which can digest 1 ml. of 0.1% soluble starch solution in $\frac{1}{2}$ hour at 37°C.

(b) Its normal range is 6.7 – 33.3 and the value increases in acute pancreatitis.

Q25. Name the constituents of normal urine.

(a) **Nitrogenous constituents :** Urea, uric acid, creatinine, ammonium salts, hippuric acid, enzymes (amylase, pepsin, trypsin etc.).

(b) **Non-nitrogenous cnstituents :** Chloride, phosphate, inorganic sulphates, ethereal sulphates, sodium, potassium, calcium, iron, pentoses, lactose, lactic acid, oxalic acid, sex hormones, vitamins (B_1 and C).

Q26. Name the abnormal constituents of urine.

Reducing sugars, acetone bodies, proteins, blood pigments, bile pigments, bile salts, urobilinogen, indican, homogentisic acid.

Q27. Mention the composition of feces.

Water (70%); cellulose and very little food residues; calcium and phosphates; bacteria together with mucous and secretion; certain amount of fatty material.

Q28. Indicate the amount of feces discharged per day.

About 100 gms per day; varies with diet.

Q29. Mention the colour and odour of feces.

Colour :

(a) Brown colour due to stercobilin derived from bilirubin.

(b) Black colour due to the presence of blood and if no bile is allowed into the intestine.

Odour : Due to indole, skatole, and gases formed by carbohydrate fermentation.

Q30. What is the composition of sweat?

(a) It is the hypotonic fluid containing 0.3% NaCl with an amount of urea, lactic acid, sugar, potassium, and calcium.

(b) It P^H is 4.5 but the P^H becomes 7.0 – 7.4 in case of washed and dried skin.

Q31. Name the tests for the detection of urea, uric acid, creatinine, ammonium salts, reducing sugars, acetone bodies, proteins, blood pigments, bile salts, bile pigments, urobilinogen, indican.

(a) Specific urase test for urea.

(b) Murexide test for uric acid.

(c) Jaffe's test for creatinine.

(d) Test for ammonium salts.

(e) Benedict's test for reducing sugars.

(f) Rothera's test for acetone bodies.

(g) Heat coagulation test for albumin.

(h) Benzidine test for blood pigments.

(i) Hay's test for bile salts.

(j) Fouchet's test for bile pigments.

(k) Ehrlich's aldehyde test for urobilinogen.

(l) Obermeyer's test for indican.

Q32. Mention the calculation for inulin clearance test.

$C_{in} = \frac{U \times V}{P}$ where, U = mg of inulin/100 ml of urine.

V = ml. of urine/mt.

P = mg of inulin/dl of plasma.

Normal average : Inulin clearance in an adult = 125 ml. of plasma cleared of inulin/mt.

Normal level :

Males : 110–150 ml/min.

Females : 105–132 ml/min.

CHAPTER 23

BIOCHEMICAL ANALYSIS OF NERVE TISSUE

Q1. What is Perikaryon?

(a) Perikaryon consists of a cell membrane—cytoplasm, a large vesicular nucleus having a single prominent nucleolus, numerous rod-like or spherical mitochondria, and a golgi aparatus.

(b) It also contains certain other organelles which are not present in a usual cell of the body. These organelles include dendrites, sex-chromatin, nissl granules, and neurofibrillae etc.

Q2. What is meant by sex-chromatin?

It is a large granule often found adjacent to the nucleolus. It is found only in females. Therefore, it is termed as sex- chromatin.

Q3. Express the nature of nissl granules.

(a) The cytoplasm of a perikaryon contains the basophilic masses known as nissl ganules.

(b) They contain many thin, membrane bound cavities arranged in parallel fashion.

(c) Their numbers and sizes vary with the physiological conditions of the cell.

Q4. What are neurofibrillae?

Neurofibrillae are the aggregations of thin, thread-like structures of variable length having 60 – 100 Å in diameter forming a loose network of fibrils in the cytoplasm.

Q5. State the clinical significance of nervous system.

(a) The nervous system controls the rapid activities of the body, eg, muscular contraction, respiration, and the rates of secretion of some endocrine glands such as adrenal medula and neurohypophysis.

(b) This system has the capacity to transmit rapidly a number of information at a time from one part of the body to another.

Q6. What is perineurium?

A nerve consists of a group of nerve fibers which are enclosed in a connective tissue sheath known as perineurium.

Q7. What is meant by epineurium?

A number of bundles of nerve fibers are bounded together by another sheath of connective tissue known as epineurium to form a nerve.

Q8. What is node of Ranvier?

A small uninsulated area exists at the junction between each two successive Schwann cells along the axon where ions can flow at ease between the extracellular fluid and the axon. This area is said to be the node of ranvier.

Q9. What do you mean by saltatory conduction?

The higher permeability at the nodes of ranvier causes the impulses to be conducted from node to node by the myelinated nerve fiber rather than continually along the entire fiber. This process is said to be saltatory conduction.

Q10. Why a minute supply of blood glucose is specially important to the nervous system?

Owing to the less amount of glycogen storage in the nerve tissue.

Q11. Mention the chemical composition of the nerve tissue.

(a) Nerve tissue is composed of water (80 per cent) and solids (20 per cent).

(b) The solids consist of proteins (38 to 40 per cent of total solids), lipids (50-54 per cent), small amounts of organic extracts and inorganic salts.

(c) Potassium is highly significant in the nerve impulse.

Q12. What is meant by nerve impulse?

It is defined as an electro-chemical change which is transmitted by nerve fibers.

Q13. Enumerate the prominent nervous symptoms.

Mental confusion, dizziness, delirium etc. in hypoglycemia.

Q14. Explain the role of brain tissue on protein metabolism.

Glutamic acid is the only amino acid metabolized by brain tissue. This acid serves as a precursor of GABA, one of the chemical transmitters, and as a major acceptor of ammonia. Thus, the brain tissue is protected against the toxic effects.

Q15. What is action potential?

The sequence of rapid changes in the electrical potential is known as the action potential.

Q16. Mention the types of membrane potential.

Action potential and resting membrane potential.

Q17. Define resting membrane potential.

It is defined as the net difference in the charge between inside and outside membrane involving little or no expenditure of energy.

Q18. State hyperpolarisation.

(a) The permeability of K^+ increases due to release of inhibitory transmitter.

(b) K^+ moves out rapidly through the membrane because of excess of K^+ inside the cell leaving a greater degree of negativity inside the neurone.

19. What do you mean by diffusion electrical potential?

The interior of the axon (nerve cell body) is about – 85 mv with respect to the exterior in the resting stage of the nerve. Since this potential is dependent on diffusion of K^+ it is said to be as diffusion electrical potential.

Q20. Explain the terms "Vagusstoff" and "Acceleransstoff".

The stimulation of sympathetic nerve supplied to the first heart can cause the stimulation of the second heart. The substance so liberated by vagus stimulation is termed as **Vagusstoff** and that liberated by sympathetic stimulation is termed as acceleransstoff.

Q21. Express Nernst equation.

$$E(mV) = \frac{RT}{\eta F} \times 23 \log_{10} \frac{[Ct]}{[Co]}$$

$$= \frac{61.5}{I} \times \log_{10} \frac{[30]}{I} = -90 \text{ mV (appr.)}$$

CHAPTER 24

BIOCHEMICAL AND GENETIC BASES OF DISEASES

Q1. How are diseases caused?

Diseases are caused by the alterations of structures; or the amounts of certain biomolecules; or disturbances of important biochemical reactions or processes.

Q2. How are patients treated by the physicians?

Physicians must treat the patients considering social, psychologic, cultural, economic, and other factors, but always depending on a sound knowledge of biochemical, physiologic, and pathologic mechanisms.

Q3. How are biomolecules affected?

(a) Biomolecules are affected in genetic diseases.

(b) The primary defect resides in DNA, and the structures or functions of other biomolecules are affected secondarily.

Q4. Mention the basis on which health depend.

Health fully depends on the regulated and harmonious functioning of many biochemical reactions and processes that occur in normal cells.

Q5. State the deficiency or excess of certain biomolecules that cause diseases.

(a) Deficiency of vitamin A causes night blindness; whereas excessive intake of vitamin A results in chronic states of toxicity.

(b) Deficiency of vitamin D results in rickets, but excess causes severe hypercalcemia.

Q6. Enumerate the biochemical changes that cause disease to occur rapidly.

(a) Some diseases progress rapidly causing death within minutes after coronary thrombosis on the fact that brain and heart are particularly sensitive to lack of oxygen and fuel.

(b) Life is threatened within hours at the excessive loss of water and electrolytes in cholera.

(c) Rapid and large changes of distribution of certain electrolytes (eg, K^+) become hazardous very quickly because of the sensitivity of myocardial muscles to such changes.

Q7. Mention the major intracellular organelles that are affected by diseases.

Organelles	Diseases
Nucleus	Most genetic diseases
Mitochondrion	Mitochondrial myopathies
Endoplasmic reticulum	chemical toxicities
Plasma membrane	Metastasis of cancer cells
Lysosome	Lysosomal storage diseases
Peroxisomes	Zellweger syndrome.

Q8. How will you distinguish among diseases?

By the history, physical examination, and suitable laboratory tests.

Q9. Point out the classes in which genetic diseases may be arranged.

(a) **Major classes :** Chromosomal disorders, monogenic disorders, and multifactorial disorders.

(b) **Additional classes :** Somatic disorders and mitochondrial disorders.

Q10. What is the characteristic of monogenic disorders?

(a) Monogenic disorders involve single mutant genes.

(b) They are classified as autosomal dominant, autosomal recessive, or x-linked.

(c) The term "dominant" signifies that the mutation will be clinically evident if only one chromosome is affected.

(d) The term "recessive" denotes that both chromoses must be affected.

(e) X-linked disorders are those in which the mutation is present on the x-chromosomes. Females have two x-chromesomes, whereas males have only one.

Q11. State the characteristic of multifectorial disorders.

(a) Multifactorial disorders are associated with a number of genes and environmental factors.

(b) Ischemic heart disease and hypertension are the members of this group.

Q12. Mention the characteristic of chromosomal disorder.

In chromosomal disorders, there is an excess or loss of chromosomes. The best-known such condition is Down syndrome.

Q13. Which are highly involved in genetic disorders?

Mutations.

Q14. What are the consequences of metabolic block?

(a) Decreased formation of the product.

(b) Accumulation of the substrate.

(c) Increased formation of metabolites.

Q15. What do you know about mutant protein?

(a) A mutant protein is synthesized if a structural gene for a non- catalytic protein is affected.

(b) The mutant protein does not function properly or move very slowly through the cell.

Q16. Mention the cues to the diagnosis of an inborn error of metabolism.

(a) Family history of a genetic disease.

(b) Positive screening test.

(c) Physical findings such as enlarged orgańs.

(d) Unusual odour from breath.

(e) Low blood sugar (eg, glycogen storage disease)

(f) Low blood pH (eg, due to accumulation of organic acids in maple syrup urine disease).

Q17. Enumerate the specific tests for the diagnosis of genetic diseases.

(a) Electrophoresis (eg, for Hbs).

(b) Measurement of low amount of product or elevated amount of precursor (eg, phenylalanine in PKU).

(c) Analysis by southern blotting of restriction fragment length polymorphisms (RFLPs) and other features of DNA structure linked to or causing specific diseases (eg, Hbs, Huntington disease, Duchenne muscular dystrophy).

Q18. Mention the successful treatment for some genetic diseases.

(a) Attempt to correct the metabolic consequences of a disease by administration of a missing product or by arranging the availability of substrate.

(b) Attempts to replace the absent enzyme or protein or to increase its activity.

(c) Attempts to remove excess of stored compound.

(d) Attempts to correct the basic genetic abnormality by gene therapy.

(e) Dietary treatments of phenylketonuria and galactosemia.

(f) Replacement therapy for hemophilia.

(g) Enzyme therapy is difficult in case of the brain.

(h) Sufficient efforts are being devoted to gene therapy.

Q19. What should be the treatment for methylmalonic aciduria?

Injections of vitamin B_{12}.

Q20. Mention the methods of introducing genes into cells for gene therapy.

(a) Intranuclear injection.

(b) Transfection (eg, with calcium phosphate)

(c) Electroporation.

(d) Adenovirus (used for introduction of the cystic fibrosis gene into cells of the respiratory tract)

(e) Retroviruses (specially constructed)

(f) Plasmid-liposome complexes.

(g) Site-directed recombination.

Q21. What is a triplex or antisense molecule?

A triplex molecule is formed by the binding of oligonucleotides (about 20 bases long) through base-pairing to a specific mRNA or to double-stranded DNA.

Q22. Name the substance that can act as genetic medicine.

Specific oligonucleotides.

Q23. Mention the substance which is useful in the treatment of AIDS.

Oligonucleotide blocks the replication of HIV-1 virus by combining with its gag gene and thus prove useful in the treatment of AIDS.

Q24. State some recent major achievements relating to molecular medicine.

(a) Development of gene and antisense therapy.

(b) Production of biotechnology—derived products based on recombinant DNA technology.

(c) Development of various DNA—based methods for diagnosis and screening of disease.

(d) Isolation of oncogenes, tumor suppressor genes, and cancer susceptibility genes and proposal of specific model for the development of colorectal cancer.

(e) Discovery that mutations in mitochondrial genes can cause diseases.

Q25. How does HIV infection occur?

(a) HIV infection is caused by the exchange of semen, vaginal secretions, milk or blood which is infected by virus.

(b) HIV transmission occurs through anal or vaginal intercourse.

(c) The infection with HIV is higher with anal than with vaginal intercourse.

(d) IVDA (intravenous drug abusers) are affected by HIV infection due to practice of "needle sharing" which allows transmission of HIV infected blood from one individual to another.

Q26. Where does the disease AIDS occur?

This disease specially occurs among homosexuals, contaminated bisexuals, drug abusers, babies from affected mother's milk, people receiving blood transfusion.

AIDS is a viral disease due to HIV.

Q27. What is HIV?

(a) HIV means human immunodeficiency virus.

(b) It belongs to the subfamily of retroviruses. Retroviruses have an RNA genome and transcribe a DNA copy of the RNA genome after penetration of the host cell.

(c) It has a core consisting of the RNA genome and core protein surrounded by an envelope with high lipid content rendering it sensitive to organic compounds.

(d) HIV infects cells in the central nervous system.

Q28. How will you confirm HIV infection?

This is confirmed by detecting the presence of antibodies to HIV in the patient's serum. Such antibodies are detected by the current ELISA test which is simple and cheap.

Q29. What is the effect of bacterial infections in HIV?

(a) Bacterial infections in AIDS causes tuberculosis. The diagnosis in this case is very difficult.

(b) TB may accelerate HIV disease and the prognosis is poor.

Q30. Mention the sites of involvement of AIDS.

(a) The most common site is the skin.

(b) The favoured sites are the mouth, tip of nose, penis, and lower legs.

(c) Lymph nodes are the second most common site.

(d) Stomach or rectal involvement may present with pain, bleeding or obstruction.

Q31. Which is the disease common in HIV infected patients?

(a) Skin disease is very common.

(b) The common skin diseases are seborrhoeic dermatitis, secondary symphilis, fungal infections, drug erruptions.

(c) Skin disease consists of weight loss, diarrhea, and dermatitis.

Q32. What should be the terminal care for AIDS?

(a) Good support from a partner, family, and friends can be of great help during this distressing time.

(b) Attention is to be paid to the control of nausea, vomiting, and diarrhea.

(c) Intramuscular injections should be avoided.

CHAPTER 25

CARDIAC PROFILE

Q1. What are the major manifestations of heart disease?

(a) Chest pain.

(b) Heart failure.

Q2. Mention the indications of the severe pain of myocardial infarction.

(a) Sweating

(b) Nausea.

(c) Vomiting.

Q3. What do you mean by Angina?

(a) Angina is a choking or constricting chest pain associated with exertion. The pain becomes worse by large meals or cold wind.

(b) The pain is relieved by rest or nitrates.

Q4. How is heart failure diagnosed?

It is diagnosed when a patient shows the symptoms of low cardiac output, pulmonary congestion or systemic venous congestion.

Q5. What should be the advice in case of patients with heart failure?

These patients are advised to avoid a high dietary salt intake, excess alcohol, salt or fluid retaining drugs.

Q6. What is the use of Electrocardiography (ECG)?

(a) It is used to find out cardiac arrhythmias and conduction defects.

(b) It gives information regarding electrolyte imbalance and the toxicity of certain drugs.

Q7. Mention the use of Echo Cardiography (ECHO).

It uses reflected ultrasound to study blood flow, the structure of the heart, the movement of valves and cardiac muscles.

Q8. What are the indications of Diastolic murmurs?

(a) These are indicatives of abnormal pathology.

(b) Continuous murmurs result from or combination of systolic and diastolic flow.

Q9. Mention the other manifestations of Cardiovascular disease.

(a) Oedema

(b) Palpitations

(c) Syncope.

Q10. Why does oedema occur?

It occurs due to excessive salt and water retention.

Q11. What does oedema indicate?

(a) It is the symptom of chronic heart failure.

(b) It usually affects the ankles, legs, thighs, and lower abdomen.

(c) Oedema of nephrotic syndrome is more severe and often affects the face and arms.

Q12. What is Syncope?

(a) This is a loss of consciousness resulting from an inadequate blood supply to the brain. This happens due to sudden vasodilatation.

(b) Exertional syncope occurs under conditions of aortic stenosis.

Q13. Mention the vital factors responsible for coronary (Ischaemic) heart disease.

(a) Age and male sex

(b) Smoking

(c) Hypertension

(d) Diabetes mellitus

(e) Hypercholesterolemia.

(f) Mental stress.

(g) Excessive alcohol intake.

Q14. How is coronary disease prevented?

(a) Smoking is prohibited.
(b) Regular exercise is a must.
(c) Maintenance of ideal body weight.
(d) Mixed diet.
(e) Taking of fats containing polyunsaturated fatty acids.

Q15. How does Angina pectoris occur?

It occurs due to an imbalance between myocardial oxygen supply and demand.

Q16. Discuss the clinical symptoms of Angenia pectoris.

(a) Stable angina is indicated by left-sided or central chest pain.
(b) Tightness in the chest—"like a band round the chest."
(c) 'Choking' sensation. Breathlessness is sometimes a prominent feature.
(d) The paid radiates to the neck and discomfort in the arms.
(e) The symptoms become worse after a heavy meal, in walking uphill, and in lying flat.

Q17. What should be the instruction for the patients with angina?

(a) Smoking is strictly prohibited.
(b) Regular exercise is a must.
(c) Severe exertion and vigorous exercise after a heavy meal must be avoided.
(d) The anti-anginal drugs, eg, nitrates, beta-blockers must be taken.

Q18. How does myocardial infarction occur?

It happens due to the formation of thrombus at the site of rupture of an atheromatus plaque in a coronary artery.

Q19. Point out the clinical symptoms of myocardial infarction.

(a) Severe pain lasts longer and the pain seems to be constriction in the chest.
(b) Breathlessness and vomiting.
(c) the unrecognised myocardial infarcts are common in elderly and diabetic patients.
(d) Sudden death may occur from ventricular fibrillation.
(e) There are signs of tissue damage causing fever and leucocytosis.

Q20. What should be the early management in case of acute myocardial infarction?

(a) Bed rest is most essential.
(b) Oral aspirin is required.
(c) High flow oxygen should be administered.
(d) Consideration of beta adrenoceptor antagonist.

Q21. State the factors for essential hypertension.

(a) Genetics and familial.
(b) Social deprivation.
(c) High salt and high alcohol intake.
(d) High renin.

Q22. Mention the causes of secondary hypertension.

(a) Renal disease.
(b) Endocrine disorders (cushing's syndrome, acromegaly, hyperparathyroidism etc.)
(c) Oral contraceptives.

Q23. Enumerate the adverse effects of hypertension.

(a) Stroke may result from cerebral hemorrhage.
(b) Loss of consciousness.
(c) Arteriolar damage.
(d) A pressure load on the heart.
(e) Higher incidence of coronary artery disease.
(f) Damage to the renal vessels.

Q24. How will you manage hypertension?

(a) Weight reduction diet.
(b) Avoidance of excess salt consumption.
(c) Heavy alcohol consumption must be reduced.
(d) Stopping of smoking.
(e) Regular exercise.
(f) Intake of drug like aspirin.

CHAPTER 26

ENVIRONMENT AND POLLUTION

Q1. How can you define Environment?

Environment is composed of three components :

(a) Physical : Water, air, soil, housing wastes, radiation etc.

(b) Biologic : Bacteria, viruses, insects, rodents, and animals.

(c) Social : Culture, habits, income, occupation, religion etc.

Q2. Define environmental sanitation.

It has been defined by WHO as "the control of all those factors in man's physical environment which may exercise a deleterious effect on his physical development, health and survival".

Q3. How can you say water is polluted?

Water is polluted when it is not

(a) free from pathogenic agents.

(b) free from harmful chemical substances.

(c) free from colour and odour.

(d) Usable for domestic purposes.

Q4. Mention the uses of water.

(a) Domestic use : For drinking, cooking, washing, bathing, gardening etc.

(b) Public purposes : For cleaning streets, swimming pools, public fountains, fire protection, public parks.

(c) Industrial purposes : For processing and cooling.

(d) Agricultural purpose : For irrigation.

(e) Waste removal purpose : For removal of wastes from all kinds of establishments and institutions.

(f) Power purpose : For power production from hydropower and steam power.

Q5. What does happen in the case of eliminated gases from power plants?

The gaseous sulfur and nitrogen oxides react with atmospheric water forming dilute solutions of sulphuric acid and nitric acid. The acid rain causes serious affects on plants and on surface water quality.

Q6. Why is not surface water safe for human consumption?

Surface water originates from rain water and it is contaminated from human and animal sources. Hence, it is not safe for human consumption.

Q7. How can you explain that the river water is quite unfit for drinking without treatment?

(a) River water contains all kinds of impurities especially the bacterial count is very high.

(b) The impurities of river water are derived from sewage, industrial and trade wastes, and drainage from agricultural areas. The water is polluted by bathing, animal washing, and disposal of the dead.

Q8. How does self-purification of river water occur?

It occurs by dilution, sedimentation, aeration, oxidation, and sunlight.

Q9. Why is ground water superior to surface water?

Because ground water has an effective filtering medium. It is mostly free from pathogenic agents and does not require any treatment. It contains salts of calcium and magnesium which make the water hard.

Q10. What do you mean by springs?

Ground water when comes to the surface and flows freely under natural pressure is called a "Spring".

Q11. How is water polluted?

(a) Natural impurities : Dissolved gases (Carbon dioxide, nitrogen, H_2S etc).

(b) Suspended impurities : Clay, sand, mud etc.

(c) Dissolved minerals : Salts of calcium, magnesium, sodium etc.

(d) Sewage : Decomposable organic matter and pathogenic agents.

(e) Industrial wastes : Toxic agents.

(f) Agricultural pollutants : Fertilizers and pesticides.

(g) Physical pollutants : Heat and radioactive substances.

(h) Pipelines : Corrosion, leaky joints, cross connection between water supply pipes and sewage drainage pipes.

Q12. State the methods for the purification of water.

(a) Ground water (Wells and springs) : No treatment required except disinfection.

(b) Surface water (River water) : Required storage, filtration, disinfection.

Q13. Mention the sources of air pollution.

(a) Automobiles : Motor vehicles emit hydrocarbons, CO, lead, nitrogen oxide etc.

(b) Industries : Fuel combustion generates heat and power produces smoke, SO_2 and nitrogen oxide. Petrochemical industries generate hydrogen fluoride and organic halides.

(c) Domestic sources : Combustion of coal and wood produces smoke dust, SO_2 and nitrogen oxide.

(d) Tobacco smoke : Affects health.

(e) Others : Pesticide spraying, fungi, molds.

Q14. What are the effects of air pollution on health?

(a) Immediate effects on respiratory systems causing acute bronchitis. If the pollution is intense, immediate death may result.

(b) Damages cardio respiratory system in various ways.

(c) Lead poisoning is dangerous for brain and nervous system to children. Elevated lead levels in children causes neuropsychologic development showing loss of IQ and poor school performances.

Q15. Define "noise".

Noise is "wrong sound in the wrong place at the wrong time".

Q16. State the effects of noise.

(a) Loss of hearing.

(b) High intensities workmen become irritated, short tempered, and impationed.

(c) Rise in blood pressure.

(d) Increase in heart rate and breathing.

(e) Increase in sweating.

(f) Nausea and fatigue may also occur.

(g) Narrowing of pupils and affects colour perception.

(h) Reduces night vision.

Q17. What are the methods for the control of noise?

(a) House front should lie not less than 15 meters from the road.

(b) Heavy vehicles should not be routed into narrow streets.

(c) The use of pressure horn should be prohibited.

(d) It is wise to construct detached buildings than a single large building.

(e) Buildings should be sound-proof.

(f) Industries and railways must be in special areas outside residential areas.

(g) Legislation should be adopted for controlling noise.

Q18. State the affects of improper excreta disposal.

(a) Soil pollution and water pollution.

(b) Contamination of foods and propagation of flies.

(c) Occurence of diseases like typhoid & paratyphoid fever, dysenteries, and other intestinal infections.

Q19. What criterias are fulfilled by sanitary latrines?

(a) Excreta does not contaminate the ground or surface water.

(b) Excreta does not pollute the soil.

(c) Excreta is not available to flies and animals.

(d) Excreta can not produce odour.

Q20. How are fecal-borne diseases transmitted?

Through various channels such as water, fingers, flies, soil and food.

CHAPTER 27

FAMILY PLANNING

Q1. What are the objectives of the family planning?

(a) To avoid unwanted births.

(b) To bring about wanted births.

(c) To regulate the intervals between pregnancies.

(d) To determine the number of children in the family.

Q2. What should be the ideal contraceptive methods?

(a) Inexpensive.

(b) Safe.

(c) Simple to use.

(d) Highly effective.

(e) Widely acceptable.

Q3. What do you mean by conventional contraceptive?

The methods that require action at the time of sexual intercourse eg, condoms, spermicides etc.

Q4. State the advantage of condom.

(a) No side effects.

(b) Cheaper in case of cost.

(c) Easy to carry and simple to use.

(d) Protection against sexually transmitted diseases.

(e) Protection against pelvic inflammatory diseases.

(f) Protection against cervical cell abnormalities.

(g) Reduces the incidence of tubal infertility and ectopic pregnancy.

Q5. What are the disadvantages of condom?

(a) Inadequate sexual pleasure.

(b) May accidentally break during coitus.

(c) Disposable after one coital act.

Q6. How many bands of condoms are used for males?

Three bands : eg, Dry Nirodh, Delux Nirodh etc.

Q7. Explain Female Condom.

It is a pouch made of polyurethane which lines the vagina and also the external genitalia. Inner ring at the closed end is smaller compared to the outer ring.

Q8. State the complications in case of IUCD.

(a) Cramp like pain.

(b) Perforation.

(c) Abnormal menstrual bleeding.

(d) Pelvic inflammatory disease.

(e) Spontaneous expulsion.

(f) Syncopal attack.

Q9. Give examples of IUCD.

(a) CuT200.

(b) CuT380A.

(c) Multiload 250

(d) Multiload 375

(e) Progestasert.

(f) LNG 20.

Q10. Explain the mechanism of action of combined oral contraceptives (pills).

It prevents ovulation :

(a) By supressing LH surge.

(b) By supressing FSH.

(c) By supressing the pulsatile sensation of GnRH from hypothalamus.

(d) By making the cervical mucosa thick, teneceous, so that penetration of sperm is inhibited.

Q11. State the complication of OCP.

(a) Nausea, vomiting, headache, leg cramps.

(b) Weight gain.
(c) Depression, mood change, sleep.
(d) Hypertensions.
(e) Thromboembolic phenomenon.
(f) Disturbances.
(g) Mastalgia.

Q12. Mention the non contraceptive benefit of OCP.

(a) Decreased incidence of endometrium and ovarian carcinoma.
(b) Diminished premenstrual tension.
(c) Relief of menorrhagia, dysmenorrhoea.
(d) Relief of iron deficiency anemia.

Q13. What are the causes for the indications for withdrawal of OCP?

(a) Severe migraine.
(b) Visual disturbances.
(c) Sudden chest pain.
(d) Severe cramp and pain in leg.
(e) Excessive weight gain.
(f) Severe depression.
(g) Desire pregnancy.

Q14. What do you mean by sterilisation?

It is a surgical method by which the reproductive function of an individual male or female is purposefully and permanently destroyed.

The operation done in female is called tubectomy.

Q15. What is vasectomy?

It is the operation done in male where a segment of vas deferens of both the sides are resected and the cut ends are ligated.

Q16. State the advantage of vasectomy.

(a) It can be done in outdoor procedure or in a mass camp even in remote villages.
(b) It is simple and can be performed by one with minimal training.
(c) Failure rate is minimum.

Q17. What are the drawbacks of vasectomy.

Additional contraceptive protection is required for 2-3 months after operations i.e., till the semen becomes free of sperm.

Q18. Mention the complications in case of vasectomy.

(a) Wound sepsis.
(b) Impotency.
(c) Sperm granuloma.
(d) Scrotal hematoma.
(e) Spontaneous recanalisation.

Q19. What is tubectomy?

It is an operation where resection of a segment of both the fallopian tubes is done to achieve permanent sterilisation.

Q20. What are indications for vesectomy?

(a) Family planning purpose.
(b) Socio-economic.
(c) Medico-surgical indications.

Q21. Mention the time of tubectomy.

(a) During puerperium — 24-48 hrs following delivery.
(b) Concurrent with MTP.
(c) Complete her family.

Q22. What are the different techniques for tubectomy?

(a) Park land.
(b) Irving.
(c) Madiener.
(d) Minilap.

CHAPTER 28

ACID–BASE REGULATION

Q1. What is conjugate acid base pairs?

The stronger the acid, the weaker the base. Such pairs are termed as conjugate acid base pairs. Chloride ion (Cl^-) is the conjugate base of the acid HCl.

Q2. What do you mean by aprotes?

The cations such as Na^+, Ca^{++}, K^+, Mg^{++} cannot donate or accept protons. Hence, they are neither acids nor bases. Such cations are termed as aprotes.

Q3. How is sulphuric acid produced in the body?

This acid is formed by the oxidation of sulphur containing amino acids in the body, eg, methionine and cystine or cysteine.

Q4. Mention the factors involved in the regulation of blood pH.

(a) Buffer systems in the blood.

(b) Respiratory mechanisms.

(c) Renal mechanisms.

(d) Dilution factor.

Q5. Why is $NaHCO_3$ in the blood is represented as the alkali reserve?

Because it does combine with strong and non-volatile acid.

Q6. What is the disadvantage of bicarbonate buffer in the body?

This buffer is very weak as a chemical buffer and hence pKa is far away from the physiological pH.

Q7. Prove that the phosphate buffer system works in conjugation with the kidneys.

(a) When a strong acid enters the blood, it is fixed up by Na_2HPO_4 which is converted to acid phosphate. This acid phosphate is excreted by the kidneys showing that the urine becomes more acidic.

(b) When an alkali enters, it is buffered by the acid phosphate forming the alkaline phosphate which is excreted in the urine showing the increased alkalinity of urine.

Q8. Show the effects of imidazole group of histidine on the physiological buffering action of Hb.

Imidazole contains two groups :

(a) Fe^{++} containing group which is concerned with carriage of O_2.

(b) Imidazole N_2 group which can give up H^+ and accept H^+ depending the pH of the medium.

Q9. What do you mean by isohydric transport of CO_2?

The newly formed H^+ in tissues from H_2CO_3 does not bring about a change in pH. This circumstance in the role of Hb buffers is sometimes referred to as isohydric transport of CO_2.

Q10. What is meant by paradoxic aciduria?

Potassium ions (K^+) leave the cell in K^+ deficiency and H^+ enters the cells producing intracellar acidosis. ECF becomes alkaline. Increased excretion of H^+, NaH_2PO_4, and NH_4Cl increase the titratable acidity. ECF is though alkaline, highly acidic urine is excreted. This condition is referred to as paradoxic aciduria.

Q11. State the importance of the drug diamox clinically.

The drug diamox inhibits the enzyme carbonic anhydrase. The increased amount of $NaHCO_3$ appears in the urine. There is reduction in titratable acidity and ammonia excretion in the urine. Hence, this drug is clinically used as a diuretic to induce a loss of Na^+ and water in congestive cardiac failure and in hypertensive heart diseases.

Q12. Mention the causes for the abnormal increase in "anions" other than HCO_3^-.

(a) Endogenous excessive production of acid ions occur in diabetic acidosis, starvation, high fever, violent exercise, hemorrhage, and shock.

(b) Ingestion of excessive quantities of phosphoric acid, HCl, salicylic acid, and mandelic acid etc.

(c) Renal insufficiency causing retention of acids commonly observed in the terminal stages of nephritis.

Q13. What are the conditions causing depression of respiration?

(a) Damage to CNS :

(i) Brain damage.

(ii) Drug poisoning.

(iii) Excessive anaesthesia.

(b) Loss of ventilatory functions :

(i) Tension.

(ii) Pulmonary and mediastinal tumors.

(c) Effect of pain eg, pleurisy.

Q14. Mention the factors by which metabolic alkalosis occurs.

(a) Decreased ionic calcium in blood producing tetany.

(b) Hypokalamia due to the increased excretion of K^+ from the distal tubules.

(c) Kidney damage caused by the degenerative changes in the tubules.

(d) Ketosis and ketonuria due to excessive vomiting for the insufficient carbohydrate intake.

Q15. What is the effect of large doses of salicylate?

Large doses of salicylate cause stimulation of respiratory centre (RC) with hyperventilation showing alkalosis.

RENAL FUNCTION TESTS (RFT)

CHAPTER 29

Q1. Classify renal function tests.

(a) Tests based on Glomerular filtration.

(b) Tests based on Renal Plasma Flow (RPF).

(c) Tests based on Tubular Function.

(d) Certain Miscellaneous Tests.

Q2. Mention the Glomerular Filtration Tests.

(a) Urea clearance test.

(b) Endogenous Creatinine clearance test.

(c) Inulin clearance test.

(d) Cr^{51}—EDTA clearance test.

Q3. Define clearance of a substance.

It is defined as the volume of blood/plasma which contains the amount of the substance which is excreted in the urine in one minute.

Q4. Explain blood urea clearance.

It is expressed as the number of millilitre (ml) of blood/plasma which are completely cleared of urea by the kidney per minute.

Q5. What is meant by maximum urea clearance (cm)?

The clearance of urea which occurs when the urinary volume exceeds 2 ml per minute is termed as maximum urea clearance (cm).

Q6. Clearify standard urea clearance (Cs).

The rate of urea elimination is reduced when the urinary volume is less than 2 ml. per minute. Such clearance is termed as standard urea clearance.

Q7. When does terminal uremia occur?

It appears when the urea clearance falls to about 50 per cent of the normal values.

Q8. What is the importance of determination of Cr^{51}—EDTA clearance?

This clearance is widely used for routine assessment of glomerular filtration rate (GFR) in adults as well as in children. It is especially convenient in children where it is not easy to collect 24 hour urine sample. It is used for children younger than one year old.

Q9. Clearify filtration fraction (FF).

It is the fraction of plasma passing through the kidney which is filtered at the glomerulus. This is obtained by dividing the inulin clearance by the PAH clearance.

Q10. In which cases pyelography should not be done?

In case of patients suffering from acute nephritis, severely impaired liver function, congestive cardiac failure, hypersensitivity etc.

Q11. Explain the importance of radio-active scanning.

(a) This technique is useful for the detection of abnormalities in size, shape, and position of the kidneys.

(b) This technique also shows renal tumors and renal infarcts which are not detected in pyelography.

Q12. Name the recently used substance which gives better result in the indication of renal function.

Diodone (30 per cent) which is available as pyelosil (Glaxo) gives better result. The excretion of diodone is used as a rough indication of renal function.

CHAPTER 30

GASTRIC FUNCTION TESTS

Q1. Mention the purposes for performing gastric analysis.

(a) Diagnosis of gastric ulcer.
(b) Diagnosis of pulmonary tuberculosis.
(c) Diagnosis of Zollinger-Ellison syndrome.
(d) Confirming surgical vagotomy.

Q2. What does it indicate in the presence of food residue in the gastric juice?

Carcinoma of the stomach.

Q3. What will be the indication if there is brown or reddish-brown blood in the gastric juice?

Gastric ulcer as well as in gastric carcinoma.

Q4. In which states increased amount of mucus are present in gastric juice?

(a) In gastritis and in gastric carcinoma.
(b) In swallowed saliva.

Q5. Mention the normal values of free acid and total acid.

Free acid : 0 - 30 mEq/L.
Total acid : 10 - 40 mEq/L.

Q6. In which case will you find achlorhydria?

Carcinoma of stomach.

Q7. When does gastric juice show low acidities and absence of free HCl?

(a) Carcinoma of stomach in case of low acidities.
(b) Pernicious anemia in case of absence of free HCl.

Q8. State the conditions in Achylia gastrica.

Gastric secretion is completely absent due to advanced cases of cancer of stomach, advanced cases of gastritis, and acute pernicious anemia.

Q9. Express the advantages of alcohol test meal.

(a) More easily administered.
(b) Consumed better.
(c) Quick emptying of the stomach.
(d) Specimens are cleared and easily analysed.

Q10 What does happen in case of insulin stimulation test?

(a) Acid level is more in duodenal ulcer.
(b) The concentration of free acid may be over 100 mEq/L.

Q11. What is pentagastrin?

(a) Synthetic peptide composed of butyloxycarbonyl β- alanine.
(b) Active stimulator.

Q12. Express the findings after pentagastrin test.

(a) In duodenal ulcer, the gastric secretion is 15 to 83 mEq/hour.
(b) Little value in gastric ulcer.
(c) The "true" achlorhydria is found in cancer of the stomach and in pernicious anemia.
(d) Reduced acid level in found in acute gastritis.
(e) The secretion of pepsin occurs after stimulation with pentagastrin.

Q13. State the characteristics of Zollinger-Ellison Syndrome.

(a) This syndrome is characterised by a high basal secretion usually above 10 mEq/hour.
(b) It is also characterised by a peptic ulcer, gastric hypersecretion, and diarrhoea in patient with "gastrin".
(c) This is also accompanied by parathyroid adenomas with hyperparathyroidism.

Q14. Name the valuable test which is used as "screening test" only.

"Diagnex Blue" test.

Q15. Explain the significance of "Diagnex Blue" test.

(a) A positive result indicates the secretion of acid by the stomach.

(b) A negative result is an unreliable indicator of "true achlorhydria".

(c) This test is not reliable in patients suffering from renal diseases, urinary retention, malabsorption, pyloric obstruction.

CHAPTER 31

LIVER FUNCTION TESTS

Q1. State the important metablic functions of liver.

(a) Carbohydrate, fat, and protein metabolism.

(b) Storage of glycogen, vitamins, and iron.

(c) Destruction of red blood cells, formation and excretion of bile pigments.

(d) Synthesis of plasma proteins, fibronogen, and prothrombin.

(e) Detoxifying functions.

Q2. Explain the fate of bilirubin as a result of bacterial action.

Bilirubin is converted into mesobilirubin which is then reduced to mesobilirubinogen in the intestine. This compound is further reduced to stercobilinogen (urobilinogen) which is highly soluble. About 5 percent of it is excreted into the urine. The urobilinogen is then oxidized to urobilin. The stercobilinogen in the feces is oxidized to stercobilin.

Q3. Define jaundice.

(a) It means a yellowish tint to the body tissues including yellowness of the skin and also of the deep tissues.

(b) The usual cause of jaundice is the increased bilirubin content in the extracellular fluids.

(c) The skin becomes yellow when the concentration rises to about three times the normal.

Q4. Express the causes of jaundice.

(a) Increased destruction of red blood cells.

(b) Excessive production of bilirubin.

(c) Dysfunction of liver cells to convert bilirubin into bilirubin glucuronide form in the liver and excrete it in the bile.

(d) Obstruction of bile ducts to make usual flow of the bile to the duodenum.

Q5. What is the normal bilirubin level in the serum?

0.3 - 1.1 mg/dL.

Q6. Name the diagnostic process to differentiate between hemolytic and obstructive jaundice.

Van den Berg reaction.

Q7. Mention the biochemical tests for jaundice.

(a) Serum bilirubin concentration and nature of Van den Berg reaction.

(b) Urine test for urobilinogen and bilirubin.

(c) Serum alkaline phosphatase activity and SGPT acitivty.

(d) Thymol flocculation test and colloidal gold test.

(e) Feces colour.

Q8. Explain the causes of hepatic jaundice.

(a) It is caused by infection, toxins, and liver poisons.

(b) Infection with virus is the common cause. The infective organisms cause damage to the liver parenchymal cells. Hence, the conjugation of bilirubin in the liver is affected.

Q9. What do you mean by Gilbert's syndrome?

(a) There is a defective intracellular transport of bilirubin. In some cases, there is defective hepatic conversion of bilirubin to bilirubin diglucuronide.

(b) As a result, the serum bilirubin concentration level ranges from 4 to 6 mg per 100 ml. Sometimes, it may rise to 12 mg.

Q10. What is meant by Crigler-Najjar syndrome?

(a) This syndrome appears in new born infants.

(b) The serum bilirubin level rises to 20 mg/100 ml or more within a few days after birth.

(c) Bilirubin is not converted to bilirubin diglucuronide due to the deficiency of the enzyme of **glucuronyl transferase** in the liver.

(d) The serum bilirubin level comes down to normal level in those infants who survive.

Q11. Mention the results obtained in liver diseases.

(a) γ-globulin is increased in chronic infective hepatitis.

(b) Albumin content is decreased in cirrhosis of the liver.

(c) Fibronogen content is decreased in necrosis of the liver.

Q12. State the removal or excretion of drugs, hormones, and other substances of liver.

(a) Different drugs like sulfonamides, penicillin, ampicillin, and erythromycin.

(b) Hormones like thyroxine, all of the steroid hormones.

(c) Other substances like calcium from the blood into the bile and then into the feces.

CHAPTER 32

NORMAL LABORATORY VALUES

HEMATOLOGY

Bleeding time : 1–7 minutes (Ivy method).

Coagulation time : 4–8 minutes (Lee and white)

Clot retraction : 40–65%
Starts in 1–3 hours.
Completes in 6–24 hours.
No clot lysis in 24 hours.

Fibrinogen : 200–400 mg/dL

Activated partial thromboplastin time : 24–35 seconds.

Thrombin time : 9–10 seconds (control + 3 seconds)

Hemoglobin (In blood) : Males = 14.5–16.5 gms/dL
Females = 12.0–14.5 gms/dL.

Hemoglobin (In serum) : 2–3 mg/dL.

Packed cell volume : Males = 38–50%
Females = 34–43%

Mean corpuscular volume : 78–94 cμ.

Erythrocyte Sedimentation Rate (ESR) [Westergren method] : Males = 0–15 mm at the end of 1 hour.
Females = 0–20 mm at the end of 1 hour.

Cellular measurement of red cells : Average diameter = 7.3 μm.
Mean corpuscular volume (MCV) : Males = 80–94 fL
Females = 81–99 fL
Mean corpuscular hemoglobin (MCH) = 27–32 Pg
Mean corpuscular hemoglobin concentration (MCHC) = 32–36%

Fragility of red cells : Starts at 0.5% NaCl.
Completes at 0.3% NaCl.

Red blood count (RBC) : Males = 4.5–6.2 million/μL.
Females = 4.0–5.5 million/μL.

Average plasma volume : 45 ≠ 5 ml/kg body weight.
1487 ≠ 112 ml/sq meter body surface.

Platelets : 1,50,000–4,50,000 /cmm.

Reticulocytes count : 0.5–2% of erythrocytes.

Total leucocyte count : 4,000–11,000/cmm.

Differential Count of WBC or Leucocyte:

Myelocytes	0%
Juvenile neutrophils	0%
Band neutrophils	0-5%
Segmented neutrophils	60-70%
Lymphocytes	20-30%
Eosinophils	1-7%
Monocytes	1-6%
Basophils	0-1%
Lymphocytes (Total):	1500-4000 / µL
B cell	5-25%
T cell	60-88%
Suppressor	10-43%
Helper	32-66%
H : S	> 1.

BODY FLUIDS

Body Fluids

Total volume	: 50 per cent (in obese) to 70 per cent (Lean) of body weight.
Intracellular	: 0.3 – 0.4 of body weight.
Extracellular	: 0.2 – 0.3 of body weight.

Blood

Total volume	: Males : 69 ml/kg. body weight Females : 65 ml/kg. body weight
Plasma volume	: Males : 39 ml/kg. body weight. Females : 40 ml/kg. body weight
Red blood cell volume	: Males : 30 ml/kg. body weight (1.15–1.21L/m^2 body surface area) Females : 25 ml/kg. body weight (0.95 – 1.00 L/m^2 body surface area)

CEREBROSPINAL FLUID (CSF)

Appearance	: Clear and colourless
Specific gravity	: 1.003 – 1.008
Cells	: < 5/mm^3, all lymphocytes
pH	: 7.34 – 7.43
Pressure	: 50 – 180 mm H_2O
Glucose	: 40 – 70 mg/dL or 60% of serum glucose
Chloride	: 118 – 132 mEq/L
Protein, Lumber	: 15 – 45 mg/dL.
Albumin	: 58%
α_1-globulins	: 9%
β-globulins	: 10%
γ-globulins	: 10%
Protein, Cisternal	: 15 – 25 mg/dL
Protein, Ventricular	: 5 – 15 mg/dL

Blood Acid-base

		Normal range	
		Arterial	Venous
Actual pH	:	7.35 – 7.45	7.32 – 7.42
PCO_2 (mm Hg)	:	35 – 45	41 – 51
Total CO_2 (mmol/L. plasma)	:	23 – 27	26 – 30
Actual bicarbonate (mmol/L. plasma)	:	22 – 26	24 – 28
Oxygen saturation (%)	:	96 – 97	40 – 70
Oxygen tension PO_2 (mm Hg)	:	80 – 100	25 – 40

SYNOVIAL FLUID

Cells	: 200 cells/mm^3
Polymorphonuclear cells	: 25%
Fibrin clot	: None
Crystals	: None
Glucose	: Same as serum
Hyaluronic acid	: 2.45 – 3.97 g/L
pH	: 7.32 – 7.64
Protein	: 2.5 g/dL
Albumin	: 63%
α_1-globulins	: 7%
α_2-globulins	: 7%
β-globulins	: 9%
γ-globulins	: 14%
Uric acid	: Same as serum

SEMEN

Liquefaction	: Complete in 15 minutes
Morphology	: 60% normal forms.
Motility	: 75% motile forms.
pH	: 7.2 – 8.0
Volume	: 2.0 – 6.6 mL.
Spermatocrit	: 10%
Spermatocyte count	: 50 million/dL

STOOL

Bulk	
Dry weight	: < 66.4 g/24 hours.
Wet weight	: < 197 g/24 hours.
Water	: Approximately 65%
Fat (on a diet containing 30 g. fat/day)	: < 7.2 g/24 hours or 30% dry weight
Nitrogen	: < 2.2 g/24 hours.
Coproporphyrin	: 12 – 832 mg/24 hours.
Urobilinogen	: 40 – 280 mg/24 hours.

CLINICAL CHEMISTRY

Constituents	Blood	Serum	Plasma	Urine
Glucose (fasting)			70–115 mg/dL	Qualitative – Nothing detected
Glucose (2 hrs after meal)			115–160 mg/dL	Quantitative – 16-300 mg/24 hours
pH	7.35–7.45			4.6–8.0
Specific gravity	1.056 (varies with Hb and protein concentration)			1.003–1.030
Cholesterol (Total)		140–220 mg/dL	140–220 mg/dL	
HDL cholesterol				
Male		3.5 – 5.5 mg/dL		
Female		4.5 – 6.5 mg/dL		
Triglycerides		40–160 mg/dL		
Phospholipids				
Male		125–275 mg/dL		
Female		35–135 mg/dL		
Free fatty acids		8–25 mg/dL		
Protein (Total):		6.3–7.9 g/dL		Qualitative – None detected
Albumin		3.5–5.3 g/dL		
Globulin		1.8–3.6 g/dL		Quantitative – 10-150 mg/24 hrs.
Fibrinogen			0.2–0.6 g/dL	
Protein electrophoresis				
Alpha - 1		0.2–0.4 g/dL		
Alpha - 2		0.5–0.9 g/dL		
Beta		0.6–1.2 g/dL		
Gamma		0.7–1.7 g/dL		
Chylomicrons		Absent.		
Alpha lipoprotein		13–24%		
Beta lipoprotein		60–75%		
Pre-beta lipoprotein		8–20%		
Urea		15–45 mg/dL	15–45 mg/dL	
Urea nitrogen		7–21 mg/dL	7–21 mg/dL	
Uric acid		Males : 2.5–7.0 mg/dL Females : 2.5–5.5 mg/dL	Males : 2.5–7.0 mg/dL Females : 2.5–5.5 mg/dL	600–700 mg/24 hrs.
Bilirubin (Total)				
Adult		0.3–1.1 mg/dL		None detected
New born		0.4–4.0 mg/dL		
At 3 days		1.0–10.0 mg/dL		
Direct (Conjugated)		0.2–0.3 mg/dL		
Indirect		0.3–0.6 mg/dL		
Creatinine		0.6–1.2 mg/dL (Male/Female)	0.6–1.2 mg/dL (Male/Female)	0–200 mg/24 hours (Adult) 7–11 mg/24 hours (Children)
Ascorbic acid			0.4–1.5 mg/dL	
Sulfate		0.5–1.5 mg/dL		
Viscosity		1.4–1.8 mg/dL (Serum compared to H_2O)		
Lactate	4.5–19.8 mg/dL			
Pyruvate	0.6–1.0 mg/dL			
Chloride		98–110 mEq/L	98–110 mEq/L	120–240 mEq/24 hours (Varies with dietary intake)

Constituents	Blood	Serum	Plasma	Urine
Bicarbonate		24–28 mEq/L		
Ammonia			19–80 µg/dL	20–50 mEq/24 hrs.
Vitamin A		0.15–0.60 µg/dL		
Vitamin D :				
25-Hydroxycholecal-ciferol		8–55 ng/ml		
1, 25-dihydroxycho-lecalciferol		26–65 Pg/ml		
24, 25-dihydroxycho-lecalciferol		1–5 ng/ml		
Vitamin B12		200–850 µg/dL		
Porphobilinogen				Qualitative : None decteted
Urobilinogen				Quantitative : 0–2.4 mg/24 hrs.
Coproporphyrin				1–3.5 mg/24 hrs.
Uroporphyrin				50–250 µg/24 hrs.
Acetone and Acetoacetate		0.2–2 mg/dL		Less than 1 mg/day
Sodium		136–146 mEq/L	136–146 mEq/L	10–150 mEq/24 hours (Varies with dietary sodium intake)
Potassium		3.5–4.5 mEq/L	3.5–4.5 mEq/L	25–100 mEq/24 hours
Calcium		8.4–10.2 mg/dL		0.05–0.31 mEq/kg body wt./24 hour in normal diet
Calcium ionized		4.15–5.20 mg/dL		
Copper		75–160 µg/dL	75–160 µg/dL	0–32 µg/24 hours
Zinc		50–150 µg/dL		
Iron		65–170 µg/dL		
Iron binding capacity		250–450 µg/dL		
Lead		96–106 mEq/L	96–106 mEq/L	0–120 µg/24 hours
Magnesium		1.7–2.6 mg/dL	1.7–2.6 mg/dL	
Ferritin		Male : 30–300 ng/mL Female : 20–120 ng/mL Children : 7–140 ng/mL		
Hepatoglobin		40–170 mg of Hb binding capacity		
Phenylalanine		3 mg/dL		
Phosphorus				
Adult		3–5.0 mg/dL		0.8–2.0 g/24 hours
Females		3.5–5.0 mg/dL		
Ceruloplasmin				
Males		36 ± 5.6 mg/dL		
Females		40.9 ± 6.8 mg/dL		
Serotonin	5–20 µg/dL			
Transferrin		200–400 mg/dL		
Aldolase		Up to 7.6 U/L		
Amylase		25–125 U/L		25-260 Somogyl Unit/h
Acid Phosphatase		< 4 U/L		

Constituents	Blood	Serum	Plasma	Urine
Acid Phosphatase, Prostatic		< 1.2 U/L		
Alkaline Phosphatase				
Adult		20–70 U/L		
Child		20–150 U/L		
Glucose-6-Phosphate dehydrogenase		12.1 ± 2.09 U/g Hb		
Lipase		0–160 U/L		
Creatinine Phosphokinase (total)		170 U/L		
Creatinine Phosphokinase isoenzymes				
MM fraction		94–95%		
MB fraction		0–5%		
BB fraction		0–2%		
γ-glutamyl transpeptidase				
Male		9–50 U/L		
Female		8–40 U/L		
SGOT		8–40 Karmen units/dL		
SGPT		0–40 Karmen units/dL		
LDH isoenzymes				
LDH_1		20–34%		
LDH_2		28–41%		
LDH_3		15–25%		
LDH_4		8–16%		
LDH_5		6–15%		
Gastrin Immunoglobulin		< 200 Pg/ml		
Quantitative (Adults)				
IgG		570–1920 mg/dL		
IgE		< 0.025 mg/dL		
IgA		70–400 mg/dL		
IgM				
Males		30–250 mg/dL		
Females		30–300 mg/dL		
IgD		0–8 mg/dL		
Protein electrophoresis				
α_1		0.2–0.4 g/dL		
α_2		0.5–0.9 g/dL		
β		0.6–1.2 g/dL		

Hormones	Serum	Plasma	Urine
ACTH	15–70 Pg/ml		
Insulin (Fasting)		6–126 μU/ml	
Hypoglycemia		< 5 μU/ml	
Glucagon (fasting)	20–100 Pg/ml		
Cortisol	5–25 μg/dL		20–100 μg/24 hrs.
FSH	Males : 6–18 μU/dL Female : 5–20 μU/dL (Follicular phase) 5–15 μU/dL (Luteal phase) 50 μU/dL (Potmenopausal)		
Parathyroid Hormone	150–300 Pg/ml		
Calcitonin	Not detectable		
Catecholamines			110 μg/24 hours (Total) Epinephrine : 10 μg/24 hours Norepinephrine : 100 μg/24 hours
Metanephrine			1.3 mg/24 hours.
Thyroid stimulating Hormone (TSH)	< 10 μU/ml		
Thyroxine (T_4)	Total : 4.5–11.5 μg/dL Free : 0.8–2.4 μ/dL		
Thyroxine index (free)	1–4 mg/dL		
Tri-iodothyronine (T3)	70-190 mg/dL		
Thyroxin-binding globulin capacity	15–25 μg T_4/dL		
Radioactive iodine 131 I uptake (RAIU)	5–25% at 24 hours (Varies with iodine intake)		
Renin activity		Supine : 1.1 ± 0.8 mg/dL	
Normal diet		Upright : 1.9 ± 1.7 mg/dL	
Low-Sodium diet		Supine : 2.7 ± 1.8 mg/dL	
Diuretic and low-sodium		Upright : 6.6 ± 2.5 mg/dL	
Testosterone			
Total plasma bound			
Adolescent males		100 mg/dL	
Adult males		30–100 mg/dL	
Adult females		25–90 mg/dL	
Unbound			
Adult males		3–24 mg/dL	
Adult females		0.09–1.30 mg/dL	
17-Ketogenic steroids			
Under 8 years			0–2 mg/24 hours
Adolescent			0–18 mg/24 hours
Adult males			8–18 mg/24 hours
Adult females			5–15 mg/24 hours

Hormones	Serum	Plasma	Urine
Oestrogens			
Total : Males			4–25 μg/24 hours
Females			5–100 μg/24 hours
Oestriol : Males			1–11 μg/24 hours
Females			0–65 μg/24 hours
Oestrodiol : Males			0–6 μg/24 hours
Females			0-14 μg/24 hours
Oestrone : Males			3-8 μg/24 hours
Females			4-31 μg/24 hours
Oestriocholanolone (Female)			1.2 μg/24 hours
Norepinephrine		< 0.5 μg/L	100 μg/24 hours
Epinephrine		< 0.1 ng/ml	
Leuteinizing hromone (LH)	Male : 1–15 mIU/ml Female : < 1–30 mIU/ml Castrate : > 30 mIU/ml		
Antidiuretic hormone (ADH, Vasopressin)		0–2 Pg/ml	
Insulin — like growth factor-1 (IGF–1)		400–2000 units/L	
Progesterone			
Follicular phase	0.2–1.5 mg/ml		
Luteal phase	6–32 ng/ml		
Pregnancy	> 24 ng/ml		

RENAL FUNCTION TESTS

Specific gravity of urine	:	1.003–1.030
Maximal glucose reabsorptive capacity	:	Males : 300–450 mg/min. Females : 250–350 mg/min.
Inuline clearance (GFR)	:	Males : 110–150 mL/min
	:	Females : 105–132 mL/min
Creatinine clearance $\left(\frac{U_{Cr} \times V}{P_{Cr}}\right)$	:	Males : 130 ± 20 ml/min. Females : 120 ± 15 ml/min.

U_{Cr} = Urine creatinine (mg/dL)
P_{Cr} = Plasma creatinine (mg/dL)
V = Urine volume (ml/24 h)

P-Aminohippurate (PAH) clearance (RPF)	:	Males : 560–830 mL/min
	:	Females : 490–700 mL/min
Maximal PAH excretory capacity (Tm PAH)	:	80–90 mg/min
Filtration fraction (FF)	:	Males : 17–21%
[FF = GFR/RPF]	:	Females : 17–23%,
Anion gap $Na^+ - (HCO_3^- + Cl^-)$	:	10 ± 2 mEq/dL
Osmality	:	292–297 mosm/L

$$\text{Osmality (Serum)} = 2Na^+(\text{mEq/L}) + \frac{\text{BUN (mg/dL)}}{2.8} + \frac{\text{Glucose (mg/dL)}}{18}$$

$$\text{Bicarbonate } (HCO_3^-) \text{ deficit} = \text{Body weight (Kg)} \times 0.4 \text{ (desired } HCO_3^- - \text{observed } HCO_3^-)$$

$$\text{Renal plasma flow (RPF)} = \frac{U_{pah} \times V}{PPah}$$: Males : 700 ± 100 ml/min. Females : 600 ± 100 ml/min.

Where, U_{pah} = Urine Para–aminohippuric acid
V = Urine volume /24 h.
PPah = Plasma Para–aminohippuric acid

$$\text{Fractional excretion of sodium : (FE } Na^+) = \frac{(U/P)\,Na^+}{(U/P)\,Cr} \times 100$$

[Adapted from current Medical Diagnosis and Treatment, 2003]

$$\text{Milliequivalent per litre (mEq/L)} = \frac{\text{mg per 100 ml} \times 10}{\text{equivalent weight}}$$

APPENDIX

MODEL QUESTIONS AND ANSWERS

MODEL 1

Q1. Name one monosaccharide.

Glucose.

Q2. Mention the isomeric and epimeric form of glucose.

Isomeric form : Fructose.

Epimeric form : Mannose and galactose.

Q3. What do you mean by isomerisation and epimerisation?

Isomerisation : It is the process by which some compounds which are identical in composition but differ only in configuration.

Epimerisation : It is the process by which two sugars which differ from one another only in the configuration around a single carbon atom.

On the body by epimerase enzyme.

Q4. State the formula by which the isomers of glucose are determined.

2^n = 16 isomers. n = 4, the asymmetric carbon atom.

Q5. What do you mean by asymmetric carbon atom?

The carbon atom which is attached to four different groups or atoms is called the asymmetric carbon atom.

Q6. What is the chemical nature of glucose?

(a) It is a hexose aldose sugar.

(b) It is a reducing sugar since it has a free aldehyde group.

(c) It can reduce Benedict's solution and can form osazone with phenyl hydrazine solution.

(d) On oxidation it produces glucuronic acid and on reduction it forms sorbitol.

Q7. How will you perform Benedict's test?

Added 8 drops of sugar solution to 5 ml of Benedict's qualitative solution in a test tube. Boiled for 2 minutes. A brick red precipitate of Cu_2O is formed.

The colour of the precipitate may be light green, green, yellow, and brick red depending on the concentration of sugar in the solution.

Light green ppt.	⟶	0.1—0.5% of reducing sugar.
Green ppt.	⟶	0.5 — 1.0% of reducing sugar.
Yellow ppt.	⟶	1.0 — 2.0% of reducing sugar.
Brick red ppt.	⟶	Above 2% of reducing sugar.

Q8. What is the composition of Benedict's qualitative solution?

Copper sulphate + Sodium carbonate + Sodium citrate.

Q9. Why is Benedict's test preferred than the Fehling's test?

(a) Fehling's solution is composed of copper sulphate, potassium hydroxide and sodium potassium tartarate. Prolonged contact with hot strong alkali (KOH) may destruct the sugar present.

(b) If the sample to be tested be preserved by chloroform, a positive reaction may be obtained by Fehling's test even in the absence of sugar.

(c) Benedict's test responds to even a very small quantity of sugar (0.1%) producing a very good precipitate.

(d) Chloroform does not interfere with this test and even uric acid or creatinine do not interfere to such an extent as in case of Fehling's test.

(e) Benedict's test may also produce a white precipitate if considerable amounts of phosphates are present.

Q10. How will you explain α-D(+)-glucose?

In the structure of glucose, if – OH group is present on the right side of carbon atom no 1, it is called α- glucose. If the –OH group is present on the right side of the carbon atom adjacent to the primary alcoholic group, it is said to be D-glucose. If glucose solution rotates the polarized light towards the right, it is said to be dextrorotatary and symbolized by plus (+) sign and hence written as (+) glucose.

Q11. What do you mean by glucosides?

Glucosides are compounds formed by the condensation reaction of glucose and the hydroxyl group of a second compound or aglycone which may or may not be another sugar.

Q12. How is glucose available from starch food in the body?

(a) In the saliva, starch is hydrolyzed to maltose through dextrin and erythrodextrin by the enzyme ptyalin (salivary amylase) at the P^H 6.6 to 6.8 in presence of chloride ion (Cl^-).

(b) In the pancreas, starch is also hydrolyzed to maltose by the pancreatic amylase at the P^H 7.1.

(c) In the small intestine, maltose is hydrolyzed to glucose by the enzyme maltase at the P^H 5.8 to 6.2.

Q13. How is glucose absorbed in the body?

(a) Glucose is absorbed by simple diffusion.

(b) It is more effectively absorbed by active transport mechanism. (Explain the active transport mechanism).

Q14. Why is galactose also absorbed by active transport mechanism like that of glucose?

Because glucose and galactose have the same configuration around carbon atom no. 2.

Q15. Why does glucose not form the same osazone as that of galactose?

Because glucose and galactose have not the same configuration around carbon atom no. 4 although both are reducing sugars having free aldehyde group.

Q16. How is glucose metabolized in the body?

(a) Glucose forms glycogen in the liver and muscle by glycogenesis (anabolism).

(b) Glucose is broken down to 2 molecules of pyruvic acid or lactic acid by glycolysis under anaerobic condition (by catabolism) producing 2 ATP only.

(c) Under aerobic conditions 2 molecules of pyruvic acid reoxdized to CO_2 and H_2O producing 30 ATP.

Q17. Where and how are ATP formed?

In the mitochondria by respiratory chain.

Q18. Define respiratory chain.

The sequence of carriers (coenzymes and cytochromes) responsible for the transport of reducing equivalents from substrates of first end of the chain to molecular oxygen to the last end of the chain producing H_2O and ATP is known as the respiratory chain.

Q19. Name the inhibitors of the respiratory chain.

Site I is inhibited by the antibiotic piericidin A.

Site II is inhibited by dimercaprol and antimycin A.

Site III is inhibited by dimercaprol and antimycin A.

Q20. What do you mean by oxidative phosphorylation?

The process by which ADP is phosphorylated to ATP by Pi in the respiratory chain is called oxidative phosphorylation.

Q21. Name the uncoupling agents.

Thyroxine, oligomycin, DNP, arsenite, dicoumarol, aureomycin, gramicidin.

Q22. Name the different steps of glycogenesis, glycolysis, TCA cycle with their enzymes coenzymes, and cofactors.

See books for glycogenesis, glycolysis, and TCA cycle.

Q23. What is the normal level of glucose in the blood?

Fasting : 70–115 mg/100 ml of blood.

P.P. (after 2 hrs of 50 gms of glucose intake) : 115 – 160 mg/100 ml of blood.

Q24. What do you mean by renal threshold for glucose?

If the blood glucose level rises above 180 mg. per 100 ml. of blood, sugar spills off in the urine. This value (180 mg) is called renal threshold for glucose.

Q25. What is G : N or D : N ratio?

The reabsorption of glucose becomes less producing hypoglycemia when the kidney becomes infected. This hypoglycemia causes liberation of epinephrine which causes glycogenolysis. This glucose is also excreted out. The tissue protein then starts breaking down to supply glucose to maintain blood glucose level causing the tissue waste. In such cases, the urine is tested for glucose as well as nitrogen. It is found that the ratio of glucose and nitrogen (G : N) becomes constant.

Q26. What do you mean by diabetes mellitus?

Diabetes = Lack of insulin, Mellitus = Sweet.

Q27. Why is not insulin given orally in the treatment of diabetes mellitus?

Insulin is a protein. It is digested by the enzyme insulinase to amino acids. Hence its activity is totally lost.

Q28. What is meant by juvenile diabetes?

Glucose is detected in the urine of individual aged even 14 due to the inability of the pancreas to secrete insulin.

Q29. What is glycosuria?

It is the condition in which abnormal quantities of glucose are excreted in urine.

Q30. Mention the types of glycosuria.

(a) **Alimentary glycosuria :** Presence of sugar in urine due to more intake of carbohydrate diet and detection is done just after meals.

(b) **Diabetic glycosuria :** Sugar in urine due to diabetes mellitus.

(c) **Renal glycosuria :** Sugar in urine due to renal tubular defects in spite of normal blood sugar level.

(d) **Adrenaline Glycosuria :** Sugar in urine during fear, anger, and anxiety owing to the liberation of more epinephrine which causes glycogenolysis resulting in hyperglycemia.

MODEL 2

Q1. Name one highly nutritive food.

Egg.

Q2. Mention the constituents of egg.

(a) It contains 30% yolk, 59% white, 11% shell.

(b) The white contains proteins and salts. The greater part of the protein is ovalbumin. The other proteins are conalbumin, ovoglobulin, and ovomucoid (a glycoprotein). The white also contains 10.7% protein, 0.1% fat, and 1.4% carbohydrate.

(c) The yolk contains 51% water, 15% protein, 33% fat of which 4% cholesterol, and 1% mineral. The proteins are vitellin (a phosphoprotein) and livetin (a globulin). The minerals are calcium, iron, and phosphate. The yolk is also rich in vitamin A, B_1, B_2 and D, but not C. The other B group vitamins and E are also present.

Q3. Why is the colour of yolk yellow?

Due to the presence of riboflavin (vitamin B_2).

Q4. How will you identify the presence of albumin and globulin in eggs?

Full staturation test for albumin : To 5 ml. of the solution added solid ammonium sulphate by vigorous shaking until saturated. A gelatinous ppt. is formed.

Half saturation test for globulin : To 5 ml. of the solution added 5 ml. of saturated ammonium sulphate. Shaked vigorously. A gelatinous ppt. is formed.

Q5. How will you identify protein?

Biuret test : To 2 ml. of the solution added 2 ml. 10% NaOH and one drop of 10% $CuSO_4$ solution. Development of violet colour indicates the presence of protein.

Q6. State the nature of cholesterol.

(a) It is a derived lipid.

(b) It is an alcohol containing the cyclopentanoperhydrophenanthrene nucleus.

(c) Its empirical formula is $C_{27}H_{46}O$.

Q7. Name another higher alcohol.

Sphingosin.

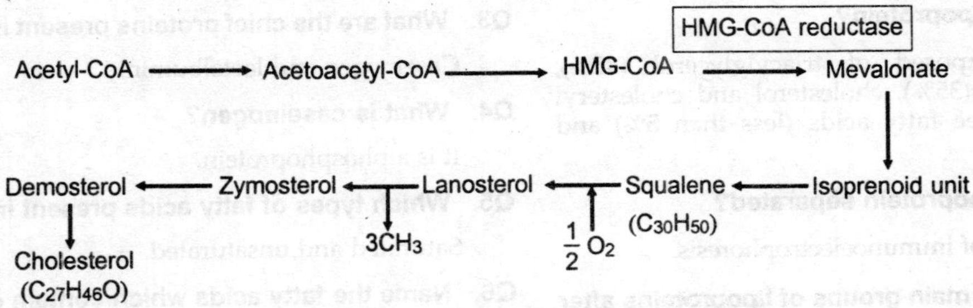

CHART 1

Q8. What is the normal cholesterol level in the body?

140—220 mg per 100 ml. of blood.

Q9. Discuss the biosynthesis of cholesterol.

See Chart 1 above.

Q10. Squalene contain 30 carbon, how does cholesterol contain 27 carbons?

In the conversion of lanosterol to zymosterol, $3CH_3$ is eliminated.

Q11. What is the synthetic condition of cholesterol during fasting?

Reduced synthesis.

Q12. What will be the rate of synthesis of cholesterol in the body on more cholesterol feeding.

More cholesterol feeding inhibits the HMG-CoA reductase for which cholesterol biosynthesis is reduced.

Q13. Name the factors that lower the cholesterol level in the blood.

Low protein intake, intake of nicotinic acid in large amounts, intake of iron salts, physical exercise, increased fiber content of diet.

Q14. What is the total amount of cholesterol in the body of a person weighing 70 kg?

140 grams.

Q15. How much of cholesterol is synthesized per day?

About one gram.

Q16. Name the substances formed from cholesterol.

Bile acids and bile salts, vitamin D, corticosteroid hormones, androgens (male sex hormones), estrogens and progesterone.

Q17. Name the bile acids and bile salts.

Bile acids : Cholic acid, deoxycholic acid, chenodeoxycholic acid, lithocholic acid.

Bile salts : Sodium or potassium glycocholate. Sodium or potassium taurocholate.

Q18. Mention the effects of drugs on cholesterol.

Sitosterol : Reduces the absorption of cholesterol.

Neomycin : Causes the increased fecal excretion of cholesterol.

Q19. How will you test for cholesterol?

Salkowski's test : A little cholesterol dissolved in 2 ml. chloroform + Conc. H_2SO_4. Shaken gently. Upper layer is turned red indicating the presence of cholesterol.

Libermann-Burchard reaction : A little cholesterol dissolved in 2 ml. choloroform + 10 drops acetic anhydride and 2 drops conc. H_2SO_4. Mixed well. A red rose colour develops which quickly changes through blue to green.

Q20. Name the conditions causing hypercholesterolemia.

Uncontrolled diabetes mellitus, obstructive jaundice, hypothyroidism, and nephrosis.

Q21. What is meant by atherosclerosis?

It is characterized by the deposition of cholesteryl ester and other lipids in the connective tissue of the arterial walls.

Q22. What is vitellin?

It is a conjugated protein, a phosphoprotein.

Q23. Name some other conjugated protein.

Glycoprotein, lipoprotein, chromoprotein, metalloprotein.

Q24. Give examples of metalloprotein and chromoprotein.

Metalloprotein : Siderophilin, Ceruloplasmin.

Chromoprotein : Hemoglobin, catalase, cytochromes.

Q25. What is cytochrome oxidase?

Cytochrome aa_3.

Q26. What is lipoprotein?

It is composed of triacylglycerol (45%), phospholipids (35%), cholesterol and cholesteryl ester (15%), free fatty acids (less than 5%) and protein.

Q27. How is lipoprotein separated?

By means of immunoelectrophoresis.

Q28. Name the main groups of lipoproteins after separation.

Chylomicrons, VLDL, LDL, HDL.

Q29. What is the clinical importance of lipoproteins?

(a) LDL concentration increases in severe diabetes mellitus, atherosclerosis etc.
(b) Serum LDL is also increased in hypothyroidism, obstructive jaundice, glycogen storage disease.

Q30. What is meant by lipotropic factors?

(a) The substances that prevent the accumulation of fat in the liver are known as lipotropic factors.
(b) These substances are choline, methionine, betaine.

MODEL 3

Q1. What have you taken today before coming to the examination centre?

Milk.

Q2. Why is milk said to be almost a complete food?

Because it contains almost all the nutritive substances, namely protein, fat, carbohydrate, minerals, and vitamins.

Q3. What are the chief proteins present in milk?

Caseinogen and lactalbumin.

Q4. What is caseinogen?

It is a phosphoprotein.

Q5. Which types of fatty acids present in milk?

Saturated and unsaturated.

Q6. Name the fatty acids which contain carbon atoms 2, 16, and 24.

Acetic acid, palmitic acid, and lignoceric acid.

Q7. How many ATP will be formed if stearic acid is completely oxidized?

(a) Stearic acid contains 18 carbon atoms.
(b) It is oxidized by β-oxidation.
(c) It undergoes 8 times β-oxidation and 9 molecules of acetyl-CoA are formed.
(d) Each turn of β-oxidation produces 5 ATP and each molecule of acetyl-CoA forms 12 ATP after complete oxidation by TCA cycle.
(e) Therefore, the number of ATP are formed as below.

$$5 \times 8 = 40 \text{ ATP}$$
$$12 \times 9 = 108 \text{ ATP}$$
$$148 \text{ ATP}$$

2 ATP are utlized in the initial activation of fatty acids.

∴ The net total number of ATP = 148 − 2 = 146.

Q8. Name the carbohydrate present in milk.

Lactose.

Q9. How is lactose synthesized in the mammary gland?

See Chart 2 below.

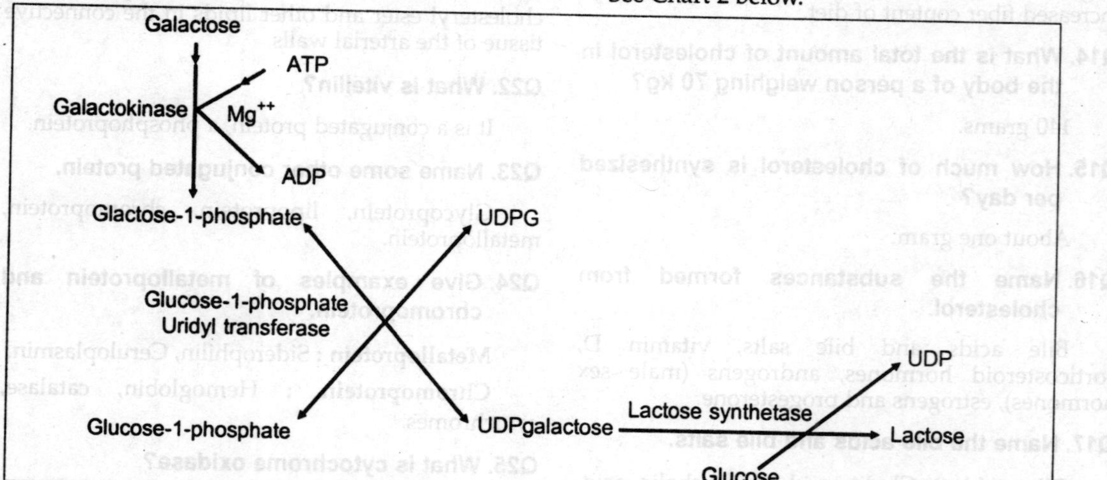

CHART 2

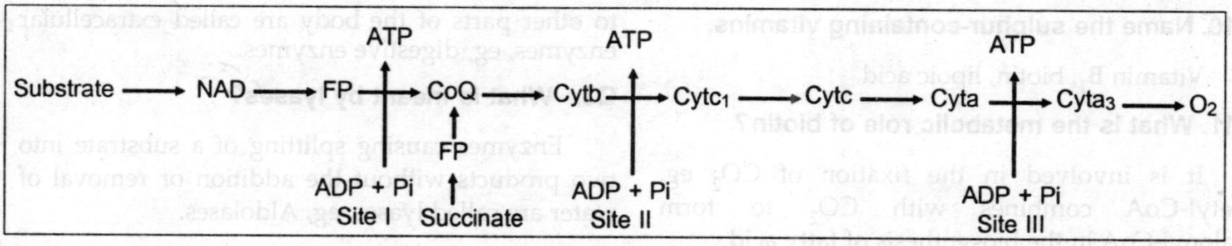

CHART 3

Q10. What is the chemical characteristic of lactose and why?

It is a reducing sugar since it has a free aldehyde group.

Q11. Where and how ATP is synthesized in the body?

In mitochondria by the respiratory chain.

See Chart 3 above.

Q12. Name some uncoupling agents.

Methylene blue, arsenite, dicaumerol, thyroxine, gramicidin.

Q13. Name the inhibitors of Site II of the respiratory chain.

Dimercaprol and antimycin A.

Q14. Which mineral of the milk is insufficient for body?

Iron.

Q15. Name the vitamins which are present in milk in low concentration.

Vitamins C and D.

Q16. What does happen when milk is kept unsterilised?

(a) Milk becomes sour due to the formation of lactic acid from lactose by bacteria present in milk.

(b) This lactic acid is quite sufficient to precipitate (curdle) the caseinogen.

Q17. Why does milk become clotted?

(a) When milk comes in contact with rennin or proteolytic enzymes.

(b) The proteins of milk neutralise gastric HCl to allow clotting.

MODEL 4

Q1. Define vitamins.

The accessory food factors which are organic in nature and necessary for the normal functioning of the tissues are called vitamins.

Q2. Name one vitamin which has got reducing property.

Vitamin C.

Q3. Why does anemia occur in the deficiency of vitamin C?

Iron of the food exists in the form of $Fe(OH)_3$. This is splitted into Fe^{+++} by HCl of the stomach. This Fe^{+++} is reduced to Fe^{++} by vitamin C and the ferrous ion is absorbed in the chelate form being combined with protein of low molecular weight. Iron in the ferrous form forms hemoglobin being transported to the plasma. Therefore, vitamin C is required in the formation of hemoglobin by the above process.

Q4. What is the percentage of iron absorbed?

Less than ten per cent.

Q5. Name the sources of iron.

Rich soures : Liver, kidney, heart.

Good sources : Apples, bananas, egg yolk, fish, beans.

Q6. What is the normal level of iron in the blood?

75–175 mcg/100 ml. of blood.

Q7. What is siderophilin?

(a) It is a iron binding protein. Iron is in the ferric state.

(b) It is a glycoprotein of molecular weight 76,000 containing 5.3 per cent carbohydrate.

Q8. What is siderosis?

It is the condition in which iron is deposited in the liver in excessive amounts.

Q9. What is ceruloplasmin?

(a) It is a copper binding plasma protein.

(b) It has feroxidase activity which convert Fe^{++} to Fe^{+++}.

Q10. Name the sulphur-containing vitamins.

Vitamin B_1, biotin, lipoic acid.

Q11. What is the metabolic role of biotin?

It is involved in the fixation of CO_2 eg. acetyl-CoA combines with CO_2 to form malonyl-CoA in the biosynthesis of fatty acid.

12. State the coenzyme activities of thiamine.

(a) The coenzyme form of thiamine is TPP.

(b) TPP is involved in the conversion of pyruvate to acetyl- CoA, α-keto glutarate to succinyl-CoA.

(c) It also functions along with transketolase enzyme in the conversion of ribose-5-phosphate and xylulose-5-phosphate to sedoheptulose-7-phosphate.

Q13. What do you mean by coenzymes?

These are heat-stable, dialyzable, non-protein organic molecules and the prosthetic groups of enzymes.

Q14. Name some nutritional conezymes.

TPP, CoA, FAD, NAD, NADP, B_6-PO_4.

Q15. What is meant by intrinsic factor?

HCl and a constituent of gastric juice which is a mucopolysaccharide secreted by the parietal cells of the gastric mucosa.

MODEL 5

Q1. Define enzymes.

Enzymes are soluble, colloidal organic catalysts formed by living cells, specific in action, protein in nature, inactive at O°C and destroyed by moist heat at 100°C.

Q2. What do you mean by intracellular and extacellular enzymes?

Enzymes which are used in the cells which make them are said to be intracellular enzymes, eg, enzymes responsible for metabolism. Enzymes which are produced by other cells and are secreted to other parts of the body are called extracellular enzymes, eg, digestive enzymes.

Q3. What is meant by lyases?

Enzymes causing splitting of a substrate into two products without the addition or removal of water are called lyases, eg, Aldolases.

Q4. What are ligases?

Enzymes causing linking together of two compounds couple to the breaking of a pyrophosphate bond in ATP or a similar compound, eg,

ATP + acetyl-CoA + CO_2 → ADP + Pi + malonyl-CoA.

Q5. Define isozymes with example.

The enzymes that occur in a number of different forms and differ from each other chemically, immunologically, and electrophoretically are called isozymes, eg, lactate dehydrogenases, transaminases, aldolases etc.

Q6. State the diagnostic importance of lactate dehydrogenases.

(a) In certain solid tumors, the levels of LD_1 and LD_2 are increased in blood.

(b) Increased level of LD_5 in serum is found in tumors.

(c) In acute leukemia, the isoenzyme levels of lactate dehydrogenase are increased in the blood.

(d) Alkaline phosphatase isozymes can distinguish liver lesions from bone lesions in metastatic carcinoma.

Q7. Give some examples of coenzymes.

NAD, NADP, CoA, ATP, B_6-PO_4, TPP etc.

Q8. Is there any coenzyme form of fat-soluble vitamins?

No.

Q9. Show some metabolic reactions where NADP acts as coenzyme.

In HMP shunt pathway :

See chart below & next page.

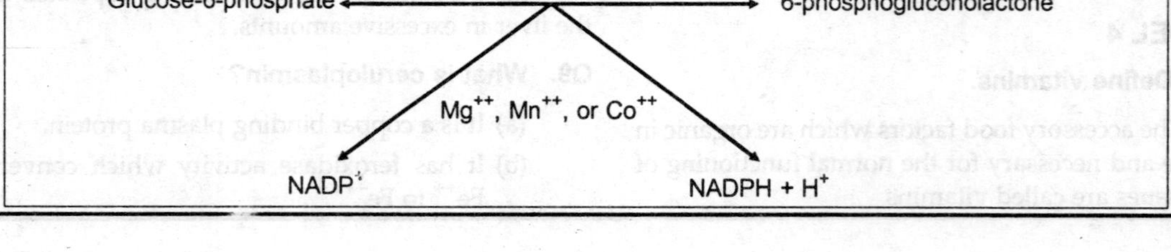

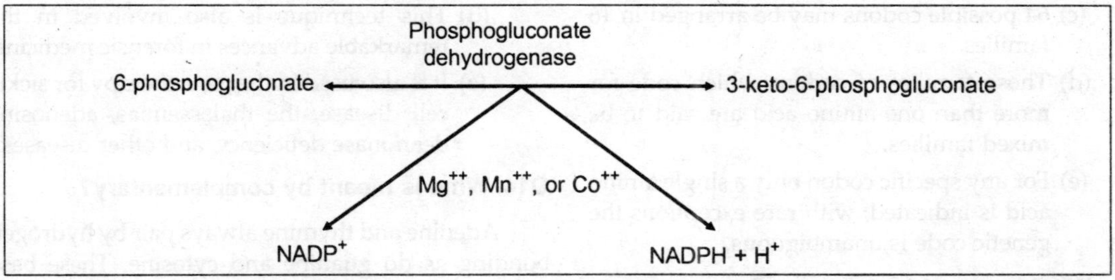

Q10. State the role of reduced NADP.

(a) It is required for the biosynthesis of fatty acids.

(b) It is also required for the biosynthesis of cholesterol.

(c) It is required for the biosynthesis of steroid hormones.

(d) It is also required for the biosynthesis of xylitol and ascorbic acid in the uronic acid pathway.

Q11. What do you mean by competitive inhibition?

The specific inhibitor is structurally similar to the substrate for which the inhibitor can combine with the enzyme in stead of the substrate. For example, the malonic acid (the inhibitor) is almost structurally similar to succinic acid (the substrate) on which the enzyme succinate dehydrogenase acts producing fumaric acid. If succinate dehydrogenase combines with the inhibitor, malonic acid, fumaric acid can be formed.

Sulphanilamide is also structurally similar to P-aminobenzoic acid which is essential for the synthesis of folic acid by the enzyme of the bacteria. If the enzyme combines with sulphanilamide, no folic is formed.

MODEL 6

Q1. What do you mean by codons?

The letters A, G, T, and C found in DNA organized into three-letter code words are called codons.

Q2. What is genetic code?

The collection of the codons constitutes the genetic code.

Q3. How are genes involved in protein synthesis?

A linear array of codons (a gene) specifies the synthesis of various RNA molecules, most of which are involved in some aspect of portein synthesis.

Q4. What is translation?

The cell possesses the machinery necessary to translate accurately and efficiently from the nucleotide sequence of an mRNA into the sequence of amino acids of the corresponding specific protein. The complete understanding of this process is termed translation.

Q5. What are nonsense codons?

(a) The codons that do not code for specific amino acids are termed nonsense codons.

(b) At least two of these nonsense codons are used in the cell as **termination signals.** They specify where the polymerization of amino acids into a protein molecule is to stop.

(c) The remaining 61 codons code for 20 amino acids. Thus, there must be **"degeneracy"** in the genetic code.

Q6. What is wobble?

The degeneracy of the genetic code resides mostly in the last nucleotide of the codon triplet, suggesting that the base pairing between the last nucleotide and the corresponding nucleotide of the anticodon is not strict. This is called wobble.

Q7. What is cistron?

It is the genetic unit coding for the structure of the subunit of a protein molecule acting as it does as the smallest unit of genetic expression.

Q8. What do you mean by the family of codons?

(a) A family of codons occupies a single column between the horizontal lines, eg, the codons CCN where N can be U, C, A or G, define a family located in the second column of the second box from the top.

(b) In some families, all 4 codons code for the same amino acids as do members of the CC family. These are referred to as unmixed families. 8 of the 16 families of codons are unmixed.

(c) 64 possible codons may be arranged in 16 families.

(d) Those families of codons which code for more than one amino acid are said to be mixed families.

(e) For any specific codon only a single amino acid is indicated; with rare exceptions the genetic code is unambiguous.

Q9. What is meant by missense?

(a) A missense effect occurs when a different amino acid is incorporated at the corresponding site in the protein molecule.

(b) This mistaken amino acid or missense may be acceptable, partially acceptable, or unacceptable to the function of that protein molecule depending upon its location in the specific protein.

Q10. Show the effect of mutation in hemoglobin structural gene.

(a) Codon for valine at position 67 of the β-chain of hemoglobin is not identical in all persons possessing the normal β-chain of hemoglobin.

(b) Hemoglobin Milwaukee has at position 67 a glutamic acid.

(c) Hemoglobin Bristol contains aspartic acid at position 67.

Q11. What is the effect of missense mutations?

Hemoglobin S, Sickle hemoglobin is the best example for this in which the normal amino acid in position 6 of the β- chain, glutamic acid, is replaced by valine.

Q12. What is the chemical nature of different factors (IF-1, IF-2 etc.) in protein synthesis?

They are protein in nature.

Q13. State the clinical application of recombinant DNA technology.

(a) Using recombinant DNA technology, human proteins can be produced in abundance for therapy eg, insulin, growth hormone, and plasminogen activator.

(b) By this technology proteins for vaccines (eg, hepatitis B) and for diagnostic tests (eg, AIDS test) can be obtained.

(c) It is also used to diagnose existing diseases and predict the risk of developing a given disease.

(d) This technique is also involved in the remarkable advances in forensic medicine.

(e) It is also used in the gene therapy for sickle cell disease, the thalassemias, adenosine deaminase deficiency, and other diseases.

Q14. What is meant by complementary?

Adenine and thymine always pair by hydrogen bonding as do guanine and cytosine. These base pairs are said to be complementary.

Q15. What do you mean by exons and introns?

The coding regions (those regions that appear in the mature RNA species) are called exons.

The noncoding regions which interpose between the exons are called introns.

Q16. What are enhancers and silencers?

Many eukaryotic genes (Q and some viruses that replicate in memmalian cells) have special regions called enhancers that increase the rate of transcription. Some genes also have DNA sequences known as silencers that diminish transcription.

Q17. What is the key tool in recombinant research?

Certain endonucleases, enzymes that cut DNA at specific DNA sequences within the molecule (as opposed to exonucleases which digest from the ends of DNA molecules) are a keytool in recombinant DNA research.

Q18. What are Phages and Cosmids?

Phages have linear DNA molecules into which foreign DNA can be inserted at several restriction enzyme sites.

Larger fragments of DNA can be cloned in **cosmids** which combine the best features of plasmids and phages.

MODEL 7

Q1. Name one acidic amino acid.

Glutamic acid.

Q2. What does happen in glutamic acid when it undergoes transamination and decarboxylation?

By transamination : α-Ketoglutarate.
By decarboxylation : Glutamine

Q3. Enumerate the fate of glutamic acid.

(a) Being transaminated to α-ketoglutarate it enters krebs cycle for complete oxidation and also can form glucose by gluconeo-genesis.

(b) Glutamine is used in the pathway of de novo purine biosynthesis.

(c) Glutamic acid is also involved as a constituent in the formation of glutathione which is a reducing substance. These glutathions conjugate with electrophilic xenobiotics.

(d) This is also used in urea formation in the liver.

Q4. Name one sulphur containing amino acid.

Methionine.

Q5. What is the active form of methionine.

S-adenosylmethionine.

Q6. State the role of methionine in the body.

(a) It acts as a lipotropic factor.

(b) It, in the active form, helps in the formation of creatine in the body.

(c) It is metabolized to succinyl-CoA which on entering into Kreb's cycle forms energy and also forms glucose by gluconeogenesis.

(d) It is converted to propionyl-CoA which helps in the formation of fat in the body.

(e) A few xenobiotics are subject to methylation by methyltransferases employing active methionine (S-adenosyl methionine) as the methyl donor.

MODEL 8

Q1. Why is sunflower oil more valuable than mustard oil?

Because sunflower oil contains sufficient quantities of essential fatty acids which prevent atherosclerosis.

Q2. What is atherosclerosis?

It is the disease caused by the deposition of cholesteryl ester and other lipids in the connective tissue of the arterial walls. The disease is accompanied by high blood pressure, obesity, and lack of exercise.

Q3. Name the essential fatty acids showing the number of carbon atoms with the double bonds.

Name	Number of carbon atoms	Double bonds
(a) Linoleic acid	18	2 (Δ^9, Δ^{12})
(b) Linolenic acid	18	3 ($\Delta^9, \Delta^{12}, \Delta^{15}$)
(c) Arachidonic acid	20	4 ($\Delta^5, \Delta^8, \Delta^{11}, \Delta^{14}$)

Q4. State the clinical importance of essential fatty acids.

(a) Prostaglandins are synthesized from arachidonic acids.

(b) The deficiency of EFA causes swelling of mitochondria.

(c) EFA combine with cholesterol forming cholesteryl ester which is oxidized in the liver to produce bile acids and bile salts. Thus, these acids lower serum cholesterol level.

(d) Linoleic acid is essential in the diet for optimal vision.

(e) Clotting time is prolonged by the ingestion of fats rich in EFA.

(f) Deficiency of EFA produces fatty liver.

(g) Deficiency of EFA causes skin lesion, cessation of growth, kidney damage, abnormalities of pregnancy and lactation in adult females, dermatitis.

(h) Arachidonic acid form leukotrienes.

Q5. Name the bile acids and bile salts.

Bile acids : Cholic acid, deoxycholic acid, chenodeoxycholic acid, and lithocholic acid

Bile salts : Sodium or potassium glycocholate (Naork + Cholic acid + glycine)
Sodium or potassium taurocholate (Naork + cholic acid + taurine)

Q6. What are prostaglandins?

(a) They act as local hormones and present in every mammalian tissue.

(b) They are synthesized from arachidonic acid to form a cyclopentane ring.

Q7. State the clinical importance of prostaglandins.

(a) They cause vasodilatation and decrease blood pressure.

(b) They are used as contraceptives to prevent conception.

(c) They effectively induce labor.

(d) They are used to prevent gastric ulcers and control inflammation.

(e) They are the effective nasal decongestant.

(f) They increase cAMP in platelets, fetal bone, and lung but lower cAMP in renal tubule cells and adipose tissue.

Q8. How are they metabolized?

Prostalandins are quickly metabolized by the enzyme 15-hydroxy prostaglandin dehydrogenase.

Q9. Mention the side-effects of prostaglandins.

(a) PGE_2 causes gastrointestinal smooth muscle to contract which leads to cramping and diarrhoea.

(b) PGE_2 also irritates the mucosa lining of the throat causing pain and coughing.

Q10. Name the inhibitors of prostaglandins.

Aspirin, Indomethacin, Phenanzine.

Q11. How many prostaglandins are there?

14 prostaglandins have so far been isolated from male reproductive tract.

Q12. Discuss the method of unsaturation test.

I drops of Hubl's iodine reagent is added to 9 ml. of chloroform. Pink colour develops. Divide it into three test tubes. To each test tube oil drop is added at an interval of 30 seconds shaking vigorouly to discharge the pink colour. The number of drops is counted for each oil in each test tube. The less the number of drops required, the more is the unsaturation of fat.

Q13. Why are vegetable fats preserved for longer periods than animal fats?

Vegetable fats contain vitamin E, phenols, hydroquinones, and tannins etc. which are anti-oxidants that prevent development of rancidity.

Q14. Why are young ladies protected against myocardial infarction?

Young ladies have higher levels of HDL which is the effect of increased estrogen level in young ladies. This raised HDL level protects them from myocardial infarction.

Q15. Name the factors involved in coronary heart disease.

(a) High blood pressure
(b) Smoking
(c) Lack of exercise
(d) male gender
(e) Obesity
(f) Drinking soft water.

Q16. What are the factors leading to higher levels of free fatty acids?

(a) Emotional stress
(b) Cigarette smoking
(c) Coffee drinking
(d) Partaking of a few large meals.

MODEL 9

Q1. Have you taken anything before coming to appear at the examination.

Nothing.

Q2. If you are in fasting condition, what will be the blood sugar level?

Hypoglycemia.

Q3. How will you then get energy?

Fats will be oxidized to supply energy.

Q4. What will happen ultimately if your fasting condition is continued?

There will be the excessive β-oxidation of fatty acids for which more acetyl-CoA will be formed. Since TCA cycle will not be functioning properly, all the molecules of acetyl-CoA formed will not be oxidized by TCA cycle. Therefore, there will be accumulation of acetyl-CoA forming more ketone bodies. As a result, there will be ketonemia and ketonuria.

Q5. Name the ketone bodies.

Acetoacetic acid, β-hydroxybutyric acid, and acetone.

Q6. Mention the principal ketone bodies.

Acetoacetic acid.

Q7. What is the normal concentration of ketone bodies in the blood and in the urine?

0.2–2.0 mg / 100 ml. of blood.

Less than 1 mg / day in the urine.

Q8. State the other conditions in which there is ketosis.

(a) Severe diabetes mellitus.
(b) Glycogen storage disease.
(c) Toxemia of pregnancy
(d) Infective hepatic disease.
(e) Continued fever.

Q9. What is ketosis?

The condition in which there is a high concentration of ketone bodies in tissues and blood is termed ketosis.

Q10. How are ketone bodies formed?

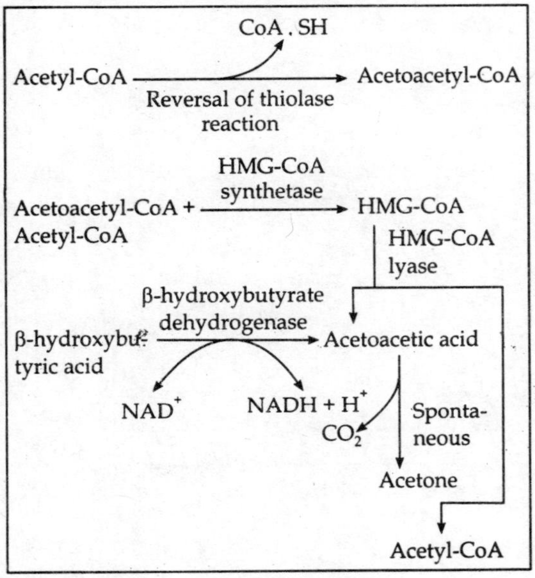

Q11. Name another lyase enzyme.

Aldolase.

Q12. Where and how the ketone bodies are utilized?

In extrahepatic tissue

1st reaction :

Transferase

Acetoacetate + Succinyl-CoA ↔ Acetoacetyl-CoA + Succinate

Second reaction :

Thiolase

Acetoacetate + ATP + CoA . SH → Acetoacetyl-CoA + AMP + PPi

Thiolase

Acetoacetyl-CoA → 2 Acetyl-CoA → TCA cycle.

Q13. How does ketoacedosis occur ?

Acetoacetic acid and β-hydroxybutyric acid are moderately strong acids and are buffered in blood or tissues. Their excretion in large quantities causes some loss of buffer cation which depletes the alkali reserve developing ketoacidosis.

Q14. How is ketosis prevented?

(a) In case of **diabetic ketosis**, carbohydrate diet, intramuscular injection of insulin and antiketogenic substance (aspartic acid) should be administered.

(b) In case of **prolonged starvation ketosis**, carbohydrate diet and antiketogenic substance (aspartic acid) should be given.

(c) Intravenous injection of isotonic solution of sodium salts should be given to restore the electrolytes and the fluids of the body.

Q15. How will you detect ketonebodies in the urine?

By performing Rothera's test.

Procedure : 5 ml. of urine is saturated with solid ammonium sulphate after vigorous shaking. Then added 2 drops of freshly prepared 5% solution of sodium nitroprusside and 1 ml. of ammonium hydroxide. Allowed to stand in the rack for a while. A permanganate colour is developed just above the layer of the undissolved ammonium sulphate crystals indicating the presence of ketone bodies.

Q16. What happens when you suffer from high tension during examination?

(a) The hormone glucagone is extensively secreted and causes glycogenolysis by increasing the formation of cAMP from ATP.

(b) There is depletion of liver glycogen and excess formation of glucose and lactic acid resulting very weakness.

(c) There may be emotional glycosuria.

(d) There may be forgetfulness.

Q17. Mention the different types of glycosuria.

(a) Alimentary glycosuria.

(b) Emotional glucosuria.

(c) Glycosuria in pregnancy and lactation.

(d) Renal glycosuria

(e) Glycosuria in hyperglycemia.

Q18. If a patient comes to you for treatment with a complain that sugar is detected in his urine, what will be your line of treatment?

First, the patient will be advised for blood sugar estimation. If his blood sugar level is found to be within the normal range, it is clear that he is not suffering from diabetes mellitus. It is confirmed that there is infection in the renal tubules for which glucose can not be reabsorbed for which glucose is excreted in urine. So, antibiotics will be given for the removal of this infection.

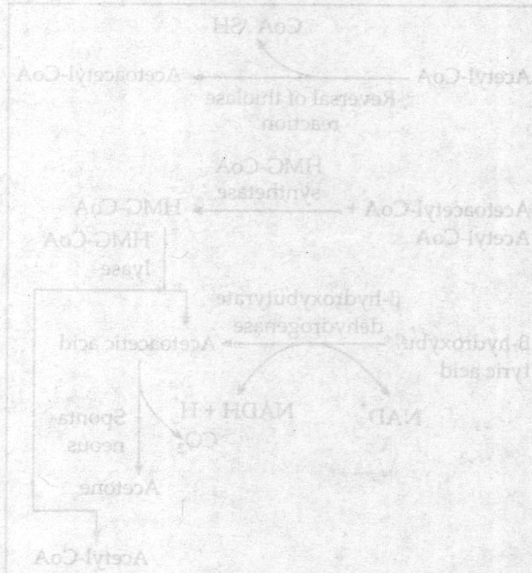

PART-II
PRACTICAL

CHAPTER 1

QUALITATIVE TESTS FOR SOME ORGANIC COMPOUNDS

UREA

1. **Biuret test :** Take about 5 ml. of urea solution in a test tube and evaporate it to dryness until a white residue with the smell of ammonia is obtained. Now, cool the residue and dissolve it in 1 ml. NaOH solution. Then add 3 drops of very diluted $CuSO_4$ solution and shake it well. A violet colour indicates the presence of urea.

2. **Test with alkaline NaOBr :** Take 2 to 3 ml of the provided solution and equal volume of hypobromite solution in a test tube. Shake the mixture. A brisk effervescence shows the presence of urea.

3. **Specific urease test :** Take about 3 ml of the provided solution in a test tube and add 1 or 2 drops of alcoholic phenolphthaline solution. Then add a pinch of soyabean powder and warm it upto 60°C. The development of pink colour indicates the presence of urea.

PHENOL

1. **Evaporation test :** Take 2 to 3 ml of the provided solution in a test tube and evaporate it to dryness by heat. It is completely evaporated leaving no residue behind indicating the presence of phenol.

2. **Test with $FeCl_3$ solution :** Take 2 to 3 ml of the provided solution in a test tube and add 3 drops of $FeCl_3$ solution to it. The appearance of deep violet colour is discharged by the addition of dilute acetic acid indicating the presence of phenol.

3. **Test with bromine water :** Take about 2 to 3 ml of the provided solution in a test tube and add equal volume of bromine water to it. The appearance of heavy white precipitate of 2, 4, 6-tribromophenol shows the presence of phenol.

4. **Test with NH_4OH and bromine water :** Take about 3 ml of the provided solution in a test tube and add ammonium hydroxide solution to it till the solution is ammoniacal. Add 2 ml bromine water and warm the mixture. The appearance of indigo blue colour shows the presence of phenol.

5. **Libermann's test :** Take about 5 ml of the provided solution in a porcelin basin. Add 1 ml NaOH solution to it and evaporate the mixture to dryness. Cool the residue and add 2 ml. conc. H_2SO_4. Add a pinch of solid $NaNO_2$ to it. The development of deep blue colour indicates the presence of phenol.

LACTIC ACID

Lactic acid is to be neutralized before performing the tests.

Process of neutralization : Take about 10 to 15 ml of the provided acid in a small beaker. Go on adding small amount of sodium carbonate to it at a constant stirring until the effervescence almost ceases. Test with a blue litmus paper which turns very light pink. In case, sodium carbonate is added in excess which is found by litmus appear blue, add a few drops of the provided solution to it which will turn the blue litmus paper light pink.

1. **Evaporation test :** Take 1 to 2 ml of the neutralized lactic acid in a test tube and evaporate it to dryness completely over a flame. No residue is formed indicating the presence of lactic acid.

2. **Test with silver nitrate solution :** Add excess of silver nitrate solution to about 3 ml of the neutral solution of lactic acid. Boil strongly for a few minutes. A slight blackening of the solution due to the reduction of $AgNO_3$ to metallic silver shows the presence of lactic acid.

3. **Test with potassium permanganate and sulphuric acid :** Add a little amount of dilute sulphuric acid and a few drops of potassium permanganate solution to about 3 ml of the neutralized solution and warm. The pink colour of potassium permanganate is discharged and there is the effervesence and smell of acetaldehyde indicating the presence of lactic acid.

4. **Indoform test :** Take about 3 ml of the neutralized solution in a test tube and 6 ml of iodine solution. Warm slightly. While still warm add dilute

sodium hydroxide solution to it dropwise till the deep brown colour almost vanishes leaving behind a pale yellow precipitate. Cool it thoroughly. The appearance of very minute dark yellow crystal on cooling having a characteristic smell indicates the presence of lactic acid.

URIC ACID

Crystalline solid uric acid is insoluble in cold water as well as in hot water. This is dissolved in Na_2CO_3 and NaOH.

1. **Schiff's test :** Pour a little solution of uric acid over a piece of filter paper previously soaked with $AgNO_3$. Black stain appears on the filter paper due to reduction of $AgNO_3$ to metallic silver by uric acid indicating the presence of uric acid.

2. **Fehling's test :** Dissolve a pinch of solid uric acid in NaOH solution mixed with Fehling reagent. Heat it. A white precipitate of copper urate first appears which on further heating converted into red precipitate showing the presence of uric acid.

3. **Murexide test :** Mix a pinch of solid uric acid with 1 or 2 drops of concentrated nitric acid and heat over a low flame in a porcelin beaker. A magenta colour appears. Cool this thoroughly and add 2 drops of liquor ammonia. The development of brilliant purple colour indicates the presence of uric acid.

ETHYL ALCOHOL

Iodoform test : To 1 ml of absolute alcohol add 0.5 gram of anhydrous Na_2CO_3 and 5 ml. iodine solution (10%). Warm in a water bath at 70° – 80°C until it is decolourised. Iodoform separates out as a yellow precipitate on cooling indicating the presence of ethyl alcohol.

ALDEHYDE

1. **Sliver mirror test :** To 1 ml. AgNO3 solution add 4 drops of NaOH and ammonia drop by drop until the precipitate dissolves. Then add 1 ml acetaldehyde and shake briskly. Bright mirror develops due to the reduction of AgNO3 on the side of the test tube showing the presence of aldehyde.

2. **Fehling's test :** Add 3 ml. Fehling's solution to 3 ml. provided solution. Heat. Brick red precipitate of cuprous oxide is formed due to the reduction of alkaline copper sulphate solution showing the presence of aldehyde.

ACETONE OR KETONE BODIES

1. **Iodoform test :** To 1 ml of the provided solution add 5 grams of Na_2CO_3. Add 5 ml iodine solution (10%). Add iodine solution till no further decolouration occurs. Crystals of iodine separates out even in cold showing the presence of acetone.

2. **Rothera's test :** Saturate 5 ml of the provided sample with solid ammonium sulphate by shaking it vigorously. Then add 2 drops of freshly prepared 5% sodium nitroprusside solution and 1 ml ammonium hydroxide solution. Allow it to stand in the reck for a while. The development of permanganate colour above the layer of the undissolved ammonium sulphate crystals indicates the presence of acetone.

FORMIC ACID

1. **Silver mirror test :** To 1 ml. $AgNO_3$ solution and 4 drops of NaOH and ammonia drop by drop until the precipitate dissolves. Then add 1 ml acetaldehyde and shake briskly. Bright mirror develops due to the reduction of $AgNO_3$ on the side of the test tube showing the presence of formic acid.

2. **Test with mercuric chloride :** To 3 ml of the provided sample add little mercuric chloride solution and warm. White precipitate is formed and the precipitate turns grey on boiling indicating the presence of formic acid.

ACETIC ACID

1. **Test with ferric chloride :** To the provided solution add a few drops of 1% $FeCl_3$ solution. The development of light red colour indicates the presence of acetic acid.

2. **Test with absolute alcohol :** To 1 ml of the provided sample add 4 drops of conc. H_2SO_4 and 1 ml absolute alcohol. Boil carefully. The fruity fragrant odour of ethyl acetate shows the presence of acetic acid.

GLYCEROL

1. **Test with copper sulphate :** Add 5 drops of the provided sample to 1 ml $CuSO_4$ solution and add 5 drops of NaOH solution. A clear blue solution is formed. No reduction occurs on boiling indicating the presence of glycerol.

2. **Test with sodium hydrogen sulphate :** Heat 2 drops of the provided sample with a pinch of sodium hydrogen sulphate. Acrolein is formed. The characteristic pungent odour shows the presence of glycerol.

3. **Test with Borax :** To 5 ml borax solution (0.5%) add 4 drops of phenolphthalein indicator. A bright pink colour is formed. Add the provided sample drop by drop until the pink colour just disappears. The pink colour reappears on boiling indicating the presence of glycerol.

CHAPTER 2

CARBOHYDRATES

1. MOLISCH'S TEST :

Principle : Alcoholic alpha naphthol forms furfural and furfural derivatives, such as hydroxymethylfurfural, by the concentrated sulphuric acid acting on the sugar. This compound forms a reddish-violet coloured ring at the junction of the two liquids.

Molisch's reagent : A 5 per cent solution of alpha naphthol in alcohol.

Procedure : Add 2 drops of of Molisch's reagent to 2 ml. of sugar solution in a test tube. Mix thoroughly. Add 2 ml. of Conc. H_2SO_4 by the side of the test tube slanting the tube. Then errect the test tuble slowly. The formation of reddish violet ring at the junction of two liquids indicates the presence of carbohydrate.

Discussion : Concentrated solution of organic compounds may give a red instead of a violet colour due to the charring action of the sulphuric acid. In case of doubt the experiment should be repeated on a more diluted solution of the substance to be tested.

2. IODINE TEST :

Principle : The composition of the blue- or red- or wine red- coloured substance is not well defined. This may be an adsorption complex of starch or dextrins or glycogen with iodine rather than a definite compound.

Iodine reagent : 0.5 ml. of iodine diluted to 5 ml. with distilled water.

Procedure : Add 1 or 2 drops of dilute iodine solution to 2-3 ml. of dilute starch or dextrin or glycogen solution. A blue, red and brown colour develop in case of starch, dextrin, and glycogen, respectively. In case of starch, the blue colour disappears on heating and reappears on cooling. But the red colour and the brown colour in cases of dextrin and glycogen, respectively, do not reappear on cooling as in the case of starch.

3. REDUCTION TESTS :

Carbohydrates with free aldehyde or ketone groups have the ability to reduce solutions of various metallic ions. These properties are mentioned below :

A. FEHLING'S TEST :

Principle : Carbohydrates with free aldehyde or ketone groups reduce copper sulphate to cuprous oxide forming a yellow- or brownish-red-coloured precipitate.

Fehling's reagent : Prepare freshly by mixing equal volumes of two stock solutions A and B.

Solution A : 6.93 grms of $CuSO_4.5H_2O$ per 100 ml. of water.

Solution B : 20 gms. of KOH and 34.6 gms. of sodium potassium tartarate (Rochelle salt) per 100 ml. solution.

Procedure : Add a few drops of sugar solution at a time to 5 ml. of Fehling's solution and heat the mixture after each addition. The production of yellow or brownish-red cuprous oxide precipitate indicates the presence of reducing sugars.

B. BENEDICT'S TEST :

Principle : Carbohydrates with free aldehyde or ketone groups reduce copper sulphate of Benedict's solution to cuprous oxide on boiling forming a red, yellow or green coloured precipitate depending on the concentration of the sugar.

Benedict's qualitative solution : Dissolve with heat 173 gms. of sodium citrate and 100 gms. of anhydrous sodium carbonate in 600 ml. of water in a beaker. Into this with constant stirring, run slowly a solution of copper sulphate ($CuSO_4.5H_2O$) containing 17.3 gms. dissolved in about 100 ml. of water. Cool and transfer to a litre flask and make to the mark with water.

Precedure : Add 8 drops of sugar solution to 5 ml. of the Benedict's qualitative reagent in a test tube. Boil vigorously for 2 minutes or place in a

water bath for 3 minutes. Allow to cool spontaneously (do not cool it by immersion in cold water). A red, yellow or green precipitate develops depending on the concentration of sugar present.

Colour	Approximate amount of reducing sugar
No change of blue colour	Absence of reducing sugar
Blue changes to green ppt.	0.1—0.5 gm% of reducing sugar
Blue changes to yellow ppt.	0.5—1.0 gm% of reducing sugar
Blue changes to orange-red ppt.	1—2.0 gm% of reducing sugar
Blue changes to Brick-red ppt.	over 2.0 gm% of reducing sugar.

C. BISMUTH REDUCTION TEST (NYLANDER) :

Principle : Carbohydrates with free aldehyde or ketone groups reduce to cause the black precipitation of metallic bismuth.

Procedure : Add 8 drops of Nylander's reagent to 5 ml. of sugar solution in a test tube and heat for 5 minutes in a boiling water bath. A black coloured precipitate develops on standing a few minutes.

D. COLE'S TEST :

Principle : Carbohydrates with free aldehyde or ketone groups reduce copper sulphate to cuprous oxide on boiling forming a yellowish precipitate.

Procedure : Add a large quantity of anhydrous sodium carbonate to 5 ml. of 0.1% sugar solution and add 3 drops of glycerol and 3 drops of 5% $CuSO_4$ soltuion. A yellowish precipitate develops on boiling.

E. REDUCTION OF METHYLENE BLUE :

Principle : Carbohydrates containing free aldehyde or ketone groups can reduce methylene blue to leuco-methylene blue with the discharge of the blue colour.

Procedure : Add one drop of 1% methylene blue, a few drops of 5% NaOH to 2 ml. of water and boil taking care that the blue colour does not disappear. Now add a few drops of 0.2% sugar solution and boil again. The colour is now discharged. Shake and note the reoxidation of leuco-methylene blue by oxygen of air to the coloured dye again.

Discussion of reduction tests :

(i) Ammonium salts interfere with Fehling's test. If present in excess, the soiution should be made alkaline with Na_2CO_3 and boiled in order to decompose the ammonium salts. Prolonged contact with hot strong alkali may destruct the sugar present. The solution under examination by Fehling's test must be made neutralized or alkaline before the commencement of the test.

If the sugar is preserved by Chloroform, a positive reaction may be obtained in the absence of sugar even. This statement is based on the fact that hot alkali produces reducing substancs from the sugar present. Uric acid and creatinine interfere this test.

(ii) Benedict's test responds to even a very small quantities of sugar (0.1%) yielding a very good precipitate.

Chloroform does not interfere with this test and even uric acid or creatinine do not interfere to such as extent as in the case of Fehling's test.

This test may produce a white precipitate if considerable amounts of phosphates are present.

On the above grounds Benedict's test is better than Fehling's test.

(iii) Cole's test is suitable for detecting a small quantity of sugar (0.1 mg).

(iv) Bismuth reduction test responds to albumin, if present. Chloroform also interferes with this test. Uric acid and creatinine do not interfere. This test can detect sugar to the extent of 0.08 per cent.

4. BARFOED'S TEST :

Principle : The monosaccharides with free aldehyde or ketone groups can cause the reduction of copper sulphate to cuprous oxide with the development of a red precipitate.

Barefoed's reagent : Dissolve 24 grams of copper acetate in 450 ml. boiling water. If a precipitate form, do not filter. Immediately add 25 ml. 8.5 per cent lactic acid to the hot solution. Shake, cool and dilute to 500 ml. and filter off the impurities.

Precedure : Add 8 drops of sugar solution to 5 ml. of Barfoed's solution. Heat to boiling for 30 seconds. A red precipitate appears indicating the presence of monosaccharides.

Discussion : This test is a reduction test for monosacchardies and the reduction is brought about in an acid solution. If the solution with the sugar is boiled for a few minutes, the disaccharides also are hydrolysed giving a positive test. This test is specific for monosaccharides.

5. BIAL'S TEST FOR PENTOSES :

Principle : Hydrochloric acid acting on pentoses yields substances which link with orcinol to form green compounds.

Bial's reagent : Dissolve 300 mg. of orcinol to 100 ml. of Conc. HCl and add 5 drops of 10 per cent ferric chloride solution.

Procedure : Mix 5 ml. of Bial's reagent with 0.5 ml. of the sugar solution. Heat to boiling. The development of green colour indicates the presence of pentoses.

Discussion : This test is sensitive to 0.1 per cent of Pentoses. Glucuronates can give a similar colour if the boiling is prolonged due to the conversion of glucuronates to pentoses.

6. TAUBER'S BENZIDINE TEST FOR PENTOSES :

Procedure : To 1 ml. of Benzidine solution add 2 drops of sugar solution. Boil and cool quickly. A violet colour indicates the presence of pentoses.

7. SELIWANOFF'S TEST :

Principle : Fructose is acted on by hydrochloric acid to form a derivative of furfuraldehyde which gives a red coloured compound when linked with resorcinol.

Seliwanoff's reagent : Dissolve 50 mg. of resorcinol in 33 ml. of concentrated hydrochloric acid and dilute to 100 ml. with water.

Procedure : Add a few drops of sugar solution (0.5 ml.) to 5 ml. of Seliwanoff's reagent in a test tube. Heat to boiling for 30 seconds. Formation of red colour indicates the presence of fructose. The test may be positive for sucrose also if it is hydrolyzed during the course of the test.

Discussion : A similar colour may also develop in case of glucose or maltose if the boiling is prolonged due to the transformation of glucose into fructose by the catalytic action of the hydrochloric acid.

This test is sensitive to 0.1 per cent of fructose if glucose is absent. But if glucose is present, it is less sensitive to fructose. Large amounts of glucose can give a similar colour.

8. HYDROLYSIS TEST FOR SUCROSE :

Principle : Sucrose on hydrolysis by HCl is converted into glucose and fructose. The presence of these two monosaccharides are detected by tests mentioned above.

Procedure : Add two drops of HCl and one drop of thymol blue to 5 ml. of sucrose solution. The development of pink colour indicates the solution acidic. Divide it into two equal parts. Boil one portion for about one minute and then cool it under the tap. Neutralize both portions by adding 2% Na_2CO_3 drop by drop. Formation of blue colour indicates the neutralization. The sucrose in the boiled portion has been hydrolyzed to form glucose and fructose which can be detected by Benedict's test and Seliwanoff's test. But the unboiled sucrose does not reduce Benedict's solution.

9. METHYLAMINE TEST FOR LACTOSE :

Procedure : Add one ml. of 0.2 per cent solution of methylamine hydrochloride in water, followed by 0.2 ml. of 10 per cent sodium hydroxide, to about 5 ml. of sugar solution. Mix by inversion, cover the tube with a glass bulb and heat at 56°C for 30 minutes. Remove from the bath and stand at room temperature. The solution will show a red colour before the end of the heating if much lactose is present. Colour increases on standing.

Discussion : The only sugars which give a red colour are the disaccharides, lactose and maltose.

Glucose, fructose, galactose, xylose and sucrose in large amounts give a yellow colour which is easily distinguished.

10. OSAZONE TEST :

Principle : A solution of reducing sugar when heated with phenylhydrazine, characteristic yellow crystalline compounds called osazone are formed. Simple sugars like glucose, fructose and mannose produce the same osazone because of the similarities in their molecular structures.

Procedure : Add 10 drops of glacial acetic acid to 5 ml. of sugar solution in test tube. Then add a knife point of phenylhydrazine hydrochloride and double the amount of sodium acetate crystals. Mix and warm a little to see that the solids are dissolved. Filter the solution in another test tube and keep the filtrate in a boiling water bath for 20 minutes. Allow the tube to cool slowly in the water bath without cooling it hurriedly under the tap to have better crystals and examine the crystals under the microscope.

Discussion : Formation of osazone crystals of different sugars depends on the time schemed below :

Osazones	Minimum time of formation of crystals	Appearance of Crystals
Glucosazone	5 minutes	Broomstick like
Fructosazone	2 minutes	Broomstick like
Galactosazone	7 minutes	Rhombic like
Lactosazone	10-12 minutes	Powderpuff like
Maltosazone	10-15 minutes	Sunflower like

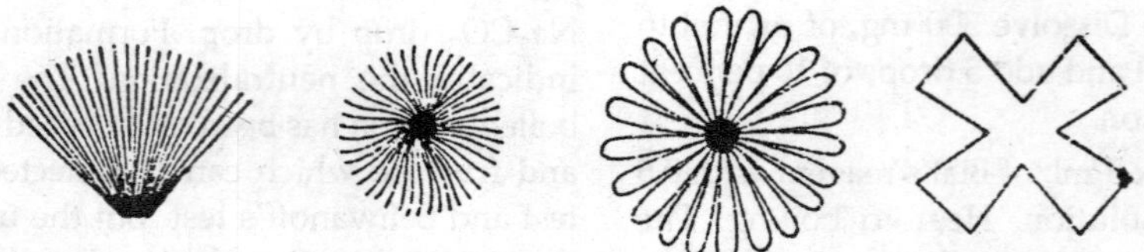

Broomstick like Powderpuff like Sunflower like Rhombic like

Sucrose does not produce osazone crystals. But it produces osazone crystals when the solution is kept in the boiling water bath for 30 minutes because of the hydrolysis of sucrose to glucose and fructose.

11. PRECIPITATION REACTION FOR POLYSACCHARIDES :

Principle : Generally, the bigger molecules are more easily precipitated by salts. The starch is precipitated when its solution is half-saturated with ammonium sulphate. Glycogen is completely precipitated only when its solution is fully saturated with ammonium sulphate; whereas dextrin is not completely precipitated even on full saturation.

Procedure : Add 5 ml. of saturated solution of ammonium sulphate to 5 ml. of starch solution to make the solution half saturated with ammonium sulphate. Mix well by inverting the test tube several times and allow it to stand for 5 minutes. Flocculent precipitate of starch will be formed. Filter. Perform iodine test with the residue as well as with the filtrate. Blue coloration with the residue indicates the presence of starch and no blue colouration with the filtrate indicates the complete precipitation of starch.

Now, take 5 ml. of glycogen or dextrin solution in a test tube and saturate it fully with solid ammonium sulphate. Wait for 5 minutes and filter. Perform iodine test with the filtrate. In case of dextrin solution, the filtrate showing red colour indicates that dextrin is not completely precipitated even on full saturation. But in case of glycogen solution, the filtrate will develop no brown coloration indicating the complete precipitation of glycogen by full saturation.

REVIEW OF THE TEST OF CARBOHYDRATES

A chart showing different tests is given below for the facility of students to put + and – signs for positive and negative reactions respectively on the basis of their practical observations.

Carbohydrates	Molisch's test	Iodine test	Benedict's test	Fehling's test	Barfoed's test	Bial's test	Tauber's test	Soliwanoff's test	Hydrolysis test	Methylamine test	Osazone test	Precipitation by half saturation	Precipitation by full saturation
Glucose													
Fructose													
Galactose													
Pentose													
Maltose													
Lactose													
Sucrose													
Starch													
Glycogen													
Dextrin													

DETECTION OF CARBOHYDRATE OF AN UNKNOWN SOLUTION

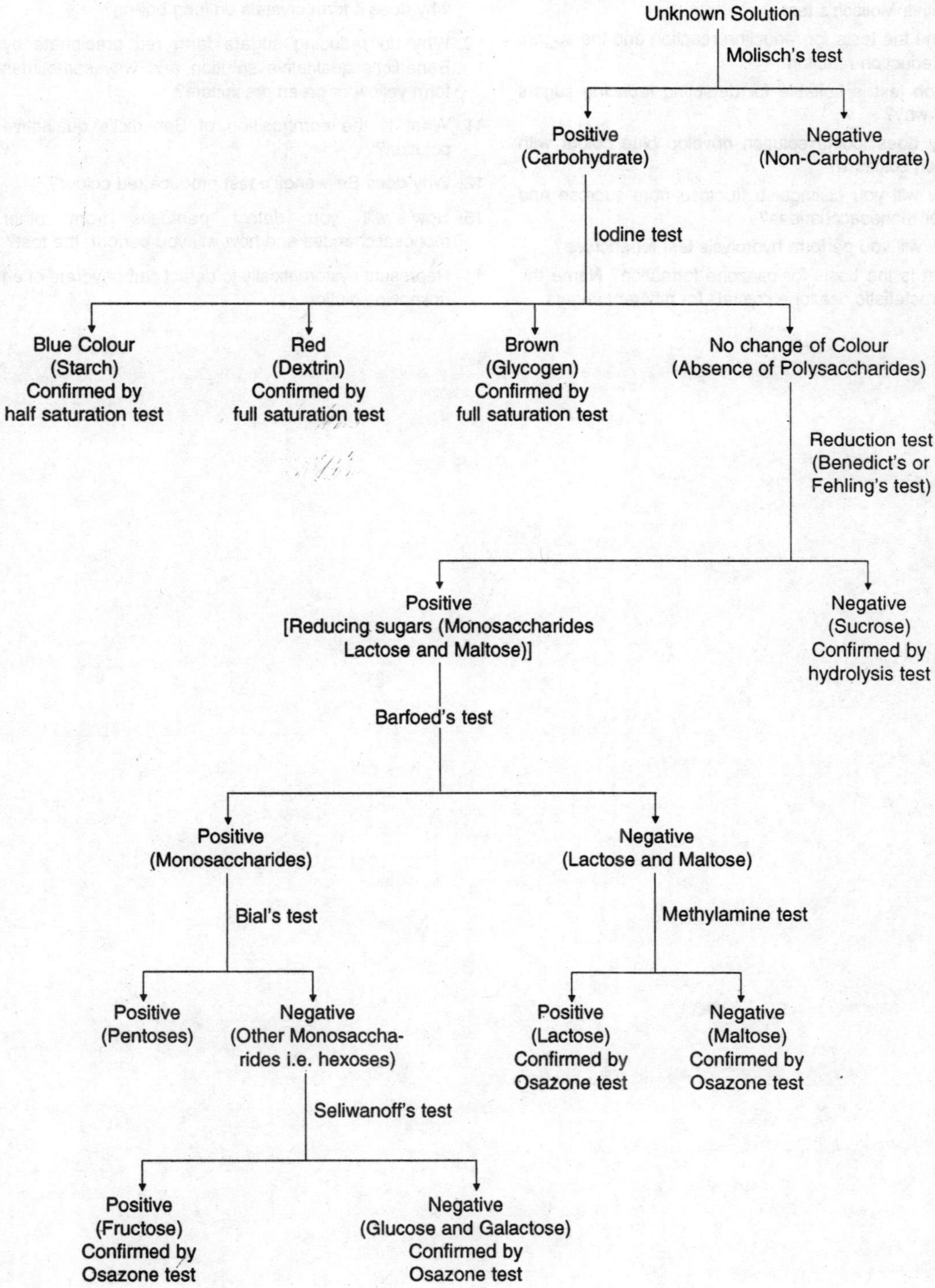

Questions to Answer

1. Discuss the principle of Molisch's test. State why some of the substance other than carbohydrates give positive Molisch's test.
2. Name the tests for reduction reaction and the sugars for reduction reaction.
3. Which test is suitable for detecting reducing sugars and why?
4. Why does iodine solution develop blue colour with starch solution?
5. How will you distinguish fructose from sucrose and other monosaccharides?
6. How will you perform hydrolysis test for sucrose?
7. What is the basis for osazone formation? Name the characteristic osazone crystals for different sugars.
8. Why do glucose, fructose and mannose form similar osazone crystals?
9. Whey sucrose does not form osazone crystals and why does it form crystals on long boiling?
10. Why do reducing sugars form red precipitate by Benedict's qualitative solution and why sometimes form yellow or green precipitate?
11. What is the composition of Benedict's qualitative solution?
12. Why does Seliwanoff's test produce red colour?
13. How will you detect pentoses from other monosaccharides and how will you perform the test?
14. Represent systematically to detect carbohydrate of an unknown solution.

CHAPTER 3

FAT

1. SOLUBILITY :

The glycerides of lower fatty acids are slightly soluble in water. All glycerides are soluble in ether, chloroform and benzene. They are slightly soluble in cold methyl alcohol, ethyl alcohol and acetone but highly soluble in hot methyl and ethyl alcohol, acetone.

Procedure : Arrange five dry test tubes in a rack. Add 2 ml. of water, ether, chloroform, benzene, and ethyl alcohol to each test tube followed by one drop of mustard oil. Shake and observe. In water, oil is broken into small droplets and float at the surface indicating that oil is insoluble in water. But in other solvents, oil disappears.

2. EMULSIFICATION :

Principle : Oil or liquid fat becomes finely divided and is dispersed in water when shaken with water to form emulsification. Emulsification is permanent and complete in the presence of emulsifying agents. The important emulsifying agents are bile salts, proteins, soaps, mono- and diglycerides. Emulsification is important in the processes of fat digestion in the intestine. Emulsifying agents lower surface tension of the liquid.

Procedure : Take 2 clean and dry test tubes. In one tube, 2 ml. water and in the other 2 ml. dilute bile salt solution. Now, to each add 2 drops of mustard oil and shake vigorously for about one minute. Allow the tubes to stand for 2 minutes and note that in water, oil is broken in small pieces and floats on the suface; whereas, in bile salt solution, the oil can be seen in minute droplets suspended in the liquid (e.g., permanent emulsification).

3. SAPONIFICATION :

Principle : Oil or liquid fat when boiled with an alkali is hydrolyzed and the liberated fatty acids form salt with alkali (soap). This process is said to be saponification. The saponification number is the number of milligrams of KOH required to saponify one gram of fat.

Procedure : Take 4 ml. 2% Sodium Carbonate Solution in a test tube and add 2 drops of mustard oil. Shake vigorouly and boil. A clean soapy solution is formed. Cool and divide it into three parts for further study on the properties of soap.

(i) In one test tube, add a few drops of Conc. HCl and observe that the fatty acid separates out and floats up. This is due to the hydrolysis of soap by the acid.

(ii) In another test tube, dissolve sufficient amount of finely powdered NaCl. White precipitates of soap separate out and float on the surface. This process is called 'salting out' of soap.

(iii) In the third test tube, add a few drops $CaCl_2$ solution. A precipitate of insoluble calcium soap is obtained.

4. GREASE-SPOT TEST :

Procedure : Put a drop of oil over a piece of ordinary writing paper (not filter paper). The translucent spot indicates the presence of fat.

5. UNSATURATION TEST :

Principles : All neutral fats contain glycerides of some unsaturated fatty acids. These unsaturated fatty acids become saturated by taking up iodine. If the fat contains more unsaturated fatty acids, it will take up more iodine.

Procedure : Add 10 drops of *Huble's iodine reagent* (alcoholic solution of iodine containing some mercuric chloride) to 10 ml. chloroform. The chloroform shows pink colour due to the presence of free iodine. Divide this solution equally into 4 test tubes.

(i) To one test tube, add mustard oil drop by drop shaking the tube vigorously for about 30 seconds after addition of each drop until the pink colour is discharged and count the number of drops. The pink colour is discharged owing to the taking up of iodine by the unsaturated fatty acids of the oil.

(ii) Repeat the above experiment with the remaining three test tubes taking fats (coconut oil, dalda and ground nut oil). Now compare their unsaturation. *It should be remembered that more the number of drops required to discharge the pink colour, the less is the unsaturation.*

Iodine value : It is the number of grams of iodine taken up by 100 grams of fat.

CHOLESTEROL

1. SALKOWSKI'S TEST (H_2SO_4 TEST) :

Procedure : Dissolve a little cholesterol in 2 ml. chloroform in a dry test tube. Add an equal volume of Conc. H_2SO_4. Shake gently. The upper layer of chloroform turns red and the sulphuric acid layer shows a yellow colour with a green fluorescence.

2. LIEBERMANN-BURCHARD REACTION (ACETIC ANHYDRIDE- H_2SO_4 TEST) :

Principle : Addition of H_2SO_4 to cholesterol in the presence of acetic anhydride gives a green (e.g, red-absorbing) chromphore.

Procedure : Dissolve a few crystals of cholesterol in 2 ml. of chloroform in a dry test tube. Now, add 10 drops of acetic anhydride and 1 to 3 drops of Conc. H_2SO_4. Mix well. A red rose colour is formed which quickly changes through blue to green.

3. FORMALDEHYDE-H_2SO_4 TEST :

Procedue : Add 2 ml. of formaldehyde—H_2SO_4 solution (1 part of 40% formaldehyde to 50 parts of the acid) to 2 ml. of chloroform solution in a dry test tube. The cherry colour is developed in the chloroform. Pour off the chloroform into another test tube and add 2 to 3 drops of acetic anhydride. The blue colour develops.

This test is said to be more delicate than Salkowski's test.

Questions to answer

1. Why is bile salt important for emulsification?
2. What is the importance of emulsification in the digestion of fat?
3. What is saponification value and what is the significance of high saponification value?
4. Why does the pink colour of chloroform disappear on addition of oil?
5. What is iodine value of fat and what is its significance?
6. Which test is important in the detection of cholesterol and how to perform the test?
7. Name the test for detecting fat and discuss its procedure.

CHAPTER 4

PROTEINS

A. COLOUR REACTIONS OF PROTEINS

Proteins respond to some colour reactions due to the presence of one or more radicals or groups of the complex protein molecule. All proteins do not contain the same amino acids and hence they do not respond to all colour reactions giving positive findings. The material under examination should, therefore, be subjected to several tests before concluding its nature.

1. BIURET TEST :

Principle : This test is positive for all compounds containing more than one peptide linkage (-CO-NH-) e.g., Proteins and their hydrolytic products (metaproteins, proteoses, peptones, polypeptides except dipeptides and amino acids). This test is also positive for substances which contain two carbamyl ($-CONH_2$) groups joined directly or through a single atom of nitrogen or carbon and similar substances which contain $-CSNH_2$, $-C(NH)NH_2$ or $-CH_2NH_2$ groups also respond to this biuret test. Hence nonproteins e.g., oxamide $\begin{bmatrix} CONH_2 \\ | \\ CONH_2 \end{bmatrix}$ and biuret $\begin{bmatrix} CONH_2 \\ | \\ NH \\ | \\ CONH_2 \end{bmatrix}$ give positive with this test.

Procedure : To 2 ml. of Protein solution add 1 ml. of 40% NaOH solution and 1 or 2 drops of 1% $CuSO_4$ solution. A violet colour indicates the presence of Peptide linkage of the molecule.

Note : Care must be taken that not more than 2 drops of dilute copper sulphate (1%) be added, otherwise blue colour will develop instead of violet colour.

2. NINHYDRIN TEST (THE TRIKETO-HYDRINDENE HYDRATE TEST) :

Priciple : This test is positive for all amino acids containing free amino and carboxylic groups. Hence, it is positive for proteins, peptones, peptides. It is also positive with other primary amines including ammonia. The triketohydrindene hydrate forms a complex with the amino or carboxylic group of the amino acids or other primary amino developing a blue colour on heating.

Procedure : To 1 ml. of Protein solution add 2 drops of freshly prepared 0.2% of ninhydrin solution and heat. A blue colour develops indicating the presence of proteins, peptides or amino acids.

3. MILLON'S TEST (Test for tyrosine) :

Principle : The neutral protein solution reacts with acidified mercuric chloride to form a yellow precipitate of Mercury-Protein complex. This complex forms a nitrite complex with sodium and nitrite solution on warming, developing a red colour. This test is given by phenols or phenolic substances such as Salicylic acid.

Procedure : Add 1 ml. of Protein solution acidified with H_2SO_4 to 1 ml. of acid mercuric sulphate solution (10% $HgSO_4$ in 10% H_2SO_4). Boil gently for about one minute. A yellow precipitate forms. Cool under the tap and add a drop of 1% of $NaNO_2$ (Sodium nitrite) solution. On gentle warming it turns red. This indicates the presence of hydroxy-phenyl group. Tyrosine is the only amino acid which responds positively to this test.

Note : Excess of chloride interferes by forming unionised $HgCl_2$. If so, more $HgSO_4$ must be added.

4. XANTHOPROTEIC REACTION (For tyrosine, phenylalanine and tryptophan) :

Principle : Proteoses and peptones do not form precipitate with HNO_3, but their solutions become yellow and then orange when made alkaline.

The white precipitate of protein after the addition of HNO_3 is due to the formation of metaproteins insoluble in HNO_3. The nitro-compounds from the proteins molecule containing *benzene ring* develop a yellow colour. These nitrocompounds in alkaline medium ionise freely and produce deep yellow or orange colour.

Procedure : To 2 ml. of protein solution add carefully 1 ml. of Conc. HNO_3. A white precipitate forms. Boil and the colour changes to yellow. Cool the test tube and add 2 ml. of 20% NaOH (or ammonia) to make it alkaline. The colour changes to orange indicating the presence of aromatic amino acids.

Note : Phenylalanine does not respond to this test as it is ordinarily performed. This test is not a satisfactory one for use in urinary examination because of the colour of the end reaction.

5. HOPKINS—COLE—ADAM—KICWICZ REACTION (Test for tryptophan) :

Principle : Tryptophan forms a condensation product with the aldehyde in presence of Conc. H_2SO_4 developing a purple ring.

Procedure : To about 1 ml. of the protein solution add one drop of very dilute (1 : 500) solution of commercial formalin (40% formalin) and one drop of mercuric sulphate solution (10% $HgSO_4$ in 10% H_2SO_4). Mix and incline the tube to add carefully at least 1 ml. Conc. H_2SO_4 by the side of the tube. A purple ring at the junction of the two liquids indicates the presence of *tryptophan*.

6. GLYOXALIC ACID REACTION FOR TRYPTOPHAN (HOPKINSCOLE) :

Principle : Tryptophan forms condensation product with glyoxalic acid in presence of pure Conc. H_2SO_4 developing a purple ring.

Procedure : About 2 ml. of Protein solution is mixded thoroughly with 2 ml. of glyoxalic acid reagent. Add this mixture to another test tube containing 4 ml. of pure Conc. H_2SO_4 in such a way that the two liquids do not mix. Rotate gently. A purple ring develops at the junction of the two liquids.

Note : This test does not respond in presence of strong oxidizing agents like nitrates and chlorates. Sulphuric acid must be highly pure; otherwise, the impurities may act as oxidizing agents.

7. SAKAGUCHI REACTION (Test for Arginine) :

Principle : Arginine in presence of alcoholic α-naphthol forms a complex with sodium hypochlorite developing a bright red colour.

Procedure : To about 3 ml. of protein solution add 1 ml. of 5% NaOH, 2 drops of pure alcoholic alpha naphthol and one drop of 10% sodium hypochlorite. Mix well. The development of bright red colour indicates the presence of *arginine.*

Note : This an extremely sensitive test and may be used as a general test for protein, because all known protein contain sufficient arginine.

8. SULPHUR REACTION (For cysteine, cystine and methionine) :

Principle : Sulphur of the sulphur-containing amino acids reacts with the sodium hydroxide forming sodium sulphide. A black or brown precipitate of lead sulphide is formed as a result of the reaction between sodium sulphide and lead acetate. This lead sulphide is insoluble in dilute HCl.

Procedure : Boil 1 ml. of Protein solution (concentrated solution of egg white) with 1 ml. of 40% NaOH for 1 minute. Add a drop of lead acetate solution. A black or brown precipitate is formed which is insoluble in dilute HCl.

9. MOLISCH'S TEST

Some proteins (glycoproteins) contain carbohydrate. Hence this test is done for detecting the carbohydrate present in glycoprotein. The method of this test is already discussed in carbohydrate chapter.

B. COAGULATION REACTION

Principle: Coagulation of Protein is caused by the denaturation of Protein structure by heat or acids.

1. HEAT COAGULATION

Procedure : Take protein solution of about two-thirds of the test tube and heat the upper portion of the solution. An opalescent appears which becomes deep on the addition of a few drops of 2% acetic acid. This indicates the presence of *albumin*.

2. COAGULATION WITH CONC. HNO3

Procedure : Take 2 ml. of Conc. HNO_3 in a test tube and add the protein solution by the side of the test tube. A white ring appears at the junction of the two liquids.

C. PRECIPITATION REACTION

Principle :

(i) Proteins are precipitated by salts of the heavy metals (e.g., $HgCl_2$, $AgNO_3$, $CuSO_4$, etc.)

(ii) These may be precipitated by certain acids some of which are called alk..loidal reagents (Picric acid, phosphotungstic acid, tannic acid and metaphosphoric acid etc.)

(iii) These may also be precipitated by concentrated solutions of ammonium sulphate, sodium sulphate and sodium chloride. These salts precipitate proteins by 'salting out' methods.

(iv) The proteins are precipitated by dehydrating agents such as alcohols and acetone. These agents convert them into suspensoids which flocculate upon the addition of a few drops of salt solution. Alcohol also causes denaturation of proteins. Alcohols bring protein solution into isoelectric point at which it is precipitated.

1. Precipitation by salts of heavy metals.

(a) **Procedure** : To 2 ml. of dilute protein solution (solution of egg white) add a few drops of dilute solution (1%) of Zinc Sulphate drop by drop. A white precipitate is formed indicating the presence of protein.

(b) **Procedure** : To 2 ml. of dilute protein solution (solution of egg white) add a few drops of 0.5 per cent ferric chloride solution drop by drop. An opalescent appears which disappears on addition of excess of ferric chloride solution.

2. Precipitation by alkaloidal reagent.

(a) **Procedure :** Add 1 ml. of Esbach's reagent (Solution of Pricric acid and citric acid) to 1 ml. of protein solution. A yellow precipitate is formed indicating the presence of protein.

(b) **Procedure:** To 2 ml. of protein solution add a drop or two of 2% solution of salicylsulphonic acid (a solution of salicylic acid in sulphuric acid). The formation of white precipitate indicates the presence of protein.

3. Precipitation of Neutral salts

(a) **Full Saturation :**

Procedure : Take 5 ml. of Protein solution in a test tube and add solid ammonium sulphate until it becomes saturated. A gelatinous precipitate forms which indicates the presence of *Albumin.*

(b) **Half Saturation :**

Procedure : Prepare a saturated solution of ammonium sulphate in a test tube. Add 3 ml. of saturated solution to 3 ml. of protein solution. Shake vigorously. A gelatinous precipitate formed indicates the presence of *globulin.*

4. Precipitation by alcohol.

(a) **Procedure** : To 1 ml. of protein solution (concentrated egg white solution) add 2 to 3 ml. of alcohol. A white opalescent forms indicating the presence of *protein*. Filter after 10 minutes. Try to dissolve the residue by water. The ppt. does not dissolve.

(b) **Procedure :** Take 1 ml. of protein solution in a test tube and 2 to 3 ml. of alcohol in another test tube. Keep both the tubes in the freezing mixture for 10 minutes and then mix the contents of the two tubes. A white opalescent appears. Filter at once. Try to dissolve the residue on the filter paper and observe that the precipitate redissolves.

CASEIN

Precipitation at isoelectric point : Add a drop of bromocresol green indicator to 3 ml. of casein solution. A blue colour develops. Now add 2% acetic acid drop by drop unitl the colour changes to green at P^H 4.7. A flocculent precipitate of casein is formed. Boil, cool and then add dil. NaOH to it. The precipitate dissolves again.

TEST FOR PHOSPHORUS IN CASEIN

Neumann's test : Take dry casein in a dry test tube. Add 2 drops of Conc. HNO_3 and 10 drops of Conc. H_2SO_4 to it. Heat the contents strongly over a flame in a fume cupboard with constant stirring to avoid spurting. The organic material chars and nitrous fumes are evolved. Now, allow the cupboard to cool in the rack and add about 2 ml. of distilled water and 2 ml. of concentrated ammonia solution. Mix. Now, add 2 ml. of ammonium molybdate solution and warm. The development of lemon yellow precipitate of ammonium phosphomolybdate indicates the presence of *phosphorus.*

Questions to answer

1. Why biuret test is done? Why violet colour is produced in that test?
2. What is the name of the test for detecting amino acid and how to perform the test?
3. If amino acid is detected, what other tests should be done for that?
4. If protein is detected, what other tests should be done for it?
5. Why does protein coagulate? By which tests you can have coagulation of protein?
6. Discuss the principle of precipitation of Protein.
7. Name the reagents for precipitating protein from a solution.
8. Which test should be performed for detecting albumin and globulin and how to perform it?
9. What is the principle of sulphur test for sulphur containing amino acids?
10. Name and perform the test for detecting aromatic amino acids.
11. Name and perform the test for the detection of Phosphorus in casein.

CHAPTER 5

ANALYSIS OF FOOD

CHEMISTRY OF MILK

Specific Gravity :

(i) The specific gravity of milk is determined by lactometer (hydrometer).

(ii) The sp. gr. of whole milk is usually around 1.30.

(iii) The sp. gr. increases with the removal of fat from milk and decreases with the addition of water to it.

Reaction :

(i) A fresh milk is practically neutral in reaction.

(ii) It becomes acidic on keeping milk at room temperature owing to the production of lactic acid by bacterial fermentation.

COMPOSITION OF MILK

1. It is a complete food as it contains almost all the elements for growth and maintenance. But it is not suitable for adults as it is poor in iron and contain high proteins and fat.

2. The composition of milk varies with the species of the animals. The composition of human and cow's milk are mentioned below.

3. Its dietary value is also progressed by the presence of inorganic salts and vitamins A, C, D and B-complex.

Constituents in gram/100 ml.	Human milk	Cow's milk
Protein	1.4	3.5
Fat	3.5	4.0
Carbohydrate	7.0	4.8
Mineral	0.2	0.7

4. The principal protein is casein (a phosphoprotein). It also contains lactalbumin and lactglobulin.

5. The fat contents present in fine droplets in emulsion form are neutral fat, cholesterol and lecithin.

6. The minerals are calcium, potassium, phosphates in large amounts; sodium, magnesium and chloride in fair amounts; but iron in poor amount.

EXPERIMENTS ON CHIEF CONSTITUENTS OF MILK

1. Dilute 10 ml. of milk by 10 ml. of distilled water. Then add 2% acetic acid drop by drop till a heavy precipitate of casein is formed. Allow it to stand for 5 minutes and filter. Residue and filtrate.

2. A small portion of the residue is allowed to dissolve in 4 ml. of 2% Na_2CO_3, and/with this solution perform (a) biuret test, (b) Ninhydrin test and (c) colour test for amino acids to justify that the residue contains *protein.*

3. Take a portion of the residue and perform Neumann's test for combined *phosphorus.*

4. A portion of the dried residue is shaken up vigorously with ether in a test tube for about 1 to 2 minutes and allow it to stand for a while. Decant the supernatant ether layer in a dry basin and allow it to evaporate spontaneously at room temperature. Rub a piece of ordinary writing paper on the portion left in the basin. A translucent grease spot on the paper indicats the presence of *fat.*

5. To 10 ml. of the filtrate add a drop of chlorphenol red indicator. Then add 2% Na_2CO_3 solution or 2% acetic acid to bring the P^H 5.4 showing the change of red to yellow colour. Boil the solution. A coagulum appears owing to the coagulation of *albumin and globulin*. Filter and keep the filtrate for the following experiments.

6. With the portion of this filtrate perform Molisch's test to know the presence of *carbohydrate* and other tests for the detection of the nature of *sugar present*.

7. Perform silver nitrate test for detecting *chloride* with a portion of the filtrate.

8. To the remaining portion of the filtrate add a few drops of NH_4OH to make it alkaline and boil. A slight gelatinous precipitate of calcium and magnesium phosphates are produced. Filter and dissolve the residue on the filter paper by pouring over it 5 ml. of hot 2% acetic acid. (a) To a portion of this solution add a clear solution of potassium oxalate. A turbidity of *Calcium oxalate* appears. (b) To the remaining solution add 2 ml. of ammonium molybdate solution. Mix and heat. Lemon yellow precipitate of phosphomolybdate indicates the presence of *phosphate*.

Schematic representation of the expriment for finding out the constituents of milk

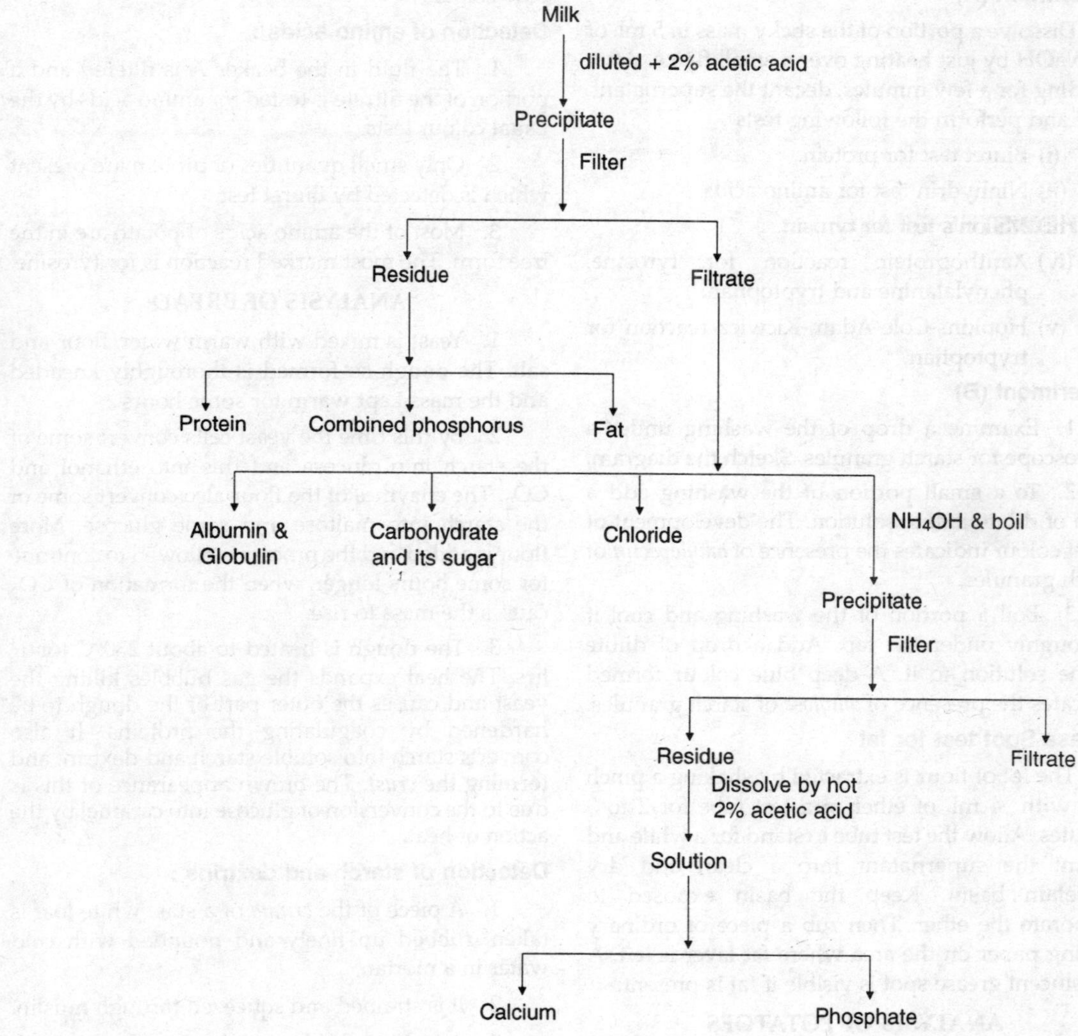

CHEMISTRY OF WHEAT FLOUR

COMPOSITION

1. Wheat flour contains about 75% starch, 8 to 10% protein and about 1% fat.

2. The remaining constituents are water, cellulose, inorganic salts and some vitamins.

3. Starch is insoluble in cold water. On boiling, cellulose layer (the covering of starch granules) breaks and starch granules go into solution.

4. The proteins are gliadins (soluble in 70-80% ethanol) and glutelin (soluble in alkali). On treatment with water these proteins form a sticky mass called *gluten*, the viscidity of which is due to gliadin. Rice and maize are poor in gliadin for which dough is not formed when mixed with water.

5. White flour is poorer in protein, fat, cellulose and vitamins.

DETECTION OF THE CONSTITUENTS OF WHEAT FLOUR

1. Mix some wheat flour with a little water to make a stiff dough and allow it to stand for 10 minutes.

2. Wrap a small amount of dough in a piece of cloth and knead it well in a small amout of water kept in a basin. Continue kneading unitl the washings are no longer turbid.

3. The yellowish sticky mass, known as gluten, remains in the muslin which is kept for peforming experiment (A) and the washings for experiment (B).

Experiment (A)

Dissolve a portion of the sticky mass in 5 ml. of 5% NaOH by just heating over a small flame. After standing for a few minutes, decant the supernatant. Cool and perform the following tests.

(i) Biuret test for protein.

(ii) Ninhydrin test for amino acids.

(iii) Millon's test for tyrosin.

(iv) Xanthoproteic reaction for tyrosine, phenylalanine and tryptophan.

(v) Hopkins-Cole-Adam-Kicwicz reaction for tryptophan.

Experiment (B)

1. Examine a drop of the washing under a microscope for starch granules. Sketch the diagram.

2. To a small portion of the washing add a drop of dilute iodine solution. The development of violet colour indicates the presence of *amylopectin* of starch granules.

3. Boil a portion of the washing and cool it thoroughly under the tap. Add a drop of dilute iodine solution to it. A deep blue colour formed indicates the presence of *amylose* of starch granules.

Grease Spot test for fat

The fat of flour is extracted by shaking a pinch of it with 4 ml. of ether in a test tube for 2 to 3 minutes. Allow the test tube to stand for a while and decant the supernatant into a clean and dry porcelain basin. Keep the basin exposed to evaporate the ether. Then rub a piece of ordinary writing paper on the area where fat layer is left. A translucent grease spot is visible if fat is present.

ANALYSIS OF POTATOES

Isolation of starch grains :

1. The clean surface of half a potato is scraped very finely with a penknife and these are placed in a beaker of water. These are stirred well and stained through fine muslin into another beaker.

2. This is allowed to stand for a few minutes and the white deposit of starch noted.

3. The supernatant fluid is poured off into another breaker labelled as A.

4. The beaker containing the starch is filled up with water, stirred well and the starch allowed to settle. Repeating this process several times—the starch can be obtained quite pure.

5. A little is examined microscopically to note the characteristic form of the grains.

6. A little is boiled with water, cooled and a drop of iodine is added to have a characteristic deep blue colour..

Detection of amino acids :

1. The fluid in the beaker A is filtered and a portion of the filtrate is tested for amino acids by the usual colour tests.

2. Only small quantities of protein are present which is detected by Biuret test.

3. Most of the amino acids of potato are in the free form. The most marked reaction is for tyrosine.

ANALYSIS OF BREAD

1. Yeast is mixed with warm water, flour and salt. The dough so formed is thoroughly kneaded and the mass kept warm for some hours.

2. By this time the yeast cells convert some of the starch into glucose and this into ethanol and CO_2. The enzymes of the flour also convert some of the starch into maltose and some glucose. More flour is added and the process is allowed to continue for some hours longer, when the formation of CO_2 causes the mass to rise.

3. The dough is heated to about 230°C for $1\frac{1}{2}$ hrs. The heat expands the gas bubbles killing the yeast and causes the outer part of the dough to be hardened by coagulating the proteins. It also converts starch into soluble starch and dextrin, and forming the *crust*. The brown appearance of this is due to the conversion of glucose into caramel by the action of heat.

Detection of starch and dextrins :

1. A piece of the *crumb* of a stale white loaf is taken, rubbed up finely and pounded with cold water in a mortar.

2. It is strained and squeezed through muslin.

3. A white fluid containing wheat starch grains is obtained. The fluid is filtered.

4. A little Fehling's solutuion is added to a portion of the filtrate and boiled; a well-marked reduction occurs. A drop or two of iodine is added to another protion; a purple colour is produced showing the presence of dextrins. If two drops of very dilute iodine is added, a pure blue colour is produced showing that a small amount of soluble starch is also present.

5. The above process is repeated using the *crust* instead of the crumb. It is noted that reducing sugar is absent or present in traces only; dextrins and starch are present.

ANALYSIS OF MEAT

Muscle tissue constitutes the most important source of proteins and essential amino acids available to man.

It mainly consists of :

(i) Proteins, (ii) Creatine, (iii) Glycogen, (iv) Lactic acid, (v) other constituents.

(i) PROTEINS

The proteins of living muscle consist of *sarcoplasmic proteins* which include many of the enzymes concerned in carbohydrate, breakdown being accompained by the contraction of the muscle; while the *contractile elements* contain myosin, actin and tropomyosin. Myosin combines with actin to form actomyosin. Myosin and actomyosin are easily denatured by heat.

Preparation and properties of muscle proteins :

1. The muscles of a rabbit are excised after killing and skannings; and then left on ice until thoroughly chilled and minced in an ice-cooled mincer. The mince is added to 5 vol. of ice-cold 0.5 M-KCl and solid $NaHCO_3$ is added to keep the P^H at 6.5 - 7.0. The mass is stirred occasionally after 20 minutes. It is then centrifuged to separate the extract from the residue.

2. The extract is diluted with 15 vol. of cold water. The precipitate of myosin is noted. The myosin is centrifuged down and a sample in water is suspended. Now, solid KCl is added a little at a time till the myosin redissolves; myosin is a globulin.

3. A little acetate buffer at P^H 5.0 is added to the solution of myosin in KCl. The protein becomes insoluble at its isoelectric P^H.

4. The solution of myosin in KCl is heated. It is easily coagulated. The coagulum is filtered off, washed and the usual amino acid colour tests applied.

5. The nitroprusside test on both the native and the coagulated myosin is carried out. It must be carefully noted that the –SH groups of the native myosin disappear on denaturation and heat coagulation.

(ii) CREATINE

Creatine is normally present in muscle as phosphagen. It is the most abundant nitrogenous extractive found in meat, in which about 0.4% is present. On boiling with mineral acids, it undergoes dehydration and conversion into creatinine. Creatinine is the normal constituent of urine but creatine—except a trace—occurs in urine under certain conditions.

Preparation of creatine from meat extract :

1. 10 grams of commercial meat extract is dissolved in 200 ml. of water.

2. A saturated solution of lead acetate is slowly added until no further precipitate is formed, care is to be taken to avoid any excess.

3. The precipitates of proteins and phosphates are filtered off.

4. The filtrate is warmed and H_2S is passed in to precipitate the lead and the resulting precipitate is filtered off.

5. The filtrate is now evaporated to a syrup. The syrup is allowed to stand in the ice chest for 2 or 3 days when rhombic crystals of creatine separate out.

6. Some of the crystals are examined under microscope to make a drawing of these.

7. The syrup is treated with 200 ml. of ethanol and stirred well with a glass rod and the crystals are collected on a small paper. The filtrate contains purine bases.

N.B. Many specimens of meat extract contain creatinine as well as creatine. Fresh rabbit muscle is a better source of creatine.

Conversion of creatine into creatinine :

Creatine is allowed to dissolve in 30 ml. of hot water and the solution is divided into two equal parts A and B.

B is treated with equal volume of 1N–HCl and heated on a boiling water-bath for 3-5 hours. The product is neutralized with NaOH A and B are tested for creatinine by the following method.

Jaffe's test for creatinine :

1. Ten ml. of the solution is treated with 15 ml. of saturated picric acid solution and 5 ml. of 10% NaOH.

2. The mixture is allowed to stand for 5 minutes and diluted to 200 ml. with water.

3. A deep orange colour develops due to the formation of a tautomeric form of creatinine picrate.

4. Sample A should give a negative response unless it is contaminated with creatinine.

(iii) GLYCOGEN

1. Fresh muscle contains about 0.5–1% of glycogen.

2. Liver glycogen is converted into glucose but muscle glycogen is converted into pyruvic acid and—under anaerobic conditions—into lactic acid.

(iv) LACTIC ACID

Lactic acid is normally found in a fatigued, excised muscle. This compound is soluble in water, ethanol and ether, crude samples of lactic acid are purified by way of the zinc salt.

Hopkins' test for lactic acid :

1. 5 ml. of Conc. H_2SO_4 and 3 drops of a saturated solution of $CuSO_4$ are added to 3 drops of a 1% solution of lactic acid in ethanol in a clean dry test tube.

2. Mixed and the tube is placed in a beaker of boiling water for about 5 minutes.

3. Cool thoroughly under the tap, add 2 drops of a 0.2% ethanolic solution of thiophene and shake.

4. The tube is replaced in a boiling water-bath.

5. A cherry-red colour develops when the mixture gets warm.

N.B. Lactic acid is oxidized in H_2SO_4 solution to formic acid and acetaldehyde which react with thiophene in the presence of excess H_2SO_4 to give a cherry red colour. Oxidation is aided by $CuSO_4$ which is inhibited by water.

(v) OTHER CONSTITUENTS OF MUSCLE

1. Other muscle constituents are the two purine bases—Hypoxanthine and xanthine.

2. The closely related to these is a third purine base—Adenine—which is also present in the muscle.

ANALYSIS OF EGGS

The hen's egg contains about 59% white, 30% yolk and 11% shell. The chemical composition of duck's egg is similar to that of hen's egg.

The *white* contains *proteins* and *salts*. The greater part of the protein is ovalbumin and the other proteins are conalbumin, ovoglobulin, ovomucoid (a glycoprotein). The white contains 10.7% *proteins*, 0.1% *fat* and 1.4% *carbohydrate*.

The pale yellow colour of yolk is due to the presence of riboflavin. The *yolk* contains 15% protein, 33% fat of which 4% cholesterol, 1% mineral and 51% water. The *proteins* are vitellin (a phosphoprotein) and livetin (a globulin). The *minerals* are mainly calcium, iron and phosphate. The yolk is also rich in *vitamins* A, B1, B2 and D. The other B group vitamins and E are also present.

A. Detection of Proteins :

The white portion of the egg is to be taken in a beaker and dilute it with water after stirring properly for a thorough mixing.

(i) Biuret test :

To 2 ml. of protein solution add 2 ml. of 40% NaOH solution and 1 or 2 drops of 1% $CuSO_4$ solution. Formation of violet colour indicates the presence of peptide linkage of *protein*.

(ii) Full saturation test :

Take 5 ml. of protein solution in a test tube and add solid ammonium sulphate until it becomes saturated. Formation of a gelatinous precipitate indicates the presence of *albumin*.

(iii) Half saturation test :

Prepare a saturated solution of ammonium sulphate in a test tube. Add 3 ml. saturated solution to 3 ml. of protein solution. Shake vigorously. Formation of a gelatinous precipitate indicates the presence of globulin.

(iv) Colour reaction test for amino acids (mainly cysteine and tryptophan).

B. Detection of Carbohydrates :

Molisch's test :

Add 2 drops of Molisch's reagent to 2 ml. of white solution of egg in a test tube. Mix thoroughly. Add 2 ml. of Conc. H_2SO_4 by the side of the test tube slanting the tube. Erect the test tube slowly. The formation of reddish-violet ring at the junction of two liquids indicates the presence of carbohydrate.

Take the yolk protion of the egg in beaker and mix it thoroughly with water after well stirring.

C. Detection of fat

(i) Grease-spot test :

Put a drop of the yolk solution in chloroform over a piece of ordinary white paper. The translucent spot indicates the presence of fat.

(ii) Salkowski's test :

Dissolve a little yolk solution in 2 ml. chloroform in a dry test tube. Add equal volume of Conc. H_2SO_4. Shake gently. The upper layer of chloroform turns red and the sulphuric acid layer shows a yellow colour with a green fluorescence indicating the presence of cholesterol.

D. Detection of minerals :

(i) Test for phosphates :

Add a little ammonium hydroxide to a test tube containing the above yolk solution to three-fourth of the test tube and then boil. A white precipitate of phosphates of calcium is formed.

Filter this precipitate and wash the residue on the filter paper with water. Dissolve the residue in about 3 ml. of hot dilute acetic acid by pouring the acid on the filter paper. Collect the solution in a clean test tube and divide it into two parts. Retain one part for the detection of calcium.

To the other part add a drop of Conc. HNO_3 and a few drops of ammonium molybdate solution. Boil. A lemon yellow precipitate indicates the presence of phosphates.

(ii) Test for calcium :

To one part of the above solution add about 1 ml. of potassium oxalate solution. A white precipitate of calcium oxalate is formed immediately.

E. Detection of Vitamins :

(i) Test for vitamin A:

Dissolve a few drops of yolk solution in 1 ml. chloroform. Add 2 ml. of the carr-price reagent (saturated antimony trichloride in chloroform) to about 0.2 ml. of the above solution. Development of blue colour indicates the presence of vitamin A.

(ii) Test for vitamin B_2 (riboflavin) :

Take 5 ml. of yolk solution in tube (A) and 5 ml. of riboflavin solution in tube (B). Add 1 ml. of glacial acetic acid and 2 ml. of pyridine to each tube. Stopper firmly and shake thoroughly. Add 2 drops of saturated solution of $KMnO_4$ to each tube to oxidize interfering substances. After 1 minute add 2 drops of 10 vol. H_2O_2 to decolorise the excess of permanganate. Add 5 gms. of anhydrous Na_2SO_4 to each tube. Shake gently. Add 10 ml. of n-butanol, stopper and shake vigorously for 2 minutes. Centrifuge the mixtures and put the clear supernatants into a clean dry test tube. Examine both under ultraviolet light and compare the intensities of the greenish-yellow fluorescence.

ANALYSIS OF CHEESE

The quantity of protein in cheese varies from 23% to 33% and fats from 7% to 20% depending on the type of cheese.

A. Detection of fat in cheese :

Boil a small amount of cheese with ethanol to extract the lipids and filter while hot. The filtrate becomes cloudy first but clears again on heating. An emulsion of fat is obtained on the addition of water.

B. Detection of protein constituents :

Rub up the residue of the above filtration with some 2% sodium carbonate in a mortar. Decant the cloudy solution and acidify it with acetic acid. A flocculent precipitate of *casein* is obtained.

If this is filtered off and the filtrate is treated with an equal volume of saturated ammonium sulphate and again filtered, the filtrate usually gives a biuret reaction when treated with an equal volume of 40% NaOH and a drop of dilute $CuSO_4$. The casein is partially broken down to proteoses, peptones etc. due to hydrolysis by bacterial proteolytic enzymes.

C. Detect different amino acids by colour reaction tests except the sulphide test.

ANALYSIS OF GELATIN

White fibrous tissues and the organic matrix of bones and teeth contain a protein known as collagen. Collagen is partly broken down by treatment with boiling water. On cooling it yields a gelatinous product free from carbohydrates. This can be dried to produce gelatin. It is a protein of low biological value because it contains little cysteine, tyrosine and phenylalanine, while tryptophan and certain other essential amino acids are altogether absent.

A. Precipitation of Gelatin

Take 2 ml. of gelatin solution and add 2 ml. of saturated ammonium sulphate solution. Gelatin is precipitated in the form of gelatinous mass.

B. Test for peptide linkages :

Perform Biuret test.

C. Tests for amino acids :

Perform colour reaction tests of different amino acids.

ANALYSIS OF PEPTONE

Two per cent solution of peptone is prepared in warm water. The P^H is adjusted to 5.4 by the addition of 1% acetic acid, a few crystals of magnesium sulphate are added and the mixture is boiled to coagulate any denatured proteins. The preparation is allowed to cool and then filtered to remove solid matter. The filtrate is used.

(i) Dilute 1 ml. of the filtrate to 4 ml. with water. Perform biuret test to have pink colour for the detection of native and denatured proteins.

(ii) Boil 1 ml. of the filtrate. No coagulum will be formed.

(iii) (a) To 1 ml. of the filtrate add a few drops of Esbach's reagent and (b) To 1 ml. of the filtrate add a few drops of lead acetate solution, precipitates will be formed in both (a) and (b) indicating the presence of albumin.

(iv) Take 5 ml. of the filtrate and saturate with finely powdered ammonium sulphate by heating with an excess of the solid and then cooling under the tap. Filter. The filtrate contains peptones which are soluble in saturated ammonium sulphate and the residues consists of proteoses. Perform biuret test for the residue and the filtrate. The biuret test is positive in both cases.

CHAPTER 6

ENZYMES

1. Enzymes are proteins having catalytic properties.

2. They are formed by living cells—both animals and plants.

3. Digestive enzymes are formed by digestive glands, such as salivary, gastric, pancreatic etc.

4. The enzymes need an optimum temperature (37°C) for most of the animal enzymes and an optimum P^H for their highest activity.

5. Their activity is lost by boiling and they become inactive at low temperature.

6. They are specific for a particular substrate for a chemical reaction to bring about.

DIGESTIVE ENZYMES
STUDY OF SALIVARY AMYLASE

1. Salivary amylase or ptyalin is the digestive enzyme of saliva which is secreted by the salivary glands.

2. This enzyme digests boiled starch into maltose. Dextrins are also formed as an intermediate product during hydrolysis of starch.

3. It works best at an optimum P^H 6.8 and optimum temperature 37°C.

4. It is activated by chloride ions.

5. During hydrolysis of starch; amylodextrin, erythrodextrin, achrodextrin and maltose formed are detected by the colours such as blue, purple, red and no colour, respectively, by the addition of dilute iodine solution.

Collection of diluted Saliva

1. Clean your mouth with water first.

2. Take about 20 ml. of 0.2% NaCl solution in your mouth and move it about in the mouth with the help of tongue for a minute or two and collect the fluid in a clean hard glass test tube.

3. Shake the contents of the tube vigorously.

4. Filter to remove any epithelial cells etc. present.

Experiment

Take 3 test tubes and mark them 1, 2 and 3 with a marking pencil.

Test tube No. 1. — 5 ml. diluted saliva + boil and cool + 5 ml. of 1% starch solution.

Test tube No. 2. — 5 ml. diluted saliva + 5 drops of dil. HCl + 5 ml. of 1% starch solution.

Test tube No. 3. — 5 ml. of diluted saliva + 5 ml. of 1% starch solution.

Place all the three test tubes in a beaker full of water at about 37°C.

Take a drop of fluid from test tube No. 3 at an interval of one minute and test it with a drop of dilute iodine till you get a faint iodine colour. The successive colours blue, purple, red and no colour are seen. Remove this test tube from the bath.

Now take a drop of fluid from each of test tube Nos. 1 and 2 at an interval of one minute and test with iodine. Only blue colour will be seen. This indicates that salivary amylase has not at all digested in these two test tubes. This explains the fact that in test tube No. 1 the activity of the enzymes is lost by boiling and in test tube No. 2 the optimum P^H of the solution is not maintained.

Perform Benedict's test in test tube No. 3. This will give positive because of the formation of maltose from starch hydrolysis by the enzyme. The enzyme has worked best owing to the optimum P^H and temperature maintained.

STUDY OF PEPSIN IN GASTRIC JUICE

1. Pepsin is secreted in the form of Pepsinogen by the peptic glands of the gastric mucosa.

2. Pepsinogen is activated to pepsin by HCl.

3. It is inactivated in alkaline medium and is destroyed by boiling.

4. To accomplish the experiment, carmine-fibrin is used as a substrate for pepsin. Carmine-fibrin is a blood fibrin stained with carmine.

Experiment

Arrange 3 numbered test tubes in a rack and put the following contents into them :

Test tube No. 1. — 1 ml. pepsin + boil and cool + 1 ml. of 0.4% HCl + 2 small flakes of carmine-fibrin.

Test tube No. 2. — 1 ml. pepsin + 1 ml. water + 2 small flakes of carmine-fibrin.

Test tube No. 3. — 1 ml. pepsin solution + 1 ml. of 0.4% HCl + 2 small flakes of carmine-fibrin.

Keep the three tubes in a water bath at about 37°C and observe the test tubes after 5, 10 and 15 minutes. It will be seen that the digestion has taken place in test tube No. 3 but not in No. 1 and 2. The digestion has been indicated by the red colour developed as a result of the liberation of carmine from the digested part of carmine-fibrin.

RENNIN

1. Milk is clotted by pepsin, trypsin and many proteolytic enzymes found in plants in the presence of a suitable concentration of calcium ions and at suitable P^H.

2. In fact, clotting is the first stage in the digestion of milk and the gastric mucosa of the infants secretes a milk-clotting enzymes—rennin.

3. Renninogen is converted to rennin being activated by HCl. The optimum P^H of rennin is higher than that of pepsin.

4. Rennet is a commercial product containing rennin and is prepared by saline extraction of the mucous membrane of the abomasum (fourth stomach) of the calf. It is used abundantly in the cheese industry.

Clotting of milk by rennin :

1. Measure 5 ml. of fresh milk into a tube and place it in a water-bath at 38°C.

2. Allow it to warm up, add 1 ml. of an active preparation of rennet. Mix, replace in the bath and examine the digest from time to time.

3. Observe that after a short time the milk sets to a semi-solid, somewhat gelatinous mass and the tube can be inverted without disturbing the contents.

4. Replace the tube in the bath and examine it again after 1 hour. Observe that the clot has been contracted and a clear whey has been formed.

EREPSIN

Intestinal erepsin consists of three enzymes—carboxypeptidases, aminopeptidases and dipeptidases. The carboxy-and aminopeptidases split off amino acids one molecule at a time so that the chain length is decreased step by step and free amino acids are liberated. This process continues until only dipeptides remain.

Action of erepsin on peptone.

1. Label two 200 ml. flasks, A and B, and into each place 100 ml. of a 2% solution of commercial peptone. To A add 10 ml. of erepsin and 2 ml. of toluene. Shake well and stopper with a cork. To B add boiled 10 ml. of erepsin, add 2 ml. of toluence and stopper. Incubate both the flasks at 38°-40°C for about 48 hours or longer.

2. Add 2 or 3 drops of strong acetic acid and bromine water drop by drop to 5 ml. of each flask. A pink colour is obtained in A but not in B.

3. To 1 ml. of A and B add 4 or 5 ml. of water, 2 drops of 1% $CuSO_4$, and 1 or 2 ml. of 5% NaOH. A strong biuret reaction is given by B but A may give none or a feeble reaction.

CATHEPSIN

An organ, after removing from the body, if incubated at 38°C in the presence of an antiseptic—such as toluene—it is found that the tissue proteins yield amino acids after slow hydrolysis. This disintegration of the tissues is known as autolysis. This happens by the action of certain proteolytic enzymes present nearly in all tissues known as kathepsin.

The kathepsins are activated by HCN and H_2S and also by reduced glutathione and cysteine. The rate of hydrolysis also depends on the condition of the tissue at the time of death and on the P^H of the medium in which the autolysis takes place. The autolysis is accelerated by previous starvation of the animal and the optimum P^H being 5 and 6.

The autolysis of OX kidney :

The ox kidney is made free from fat, minced and weighed. The pulp is treated with three times its weight with 0.2% acetic acid and shaken with toluene as preservative. The mixture is incubated at 38°C for 3 or 4 days. After filtration the filtrate is treated with more toluene and incubated for another 7 to 9 days.

10 ml. of the filtrate is boiled after adjusting the P^H 5.4 to get the complete coagulation. After filtration, bromine water is added drop by drop to

a portion of the filtrate. Formation of red or purple colour indicates the presence of free tryptophan.

10 ml. of the filtrate is cooled after boiling and 10 drops of Conc. H_2SO_4 is added and cooled again. An equal volume of $HgSO_4$ reagent is added, mixed and allowed to stand for 10-20 minutes. A precipitate consisting of mercuric compounds of tryptophan and purine bases is obtained and filtered.

To a small portion of the precipitate applied the colour reactions for tyrosine and tryptophan. Test for tyrosine is negative, while test for tryptophan is strongly positive.

The filtrate is neutralized by the addition of 40% NaOH. The precipitate after filtration is tested for tyrosine—showing markedly positive.

It is not easy to have crystals of tyrosine or leucine by direct evaporation of the original solution due to the presence of purine bases and other compounds. If these are removed by precipitation with phosphotungstic acid and 3% H_2SO_4 solution and the two acids are removed by baryta, the filtrate gives tyrosine and leucine on evaporation.

CHAPTER 7

VITAMINS

Detection of Vitamin A in cod liver oil

(i) Dissolve 1 drop of Cod liver oil in 0.5 ml. of chloroform.

(ii) Add 2 ml. of Carr-Price reagent (saturated antimony trichloride on chloroform) to it and mix.

(iii) A transient blue colour is developed indicating the presence of vitamin A and its precursor (carotene).

This test is very sensitive.

Precaution : Carr-price reagent is highly poisonous.

Detection of riboflavin in milk

(i) 2 ml. of 40% trichloracetic acid is added to 8 ml. of milk to precipitate protein, mix well and filter.

(ii) Take 5 ml. of the filtrate in a boiling tube, add to it 1 ml. of glacial acetic acid and 2 ml. of Pyridine. Shake well by the use of stopper.

(iii) 2 drops of saturated solution of potassium permanganate is added to oxidize interfering substances and after 1 minute, add 2 drops of 10% (V/V) H_2O_2 to decolorise excess of permanganate.

(iv) Add 5 grams of anhydrous sodium sulphate and shake gently.

(v) Add 10 ml. of n-butanol and shake vigorously for 2 minutes by the use of stopper. Centrifuge.

(vi) Pour the clear supernatant into a clean dry test tube and examine it by ultraviolet light. The development of greenish-yellow flourescence indicates the presence of riboflavin. The intensity of flourescence is proportional to the concentration of riboflavin.

Detection of Nicotinic acid

Principle : When nicotinic acid cyanogen at 60°C and P^H 6.5 is incubated, Pyridine ring is opened to produce an aldehyde which couples with P-aminoacetophenone in acid solution to give a yellow product.

Reagent :

Cyanogen bromide : This is prepared by adding chilled 10% KCN to chilled bromine water until the colour is just discharged. It is to be prepared in a fume cupboard.

Precaution : Cyanogen bromide solution is highly poisonous.

Procedure :

(i) Put 5 ml. of nicotinic acid solution (140 µg per ml.) in a test tube and add to it 2 ml. of cyanogen bromide solution from a burette. Mix well and incubate the tube at about 60°C in a water bath for 4 minutes.

(ii) Then cool the tube under the tap for 5 minutes avoiding strong light.

(iii) Add 1 ml. of 10% P-aminoacetophenone in ethanol and mix well.

(iv) Examine after 5 minutes. An unstable yellow colour develops.

Discussion : All these above qualitative methods can be used with known standards to determine the quantities of these vitamins.

1. DETECTION OF VITAMIN A IN COD LIVER OIL

(a) Dissolve 1 drop of cod liver oil in 0.5 ml. chloroform.

(b) Add 2 ml. carr-price reagent (saturated antimony trichloride in chloroform) to it and mix.

(c) A transient blue colour is developed.

(d) This test is positive for vitamin A and its precursor (carotene) and it is a very sensitive test.

Note : Carr-price reagent is highly poisonous.

2. DETECTION OF RIBOFLAVIN IN MILK

(a) Deproteinise 8 ml. milk (or skim milk) with 2 ml. 40% trichloracetic acid, mix well and filter.

(b) Take 5 ml. clear filtrate in a boiling tube. Add 1 ml. glacial acetic acid and 2 ml. pyridine to it. Stopper firmly and shake well.

(c) Add 2 drops of a saturated solution of potassium permanganate to oxidise interfering substances and after a minute, add 2 drops of 10% (V/V) H_2O_2 to decolorise excess of permanganate.

(d) Add 5 grams anhydrous sodium sulphate, shake gently.

(e) Add 10 ml. n-butanol, stopper and then shake vigorously for 2 minutes.

(f) Centrifuge the mixture and pour clear supernatant into clean dry test tube.

(g) Examine it by ultraviolet light and note the greenish- yellow fluorescence produced by riboflavin. The intensity of fluorescence is proportional to the amount of riboflavin.

3. DETECTION OF NICOTINIC ACID

Principle : The aldehyde, produced as a result of the opening of the pyridine ring of nicotinic acid after the incubation of nicotinic acid cyanogen bromide at 60°C and P^H 6.5, couples with P-amino-acetophenone in acid solution to give a yellow product.

Procedure :

(a) Add 5 ml. nicotinic acid solution (140 µg per ml.) and 2 ml cyanogen bromide solution from a burette (*This reagent is highly poisonous*) in a dry clean test tube. Mix well.

(b) Incubate the tube at about 60°C for 4 minutes.

(c) Cool the tube under the top for 5 minutes avoiding strong light.

(d) Add 1 ml. of 10% p-aminoacetophenone in ethanol.

(e) Mix well and examine after 5 minutes.

(f) An unstable yellow colour develops.

Preparation of cyanogen bromide :

This is prepared by adding chilled 10% potassium cyanide to chilled bromine water until the colour is just discharged. It should be prepared in a fume cupboard.

Note : All these qualitative methods can be used with known standards to determine the quantities of these vitamins.

CHAPTER 8

URINE

PHYSICAL CHARACTERISTICS

Frequency of Urination :

(i) It is generally dependent upon the amount of fluid in the bladder.

(ii) The frequency is increased in the inflammation of the urinary tract, in the affections of the spinal cord and in weakening of the sphincter.

Preservation of the Urine sample :

(i) Toluene is a very satisfactory preservative. for urine.

(ii) Formaldehyde (2 drops per 50 ml. of urine) or a bit of Camphor or thymol are also satisfactory preservative.

Volume :

(i) Normally, the volume of urine excreted by an adult individual ranges from 1,200 to 1,800 ml. per day.

(ii) The volume is influenced by the diet and by the temperature which causes the loss of water through perspiration. Hard physical exercise diminishes output of urine.

(iii) Polyuria causes the increased output of urine; but oliguria causes diminished excretion.

(iv) Certain pathological conditions such as diabetes mellitus, diabetes insipidus, contracted kidney, and amyloid degeneration of the kidney etc. cause increased volume of urine.

(v) A decreased volume is also observed in acute nephritis, fevers, diarrhea, vomiting, and diseases of the heart and lungs.

Colour :

(i) Normal human urine is yellow in colour. This colour is due to the pigment urochrome which is a compound of urobilin and urobilinogen.

(ii) The urine may be deeper yellow when it gets concentrated as in the case of low output of urine.

(iii) The urine colour may be very dark brown or balck or may assume the colour of the drugs or their degradation products on the administration of various drugs or antiseptics.

(iv) The urine colour may be greenish yellow due to the excretion of large amounts of bile pigments in the urine and may be reddish due to the presence of blood.

(v) In chyluria, the urine appears milky and in alkaptonuria, the urine becomes dark on standing.

Specific Gravity :

(i) The specific gravity of urine of normal individuals varies between 1.006 and 1.025.

(ii) Following copious water—or beer drinking the specific gravity may go down to 1.002 but in case of excessive perspiration, it may go upto 1.040.

(iii) The specific gravity depends on the amount of solutes present in urine. When large amounts of glucose or albumin is passed in urine, the specific gravity rises.

(iv) The determination of specific gravity is done by *Urinometer.* This instrument is calibrated at a specific temperature.

Appearance :

(i) Normal urine is perfectly clear and transparent when freshly voided.

(ii) On standing for a variable time, the appearance becomes turbid due to the formation of the crystals of phosphates or urates or oxalates and also for the formation of nucleoproteins or mucoid and epithelial cells.

(iii) The phosphates are deposited in alkaline urine (or sometimes in slightly acid urine). Calcium oxalate is deposited in all urines. Uric acid and Urates are deposited when the urine cools and redissolved on warming. Phosphates dissolve in dilute acetic acid.

(iv) The turbidity also appears in the presence of W.B.C., R.B.C. or bacteria. A coloured deposit is also observed in the presence of R.B.C.

Odour :

(i) Fresh normal urine has an aromatic odour.

(ii) On standing for a long time, it undergoes bacterial decomposition producing a very unpleasant ammoniacal odour.

(iii) Under normal conditions, the urine possesses a peculiar odour due to the ingestion of certain drugs or vegatables.

CHEMICAL CHARACTERISTICS

A. Reaction :

(i) Freshly voided normal human urine has a P^H 6 (acidic to litmus).

(ii) On standing for a long time, it may be alkaline due to the formation of ammonia from bacterial decomposition.

(iii) Ingestion of acid fruits (oranges, lemons etc.) causes the formation of alkaline urine because the ash of such fruits is alkaline. Bread, cereals, meat etc. produce an acid urine.

B. Urine contains normal or abnormal constituents which are detected by chemical tests. The normal constituents can be divided into nitrogenous and non-nitrogenous constituents. Urine also contains some pigments derived from blood pigments. Several enzymes are also present in normal urine e.g. amylase and trypsin. In certain abnormal conditions, some of the normal constituents of urine are excreted in abnormal amounts.

The Chief nitrogenous constituents of normal urine :

1. (a) Specific Urease test for Urea :

Principle : Soyabean powder contains the enzyme *Urease*. This *Urease* under optimum P^H and temperature decomposes Urea into ammonia and Carbon dioxide which together form ammonium carbonate (the alkaline substance) which changes the slightly acid reaction (yellow colour) to alkaline reaction (pink colour).

Procedure : Take 2 ml. urine in one test tube and 2 ml. water in another. Add a drop of phenol red indicator to each. Then add 2% Na_2CO_3 drop by drop till the pink colour develops (just alkaline). Now add 2% acetic acid to each drop by drop till the pink colour just disappears (just acidic).

Add a pinch of Soyabean powder to each and rotate the tubes between the palms or warm both the tubes to about 60°C.

The pink colour appears in the tube containing urine but not in the other tube containing water.

Precaution : Over heating should be avoided. Otherwise the enzyme will be destroyed. 60°C temperature is maintained by touch only which is likely to be pleasant. This 60°C temperature is the optimum temperature for urease for its maximum activity.

Discussion : This test is said to be the specific test for urea because the enzyme urease shows its specificity for the substrate urea.

The optimum P^H (just acidic) and temperature (60°C) must be maintained for the highest activity of the enzyme urease. Urea is formed in the liver from ammonia and carbon dioxide. Ammonia is the product in the deamination of amino acids. Therefore, urea excretion in urine is dependent on the amount of Protein ingested.

(b) Biuret test :

Principle : Urea when heated decomposes with the liberation of ammonia and the formation of biuret. The biuret is dissolved in water and develops a violet colour forming a complex with the alkaline copper sulphate solution.

Procedure : Place a small amount of urea crystals in a dry test tube and heat it in a low flame. Urea melts with the liberation of ammonia. On further heating it solidifies (In case of urine, the urine is heated to solidify). Cool the tube. Add 3 ml. of water and shake.

Add to it 1 ml. of dil NaOH and 1 or 2 drops of 1% $CuSO_4$ solution. The pink colour develops indicating the presence of urea.

Precaution : More drops of $CuSO_4$ should not be added, otherwise $CuSO_4$ will form $Cu(OH)_2$ with NaOH forming a blue colour. This is sometimes mistaken for a positive biuret test.

2. (a) Murexide test (for uric acid) :

Principle : Uric acid is oxidized to dialuric acid and alloxan which condense to form alloxantin. This reacts with ammonium hydroxide to form the purplish red compound *ammonium purpurate or murexide*.

Procedure : Take 5 ml. of the urine and saturate it fully with solid ammonium chloride. Add a few drops of strong ammonium hydroxide. A gelatinous precipitate of ammonium urate is formed. Allow the excess of ammonium chloride to settle and filter off the supernatant.

Take the residue on the filter paper in a porcelin dish. Add 2 drops of concentrated HNO_3 and heat to dryness. A reddish yellow residue remains. Cool and add a few drops of very dilute ammonium hydroxide. The development of purplish red colour is due to the formation of murexide.

(b) Benedict's test for uric acid and urates.

Principle : Uric acid is soluble in alkali. The blue colour is developed due to the reduction of phosphotungstic acid by uric acid.

Procedure : To 2 ml. of urine add a few drops of Benedict's uric acid reagent and a pinch of anhydrous soldium carbonate. Mix. A deep blue colour indicates the presence of uric acid.

Discussion : Uric acid is formed from adenine and guanine of nucleic acids. Ordinarily it occurs in urine as urates. Alkalis dissolve uric acid in the form of urates. It is insoluble in water. In gout, the kidney loses the power of eliminating Uric acid properly and it is collected in the blood in high concentration. This is accompanied by the deposition of uric acid in the joint. Uric acid is highly excreted in urine in leukemia.

3. Jaffe's test (for Creatinine) :

Principle : Creatinine, the anhydride of Creatine, forms Creatinine Picrate in alkaline medium developing a deep reddish orange colour.

Procedure : To 4 ml. of saturated solution of Picric acid add 2 ml. of 10% NaOH. Divide it equally into two test tubes.

Add 3 ml. of urine to one test tube and 3 ml. of water to the other.

The test tube containing urine produces a deep reddish orange colour due to the formation of creatinine picrate. But this does not happen in other test tube containing water.

Discussion : Creatine occurs in small amounts in the urine of normal adults. It is found in larger amounts in the urine of children and pregnant women, till it is taken to be the normal constituent.

It is increased in fasting and after high water ingestion. Increased amount are found in malnutrition, disintegration of muscular tissue and in Carcinoma of the liver.

A decrease of creatinine excretion is found in anemia, paralysis, leukemia, etc.

Under normal conditions about 1 to 1.8 grams of creatinine are excreted by an adult male per day.

4. Test for ammonium salts.

Principle : The pink colour in the urine developed by the phenolphthalein shows the urine in the alkaline medium. The ammonia evolved due to the decomposition of ammonium salts makes the phenolphthalein layer over the glass tube pink.

Procedure : To 2 ml. of urine add a drop of phenolphthalein and 2% Na_2CO_3 drop by drop till pink colour is produced.

Boil the faintly alkaline fluid and keep a glass tube previously dipped in phenolphthalein over the mouth of the test tube. This turns pink.

Precaution : Do not touch the mouth of the test tube by the glass tube previously dipped in phenolphthalein.

Discussion : The average excretion of ammonia is about 0.7 grams per day.

The kidneys manufacture ammonia in proportion to the amount of acid radicals excreted in urine. In alkaline urine, ammonium salts are absent.

Imperfect protein metabolism increases the output of ammonia in urine.

The Chief Non-Nitrogenous constituents of normal urine.

1. Chlorides :

Principle : A white precipitate of AgCl is formed when acidified urine reacts with $AgNO_3$ solution.

Procedure : Acidify 2 ml. of urine with two drops of Conc. HNO_3 and add to it 2 ml. of $AgNO_3$. A white precipitate of silver chloride indicates the presence of chlorides.

2. Sulphates :

Principle : Urine being acidified with HCl forms a white precipitate of barium sulphate by the reaction with barium chloride solution.

Procedure : Add a few drops of Conc. HCl and 1 ml. of barium chloride solution to about 3 ml. of urine. A white precipitate of barium sulphate indicates the presence of sulphates.

Note : The presence of HCl prevents precipitate of phosphates.

3. **Phosphates :**

Principle : Phosphates of calcium and magnesium are precipitated by ammonium hydroxide on boiling and these phosphates are dissolved in hot dilute acetic acid. This forms yellow precipitate of ammonium phosphomolybdate reacting with ammonium molybdate.

Procedure : Add a little ammonium hydroxide to a test tube containing urine three-fourth of the test tube and then boil. A white flaky precipitate of phosphates of calcium and magnesium is formed.

Filter this precipitate and wash the residue on the filter paper with water. Dissolve the residue in about 3 ml. of hot dilute acetic acid by pouring the acid on the filter paper. Collect the solution in a clean test tube and divide it into two parts. Retain one part for detecting calcium.

To the other part add a drop of Conc. HNO_3 and a few drops of ammonium molybdate solution. Boil. A lemon yellow precipitate indicates the presence of *phosphates.*

4. **Calcium**

Principle : Calcium forms a white precipitate of calcium oxalate on addition of Potassium oxalate to the solution for calcium detection.

Procedure : To one part of the above experiment add about 1 ml. of Potassium oxalate solution. A white precipitate of Calcium oxalate is formed immediately.

5. **Ethereal Sulphates :**

Principle : Ethereal sulphates on treatment with baryta mixture forms soluble salts with barium. On hydrolysis with strong acid and with excess baryta it forms a white precipitate of barium sulphate.

Procedure : To 10 ml. of urine add 10 ml. of baryta mixture, [Saturated solution of $Ba(NO_3)_2$ and $Ba(OH)_2$]. Phosphates and inorganic sulphates are precipitated. Filter repeatedly to have a clear filtrate.

To the filtrate add 5 ml. of Conc. HCl and boil in a dish for about 5 minutes. Allow to stand for few minutes. A faint white precipitate indicates the presence of *ethereal sulphates.*

Discussion of Non-nitrogenous constituents of urine :

(i) Sodium and potassium in the form of Na_2O and K_2O are found to the extent of 5 grams and 3 grams, respectively, in 24 hours urine.

(ii) 0.1 to 0.2 gram of calcium is excreted in urine per day. Its excretion increases in hyperparathyrodism. The excretion of magnesium is also low which approaches to 0.05 to 0.2 gram per day.

(iii) Very small quantity of iron is found in urine.

(iv) About 10-12 grams of chloride (in the form of NaCl) is excreted daily. The amount of chloride excreted depends upon the intake of food and the quantity of its loss through sweat.

(v) Phosphates are present in urine as salts of sodium, Potassium, ammonium, Calcium and magnesium. These are crystallized out in alkaline urine.

(vi) Sulphates found in urine are derived from the oxidation of sulphur containing substances. It is available in urine in three forms—(a) neutral sulphur (b) Sulphates of sodium and potassium (c) ethereal sulphates.

(vii) Ethereal sulphates in urine are the conjugated phenols, phenol-sulphuric acid and indoxysulphuric acid etc. The total output of it various forms 0.04 to 0.1 gram per 24 hours. These ethereal sulphates result from phenols produced during putrefaction of protein material in the intestine. Phenol passes to the liver where part of it is conjugated to form phenol potassium sulphate and appears in urine in this form. This indoxysulphuric acid (indican) under normal condition is excreted from 10 to 20 mg. per day. Variation also results in depending upon the protein diet. Pathologically, the indicans are excreted highly in increased putrefaction and stagnation of intestinal contents. Bacterial decomposition of body protein as in gangrene and putrid pus formation etc. result in the increased excretion of indican.

(viii) Normal human urine also contains fluorides, nitrates, silicates and hydrogen peroxide. Nitrates are available from water and food—vegetable diet provide more quantities of nitrates than meat diet.

(ix) The pigments urochrome, urobilin and uroerythrin are also found in urine. Urochrome is the principal pigment of the normal urine and the pale yellow colour of the normal urine depends on the colour of it. Urobilin is the oxidative product of urobilinogen. In fresh urine urobilinogen is present. Urobilinogen is excreted in less amount in normal urine. But its concentration in urine increases in the diseases of liver and bile passages, in fever, in malaria and in R.B.C. destruction.

The total reducing substances equivalent to about 0.05 to 0.15 per cent of glucose are present in normal human urine. Fermentable sugars are about 0.01 per cent. Most of the sugar of normal urine is not glucose but includes pentose, lactose and altered carbohydates formed through the action of bacteria in the intestinal tract.

Glucose is excreted ordinarily in traces only. Lactose is found in the urine of pregnant women and lactating mother in large quantities. Detectable amounts of pentose sugar and fructose are found in pentosuria and fructosuria, respectively. Otherwise reducing sugars are not normally found in sufficient amounts to respond to the ordinary tests.

ABNORMAL CONSTITUENTS OF URINE

Substances which are not present in easily detectable amounts in urine of normal healthy individuals but are present in urine under certain diseased conditions are said to tbe *"Abnormal"* or *"Pathological"* constituents of urine.

Before detecting the abnormal constituents of urine, note the physical characteristics (colour volume, odour, apperance, specific gravity and sediment) and chemical reaction with litmus paper.

The abnormal constituents for routine purpose are reducing sugars, acetone bodies, proteins, blood pigments, bile salts, bile pigments and pus.

1. Reducing sugars :

To detect reducing sugars in urine Benedict's test and Fehling's test are done. To be sure of the presence of glucose, lactose, pentose, fructose and glucose etc., osazone test is to be performed.

Benedict's test is preferred than Fehling's test on the fact that the strong alkalinity of Fehling's solution destroys traces of sugar; whereas Benedict's solution with its weak alkalinity responds to a little amount of sugar present in urine. Hence it is more sensitive than Fehling's test.

Presence of streptomycin and certain preservatives such as chloroform and formaldehyde give Benedict's test positive.

Benedict's test :

Principle : Copper sulphate of Benedict's qualitative solution is reduced by reducing substances on boiling to form the coloured precipitate of cuprous oxide. The light green, green, yellow and brick red precipitates of cuprous oxide depend on the concentration of reducing substances present in urine.

Benedict's qualitative reagent : Copper sulphate + sodium carbonate + sodium citrate.

Procedure : To about 5 ml. of Benedict's qualitative reagent add 0.5 ml. (8 drops) of urine and boil for 2 minutes holding the test tube firmly with a test tube holder. A light green, green, yellow and brick red precipitate indicates the presence of reducing substances in urine.

The various coloured precipitate depends on the concentration of reducing sugars in urine which gives a rough estimate of the concentration given below :

Light green precipitate	....	0.1 to 0.5 of reducing sugars.
Green precipitate	....	0.5 to 1.0% of reducing sugars.
Yellow precipitate	....	1 to 2% of reducing sugars
Brick red precipitate	....	above 2% of reducing sugars.

Note : The variation in coloured precipitate relating to the concentration of reducing sugar is found in urine only; whereas this variation does not occur in pure sugar solution in water. This difference may possibly be due to the presence of creatinine and other constituents of urine.

Precaution :

(i) Boil the urine for at least 2 *minutes*.

(ii) During boiling, the contents of the test tube get a tendency to spurt out. Hence, it is wise to keep the test tube shaking after holding it in the inclined position near the flame which does not cause overboiling.

Discussion :

(i) Normal urine also contains a trace of glucose and glucuronates; but their amount is too small to cause reduction in Benedict's test.

(ii) Benedict's test is not necessarily indicative of glucose in urine only but it too indicates the presence of other reducing sugars such as lactose (in case of pregnant women and lactating mothers), fructose (in fructosuria), glactose (in galactosuria), pentose (In pentosuria), homogentisic acid (in alkaptonuria), glucuronates and mucin. If mucin is suspected, Benedict's test is

(ii) repeated after removing mucin with kaolin. The different reducing sugars (glucose, lactose, galactose, fructose) may be identified by establishing the performance of Seliwanoff's test, osazone test and fermentation test etc.

(iii) In diabetes mellitus and in renal glycosuria, glucose is found in urine. This gives a Benedict's test positive.

2. Acetone bodies

(a) Rothera's test :

Principles : Acetoacetic acid forms a complex with sodium nitroprusside in alkaline solution developing a permanganate colour.

Procedure : Saturate 5 ml. of urine with solid ammonium sulphate by shaking it vigorously. Then add 2 drops of freshly prepared 5% solution of sodium nitroprusside and 1 ml. of ammonium hydroxide. Allow it to stand in the rack for a while. A permanganate colour develops just above the layer of the undissolved ammonium sulphate crystals indicating the presence of acetone bodies.

(b) Gerhardt test for acetoacetic acid :

Acetoacetic acid gives a red colour with ferric chloride.

Procedure : Add 5% ferric chloride solution drop by drop to about 5 ml. urine till no more precipitate of ferric phosphate is formed. Filter.

To the filtrate add some more ferric chloride. The development of red colour indicates the presence of acetoacetic acid.

Precaution : A large number of substances such as aspirin, antipyrin, salicylates etc. may develop similar red colur. If the urine is boiled, acetoacetic acid is converted into acetone; but the other substances remain unchanged. Now, if the urine gives negative test, it indicates the presence of acetoacetic acid.

Discussion :

(i) In diabetes mellitus and in prolonged starvation, fat is catabolised to produce acetoacetic acid, acetone and β-hydroxybutyric acid which are together termed as acetone or ketone bodies. These acetone boides are accumulated in blood and excreted in urine. This condition is called Ketosis. The excretion of Ketone bodies in urine is called Ketonuria.

(ii) Ketosis also occurs in the acetonemic vomiting of childhood, and frequently in pregnancy, fevers, ether and chloroform anesthesia, malnutrition, prolonged feeding of a carbohydrate poor diet high in meat and fat.

(iii) Total Ketone bodies are found in normal urine to the extent of about 20 mg in 24 hours. Pathologically, values from 0.02 to 6 grams or more per day have been observed.

(iv) β-hydroxybutyric acid, as such, does not give Rothera's or Gerhardt's test positive unless converted to acetoacetic acid and then to acetone by oxidation.

3. Albumin

(a) Sulphosalicylic acid test :

Principle : Albumin, the protein, is denatured by sulphosalicylic acid causing a coagulation.

Procedure : Add a few drops of sulphosalicylic acid to 2 ml. of clear urine (filter if not clear). A tubrbidity indicates the presence of albumin.

This test is used for routine work.

(b) Heat Coagulation test :

Principle : The urine containing albumin is coagulated after being heated.

Procedure : Fill 3/4th of the test tube by clear urine (If the urine is turbid, filter it before performing the test). Heat the upper one-third of the test tube by a small flame. A turbidity is found on the heated portion of the urine.

Add a drop of 33% acetic acid to the urine. Phsophate is dissolved but not the protein albumin.

(c) Heller's Nitric acid ring test :

Principle : Nitric acid causes the precipitation of Protein.

Procedure : To 3 ml. of nitric acid in a narrow tube add 3 ml. of urine by means of a pipette in such a way that the two liquids do not mix.

The presence of a white ring at the junction of two fluids indicates the presence of *albumin.*

Discussion :

(i) A trace of protein which is less than 250 mg. in 24 hours urine found in normal urine. This amount is so slight that it escapes detection by any of the simple tests.

(ii) Some proteins (albumin and globulin) are found in the urine under pathological conditions known as *albuminuria.*

(iii) In kidney disturbance and in altered high blood pressure, albumin is found in urine significantly.

(iv) Benign proteinuria (usually transitory), appears in severe exercise or cold baths, particularly in young poeple.

(v) Bence-Jones protein (an immunoglobulin) appear in urine in cases of multiple myeloma. This coagulates between 40-60°C.

4. Blood pigments (Benzidine test)

Principle : Hemoglobin of blood decomposes hydrogen peroxide catalytically and liberates oxygen. This oxygen oxidizes benzidine to a blue or green compound. This colour changes to brown within a few minutes on exposure to air.

Procedure : Take a pinch of solid benzidine in a test tube and 2 ml. of galcial acetic acid till the benzidine dissolves. Add 2 ml. of hydrogen peroxide (H_2O_2) to it. Divide it into two portions.

To one portion add urine drop by drop with shaking. The appearance of deep blue colour indicates the presence of blood pigments. This quickly changes to brown within a few minutes.

To the other part add water drop by drop. No blue colour appears. This part is treated as control.

Discussion :

(i) Blood occurs in urine under hematuria and hemoglobinuria. Hematuria consists of hemoglobin and unruptured corpuscles. Hematuria is brought about by blood passing into the urine because of some lesion of the kidney or of the urinary tract. Hemoglobinuria results from hemolysis. This occurs in malaria, typhoid, hemolytic jaundice, yellow fever and in transfusion with incompatible blood.

(ii) The presence of high concentration of ascorbic acid in urine is oxidized more readily than benzidine by oxygen liberated from hydrogen peroxide. The benzidine reaction then becomes negative although sufficient blood is present in urine.

(iii) The benzidine test is a very sensitive test.

5. Hay's test for bile salts :

Principle : Bile salts reduce the surface tension present in the urine for which sulphur powder sinks.

Procedure : Fill half of a test tube with urine and another test tube with water. Sprinkle gently some sulphur powder on the surface of the two liquids.

The sulphur powder spontaneouly sinks in the the test tube containing urine which indicates the presence of bile salts. But in the other test tube containing water, no sulpur powder sinks.

Discussion :

Bile duct obstruction by inflammation or gall stones in gall bladder leads to the obstruction of bile into the general circulation. The subcutaneous tissues are deeply stained resulting in the yellow complexion which is the characteristic of jaundice. In such cases, bile salts and bile pigments are detected in urine.

6. Bile pigments

A. Fouchet's test

Principle : Bilirubin is precipitated by barium chloride. This bilirubin is oxidized to green biliverdin by Fouchet's reagent (Ferric Chloride in trichloroacetic acid).

Procedure : Acidify 10 ml. of urine with a few drops of dilute acetic acid and add 5 ml. of 10% $BaCl_2$ solution. If there is not much precipitate, add 2 drops of saturated solution of magnesium sulphate. Mix and allow to stand for a few minutes.

Filter and unfold the filter paper. Add one drop of Fouchet's reagent to the precipitate. The development of green coloration indicates the presence of bile pigments.

B. Gmelin's test

Procedure : To 5 ml. of concentrated nitric acid in a test tube add an equal volume of urine carefully, as if the two fluids do not mix. The various coloured rings (green, blue, violet, red and reddish yellow) will be formed at the point of contact of the two liquids.

C. Test for urobilinogen

Add 0.5 ml. of Ehrlich's reagent [Prepared by dissolving 2 grams of P-dimethyl amino benzaldehyde in 100 ml. of 20% HCl] to 5 ml. of urine.

Discussion :

(i) A urine containing bile may be yellowish green to brown in colour and when shaken foams readily.

(ii) Bilirubin formed from hemoglobin is conjugated with glucuronic acid in the liver and then excreted.

Indican

Principle : Indoxyl is oxidized to form indigo-blue developing a blue colour which is soluble in chloroform.

Procedure : Take 5 ml. of urine and 5 ml. of Conc. HCl in a test tube. Add one drop of Potassium chlorate and 3 ml. of chloroform to it. Mix thoroughly. Allow the chloroform to settle down. The development of blue colour indicates the presence of indican. If no blue colour is developed, add another drop of Potassium chlorate and mix. If still no blue colour develops, it indicates the absence of indican.

Discussion :

(i) Tryptophan undergoes bacterial decomposition in the intestine forming indoxyl.

(ii) Indican is the potassium salt of the indoxyl sulphuric acid and hence it is an ethereal sulphate.

(iii) Sometimes in case of large abscess, indican may be formed due to putrefaction.

PUS

Microscopic identification of Pus : The urine is centrifuged and the sediment is put on the slide under microscope. In acid urine, the pus corpuscles appear as round, colourless cells, refractive, granular and nucleated. Sometimes they exhibit ameboid movements if the slide containing the sediment be warmed slightly. In alkaline urine, the pus corpuscles occur as swollen, often degenerated, transparent and non-granular.

Guaiac test : The urine is acidified with acetic acid and filtered. To the sediment on the filter paper add a few drops of tincture guaiac followed by a few drops of H_2O_2. Blue colour indicates the presence of Pus or blood or both.

Boil 5 ml. of urine in another test tube for 30 seconds and repeat the test as above. The development of blue colour indicates the presence of blood. Pus does not respond to this test after boiling the urine.

Discussion :

(i) Pus is present in urine under various inflammatory conditions. Such a condition is termed *Pyuria*. Albumin always accompanies the pus.

(ii) Pus is also present in the urine in catarrh of the urinary bladder and in inflammation of the urethra of pelvis of kidney.

(iii) Rupturing of an abscess in some part of the urinogenital tract causes a high concentration of pus in the urine.

Allow to stand for 3-5 minutes for the development of colour.

No red colour	.... Urobilinogen absent.
Faint pink	.... Urobilinogen present in small amounts.
Distinctly red colour	.... Increased amount of urobilinogen present.

Questions to answer

1. What is the volume, colour, specific gravity, odour and appearance of normal urine? How is to preserve urine?
2. What are the chief nitrogenous and non-nitrogenous constituents of normal urine?
3. Name the specific test for detecting urea in urine. What is the principle of the test and how is to perform it?
4. Name the test for detecting Uric acid in urine. Discuss the principle and performance of the test.
5. Why is red colour developed in Jaffe's test for creatinine? Discuss creatine and creatinine.
6. What do you mean by phosphaturia? How to differentiate phosphate and albumin by tests?
7. Where is ammonia manufactured in the body? How is to detect ammonium salt in the urine?
8. What do you mean by ethereal sulphates? What is the procedure for detecting it?
9. Give the discussion of the non-nitrogenous constituents of urine.
10. Name the abnormal constituent of urine. What do you mean by 'abnormal' constituents of urine?
11. Enumerate the tests for detecting abnormal constituents of urine. Discuss the principle of those tests.
12. Give the discussions of the abnormal constituents of urine.

IDENTIFICATION OF AN UNKNOWN SUBSTANCE OF PHYSIOLOGICAL IMPORTANCE

The unknown physiological substance may be given in the solid state or in solution to detect. If the substance is solid (powder), it is to be dissolved in either of the following ways :

Cold water.

Hot water.

Alcohol (95%).

2% Na_2CO_3 or 5% NaOH.

2% acetic acid or N/10 HCl.

Substances which are soluble in dilute alkali are uric acid, Protein (casein) globulin etc.

Substances which are soluble in dilute acids are calcium oxalate and alkali metaproteins.

After dissolving the substance, proceed like the substance in the solution form in the following manner :

(i) **Reaction to litmus paper :** Protein in acidic medium or free HCl will be acidic in reaction. If alkaline, it may be uric acid, casein alkalimetaprotein. If neutral, it may be carbohydrates, urea, acetone. etc.

(ii) **Appearance :** The solution of protein, polysaccharide or fat emulsion is opalescent.

(iii) **Colour :**

Reddish brown	....	Blood.
Yellow or Greenish yellow	....	Bile.
Light yellow	....	Dextrin or urine or Peptone.

(iv) **Smell :**

Peculiar smell	....	urine or bile.
Aromatic odour	....	Acetone bodies.
Characteristic odour	....	Peptone etc.

First perform test for protein, then carboydrate, fat and the nitrogenous and non-nitrogenous constituents of urine.

Tests for protein

Perform biuret test. If it is positive, perform ninhydrin test, coagulation test, Precipitation test and tests for different amino acids.

Tests for Carbohydrate

If biuret test is negative, perform Molisch's test. If Molisch's test if positive, perform other tests for individual carbohydrates given in carbohydrate chapter.

Tests for fat

If Molisch's test is negative, perform Grease-spot test. If this test is positive, perform other tests for fat mentioned in fat chapter.

Tests for nitrogenous and non-nitrogenous constituents of urine

If Gerase-spot test is negative, perform tests for nitrogenous and non-nitrogenous constituents or urine mentioned in urine chapter.

N.B. While reporting in the answer scripts, if protein is detected, one must mention all other tests although negative for a correct and complete answer for having the maximum credit.

GUIDELINE FOR WRITING PROPERLY THE DETECTED QUALITATIVE SUBSTANCES IN CLASS PRACTICAL FAIR NOTE BOOKS OR IN PRACTICAL EXAMINATION ANSWER SCRIPTS

GLUCOSE (As known sample)

	Experiment	Observation
1.	**MOLISCH'S TEST :** Added 2 drops of Molisch's reagent to 2 ml. sugar solution in a test tube. Mixed thoroughly. Added 2 ml. Conc. H_2SO_4 by the side of the test tube slanting the tube. Then erected the test tube slowly	Reddish-violet ring is formed at the junction of two liquids.
2.	**IODINE TEST :** Added 1 or 2 drops of dilute iodine solution to 2-3 ml. sugar solution.	No colour is developed.
3.	**BENEDICT'S TEST :** Added 8 drops of the sugar solution to 5 ml. Benedict's qualitative reagent in a test tube. Boiled vigorously for 2 minutes. Allowed to cool spontaneously.	A brick-red precipitate is formed.
4.	**BARFOED'S TEST :** Added 8 drops of sugar solution to 5 ml. Barfoed's reagent. Heated for 30 seconds.	A red precipitate is formed.
5.	**BIAL'S TEST :** Mixed 5 ml. Bial's reagent with 8 drops sugar solution. Heated to boiling.	No green colour is developed.
6.	**SELIWANOFF'S TEST :** Added a few drops of sugar solution to 5 ml. Seliwanoff's reagent in a test tube. Heated for 30 seconds.	No red colour is formed.
7.	**OSAZONE TEST :** Added 10 drops of glacial acetic acid to 5 ml. sugar solution in a test tube. Then added a knife point of phenylhydrazine hydrochloride and double the amount of sodium acetate crystals. Mixed and warmed a little to see that the solids are dissolved. Filtered and kept the filtrate in a boiling water bath for 5 minutes.	Broomstick—like crystals are formed.

UNKNOWN PHYSIOLOGICAL SAMPLE : (Y)

	Experiment	Observation	Inference
1.	**BIURET TEST :**		
	To 2 ml. provided solution added 1 ml. 40% NaOH solution and 1 or 2 drops of 1% $CuSO_4$ solution	No violet colour is formed	Protein is absent.
2.	**MOLISCH'S TEST :**		
	Added 2 drops of Molisch's reagent to 2 ml. Provided solution in a test tube. Mixed thoroughly. Added 2 ml. Conc. H_2SO_4 by the side of the test tube. Then errected the test tube slowly.	No red-dish-violet ring is formed at the junction of two liquids.	Carbohydrate is absent.
3.	**GREASE-SPOT TEST :**		
	A drop of provided solution is put over a piece of writing paper.	No translucent spot is formed.	Fat is absent.
4.	**SPECIFIC UREASE TEST FOR UREA :**		
	Taken 2 ml. provided solution in a test tube. Added a drop of phenol red indicator to it. Then added 2% Na_2CO_3 solution drop by drop till the pink colour develops. Added 2% acetic acid drop by drop till the pink colour just disappears. Added a pinch of soyabean powder to it and rotated the tube between the palms or warmed the tube to about 60°C.	No pink colour is developed.	Urea is absent.
5.	**BIURET TEST :**		
	To 2 ml. provided solution added 1 ml. 40% NaOH solution and 1 or 2 drops of 1% $CuSO_4$ solution.	No violet colour is formed.	Urea is absent.
6.	**MUREXIDE TEST :**		
	5 ml. provided solution is saturated fully with solid ammonium chloride. Added a few drops of strong ammonium hydroxide. A gelatinous precipitate of ammonium urate is formed. Filtered off the supernatant. The residue on the filter paper is transferred in a procelin dish. Added 2 drops Conc. HNO_3 and heated to dryness. Cooled and added a few drops of very dilute NH_4OH	Purplish red colour is developed due to the formation of murexide.	Uric acid is present.
7.	**BENEDICT'S TEST :**		
	To 2 ml. provided solution added a few drops of Benedict's uric acid reagent and a pinch of anhydrous sodium carbonate. Mixed.	A deep blue colour is developed.	Uric acid is present.
8.	**JAFFE'S TEST :**		
	To 4 ml. saturated solution of picric acid added 2 ml. 10% NaOH. Added 6 ml. Provided solution.	No deep reddish orange colour is developed.	Creatinine is absent.
9.	**TEST FOR AMMONIUM SALTS :**		
	To 2 ml. provided sample added a drop of phenolphthalein and 2% Na_2CO_3 drop by drop till the pink colour is produced. Boiled the faintly alkaline fluid and kept a glass tube previously dipped in phenolphthalein over the mouth of the test tube.	Not turned pink.	Ammonium salts are absent.
10.	**TEST FOR CHLORIDES :**		
	Acidified 2 ml. provided sample with 2 drops of Conc. HNO_3 and added 2 ml. $AgNO_3$ to it.	No white ppt. formed.	Chloride is absent.
11.	**TEST FOR PHOSPHATES & CALCIUM :**		
	Added a little NH_4OH to a test tube containing $\frac{3}{4}$th the provided sample and then boiled.	No white ppt. formed.	
	Divided it into two parts. To one part added a drop of Conc. HNO_3 and a few drops of ammonium molybdate solution boiled.	No yellow ppt. formed.	Absence of phosphates.
	To other part added 1 ml. potassium oxalate solution.	No white ppt. formed.	Calcium is absent.

Report : The provided sample no (Y) contains **uric acid** as physiological sample.

UNKNOWN PATHOLOGICAL SAMPLE (Z)

	Experiment	Observation	Inference
1.	**BENEDICT'S TEST :**		
	To about 5 ml. of Benedict's qualitative reagent added 8 drops of provided sample and boiled for 2 minutes.	A brick—red precipitate formed.	Reducing sugar is present.

	Experiment	Observation	Inference
2.	**ROTHERA'S TEST :**		
	Saturated 5 ml. of the provided sample with solid ammonium sulphate by shaking it vigorously. Then added 2 drops of freshly prepared 5% solution of sodium nitroprusside and 1 ml. of NH_4OH. Allowed it to stand for a while.	No permanganate colour developed.	Acetone bodies are absent.
3.	**(a) HEAT COAGULATION TEST :**		
	Filled $\frac{3}{4}$th test tube by the provided sample. Heated the upper $\frac{1}{3}$ of the test tube by a small flame. Added a drop of 33% acetic acid to it.	A turbidity formed. The turbidity not dissolved.	Albumin is present.
	(b) HELLER'S NITRIC ACID RING TEST :		
	To 3 ml. nitric acid in a test tube added 3 ml. of the provided sample by means of a pipette in such a way that the two liquids do not mix.	White ring formed at the junction of two liquids	Presence of albumin.
4.	**BENZIDINE TEST :**		
	Taken a pinch of solid benzidine in a test tube and 2 ml. of galcial acetic acid till the benzidine is dissolved. Added 2 ml. of H_2O_2 to it. Added the provided sample drop by drop while shaking.	Deep blue colour not appeared.	Absence of blood pigments.
5.	**HAY'S TEST :**		
	Filled half of the test tube with the provided sample. Sprinkled gently some sulphur powder on the surface of two liquids.	Sulphur powder does not sink.	Absence of bile salts.
6.	**(a) FOUCHET'S TEST :**		
	Acidified 10 ml. of the provided sample with a few drops of dilute acetic acid and added 5 ml. 10% $BaCl_2$ solution. Added two drops of saturated solution of $MgSO_4$. Mixed and allowed to stand for a few minutes. Filtered and unfolded the filter paper. Added one drop of Fouchet's reagent to the precipitate.	No development of green colour.	Absence of bile pigments.
	(b) GMELIN'S TEST :		
	To 5 ml. of Conc. HNO_3 in a test tube added 5 ml. of the provided sample carefully as if the two fluids do not mix.	No coloured ring formed.	Absence of bile pigments.
	(c) TEST FOR UROBILINOGEN :		
	Added 0.5 ml. Ehrlich's reagent to 5 ml. of the provided sample.	No red colour developed.	Urobilinogen is absent.
7.	**TEST FOR ETHEREAL SULPHATE (INDICAN) :**		
	Taken 5 ml. of the provided sample and 5 ml. Conc. HCl in a test tube. Added one drop of $KClO_3$ and 3 ml. chloroform to it. Mixed thoroughly. Allowed the chloroform to settle down.	No blue colour developed.	Absence of ethereal sulphate.

Report : The pathological provided sample (Z) contains albumin.

CHAPTER 9

INSTRUCTIONS FOR QUANTITATIVE ESTIMATION

Qualitative tests are done only to ascertain the presence of a substance. But quantitative analysis shows the exact amount of a substance in a particular sample.

The solution which contains an exact amount of a substance in a definite volume is termed as the *standard solution*. For preparing standard solution the chemical balance and the clean pipette must be used. Standard solution is required for finding out the amount of a substance in an unknown sample. Standard solution is always required in estimation by photoelectric colorimeter.

For the use of water, distilled water must always be used. All chemicals meant for use must be of great purity. Analar or C.P. quality chemicals must be used.

The titrations are generally performed in cold; sometimes performed when the solutions are hot or even boiling. Burettes with jets and pinch cocks should be used. During titration the end point should be carefully taken by the addition of a single drop of solution from burette. More than a drop at a time should not be added to have the correct end point. The first reading of titration must be taken to be approximate but not accurate. Hence, a second and a third reading must be taken. The difference between two consecutive readings must not be more than 0.1 ml. Out of three readings if two readings are almost similar than the third one, the third one should not be taken in this case. All possible errors must be avoided. While using pipettes, last drop at the tip of these pipettes should be blown out if measurement upto the tip is wanted.

The results of some of the substances are best expressed in milli equivalent (m Eq) units :

$$\text{mEq/L} = \frac{\text{mg. per 100 ml.} \times 10}{\text{Eq. Wt. of the substance}}$$

Recently, all values in clinical chemistry are expressed in SI units (Syste'me International d' Unite's) which was adopted by the International Federation for Clinical Chemistry in 1967.

PHOTOELECTRIC COLORIMETER

Photoelectric colorimeter is an instrument, which measures the intensity of colour of a solution by registering the amout of light absorbed by the given solution. Hence this instrument can be better termed *absroptiometer*.

This instrument consists of the following essential parts :

(i) A source of light.

(ii) **The monochromatic light** : The monochromatic light (a narrow wave band of light from the source) passes through the test solution. This arrangement is being done by placing selected filter in the path of light.

(iii) **A photosensitive element** : In most instrument barrier layer cells are used. The cells are provided with a layer of selenium. It is activated when light falls on the selenium and it emits electrons according to the proportion of light falling on it. The galvanometer is connected to measure the amount of electron flow and it indicates the amount of light falling on the cells.

(iv) **Cuvettes or tubes** : These are neutral glasses and are used to contain test solutions. All cuvettes, or tubes for a particular instrument should have the same absorptive capacity.

(v) **A Sensitive Galvanometer** : A certain percentage of light is absorbed when monochromatic light passes through a coloured solution. The extent of this absorption obeys two laws—Lambert and Beer. According to these laws the optical density or Extinction of coloured solution is proportional to the concentration of the coloured substance.

The optical density (O.D.) is defined as the logarithm of the ratio of the incident to transmitted light (the light emerging out of the solution).

Therefore,

$$O.D. = \frac{\text{Log incident light}}{\text{transmitted light}}$$

$$= \frac{\text{Log } 100}{T}$$ [Where incident light is taken as 100% and T is the per cent of light transmitted (transmittance)]

$= \text{Log } 100 - \text{Log } T.$

$= 2 - \text{Log } T$

When the transmittance of a solution is 50

Then, $O.D. = 2 - \text{Log } 50.$

$= 2 - 1.699.$

$= 0.3$

Since, O.D. is directly proportional to the concentration

$$\therefore \frac{\text{Concentration of unknown}}{\text{Concentration of standard}} = \frac{\text{O.D. of unknown}}{\text{O.D. of standard}}$$

$\therefore$ Concentration of unknown =

$$\frac{\text{O.D. Unknown}}{\text{O.D. Standard}} \times \text{Concentration of standard.}$$

Most instruments show calibration both in optical density (Extinction) and per cent transmission scale. If the readings are taken in per cent transmission, it is to be converted into optical density according to the formula O.T. = 2 – Log T.

Selection of Filters : In almost all cases, the correct wave length is the wavelength of light which is most strongly absorbed by the coloured solution. The colour of the filter is complementary to the colour of the solution, i.e., for blue solution red filter, for red solution green filter and for yellow or brown solution a blue filter. A filter is designated in terms of the wavelength of peak transmittance, thus a filter called "No. 540" or "No. 54" has its peak transmittance at a wave length of 540 mμ.

Importance of Blanks : The preparation of blank is essential in colorimetric analysis because of the fact that some colour comes from the reagents etc. in addition to the colour produced by the substance for estimation. Therefore, a blank is prepared using all reagents except the substance to the estimated. The optical density (O.D.) value of the blank is subtracted from the O.D. values of the unknown and the standard. Hence, the remaining values are the true O.D. values of the unknown and the standard.

Many types of photoelectric colorimeters with different designs are made available in the market. However, these photoelectric colorimeters are basically of two types—single cell type and double cell type :

(i) **Single cell type :** In this type, the light transmitted through a solution placed between the light source and the cell is measured directly by the current output of the cell in relation to that obtained with blank and standard solutions.

(ii) **Double cell type :** This type is concerned with the use of two photo cells in the same light source and balanced against each other by means of a null point instrument like a galvanometer. The solution to be estimated is placed between one cell and the light source. The extent to which the two cells are thrown out of balance as regard to each other is a measure of the light transmitted by the solution.

Operation of the instrument :

It is better to use the instruction manual of the instrument. The chief steps are as follows :

(i) The appropriate filter is advised to be inserted in the proper place.

(ii) One cuvette or tube with distilled water is to be put at the place provided for it in the machine and the instrument is to be switched on. It is to be left for warm up for about 4 to 5 minutes. The galvanometer is then to be switched on. The needle of the galvanometer is to be adjusted and brought it to position 100% transmission (zero optical density).

(iii) Another cuvettes or tubes are to be filled with the blank, standard and test solution and are put these in place of water. The optical densities of these are to be read.

(iv) Then the instrument is to be switched off and disconnected from the mains after taking all readings.

(v) The cuvettes or the tubes are to be thoroughly cleaned with running water first and then rinsed with distilled water both inside and outside of the cuvettes or tubes. The optical parts of the cuvettes or tubes should never be touched with fingers to avoid finger marks on these. Very clean and transparent cuvettes or tubes are to be required.

Calculation :

Concentration of the substance of the unknown per 100 ml.

$$= \frac{\text{O. D. of unknown} - \text{O. D. of blank}}{\text{O. D. of standard} - \text{O. D. of blank}} \times$$

Concentration of the substance per 100 ml. of the standard solution.

COLLECTION AND PRESERVATION OF URINE SAMPLES

Single specimens of urine are used for most qualitative tests and ward examinations. But for quantitative analysis of urine, a 24-hour sample of urine helps in accurate diagonsis and interpretation. For this purpose it is most convenient to collect the specimen from one morning to the next. The individual is instructed to empty the bladder immediately preceding the initiation of collection to avoid bladder catheterization. Samples collected for prompt chemical examination only need not be sterile or preserved.

Urine on keeping undergoes changes by bacterial action. Bacteria chiefly convert urea into ammonium carbonate. Such urine is unsuitable for the determination of urea, ammonia, pH, total nitrogen. Microorganisms both bacteria and yeast may act on glucose. If the urine becomes alkaline, phosphates may be precipitated. On standing, uric acid and urates are deposited because they are less soluble in cool urine. Other changes are the oxidation of urobilinogen to urobilin and the rapid oxidation of ascorbic acid.

It is advised to collect the urine in a clean, well—washed container. Still little change takes place in a twenty four-hour specimen by the time it is received in the laboratory. If urine is to be kept unchanged, preservatives are required to be added. The preservatives chloroform, toluene, light pertorleum, thymol and formalin may be used but they have disadvantages too. Toluene and light petroleum form a thin layer on the surfae and this tends to contaminate pipettes. This can be overcome by running off the lower layer of urine. They do not stop the growth of bacteria already introduced during the collection of urine. *Chloroform* is satisfactory in this respect. But it may interfere with the detection of glucose as it reduces Fehling's solution. Chloroform can be boiled off as it mixes with any deposit. 5 ml. of 10% *thymol* is found satisfactory as preservatives for many substances as sodium, potassium, chloride, bicarbonate, calcium, phosphorus, urea, ammonia, amino acids, creatinine, creatine, proteins, reducing substances Acetone bodies and amylase etc., 3 to 4 drops of *formalin* to 100 ml. of urine can be used. If more is added, it may give positive results for the test for glucose. 10 percent *acetic acid* can be added to preserve urine for determination of ascorbic acid. In addition to the use of preservatives, the urine must be kept cool.

Before carrying out estimations, any deposit must be well mixed with the urine; before sampling, the total volume of the 24-hour urine must be measured before quantitative analysis. This volume is taken as a basis for calculating output per unit of time.

Precautions to avoid hemolysis of blood

Hemolysis makes the serum unsuitable for some determinations. Therefore, it is better to obtain serum free from hemolysis. The following precautions should be taken to avoid hemolysis :

(i) The syringe and needle must be dry.

(ii) When taking blood, a minimum amout of constriction is applied to the arm. The blood should flow slowly and steadily into the syringe. The blood should be expelled slowly into the tube after removing the needle.

(iii) Tubes into which blood is kept should be clean and dry.

(iv) Any anticoagulant should be mixed with the blood by gentle rotation. Excessive amount of anticoagulant should not be used as excessive amount can cause lysis of the blood.

(v) In case of serum preparation, the blood is allowed to clot and then centrifuged or the serum is drained off when the blood is firmly clotted.

CHAPTER 10

URINE

ESTIMATION OF CREATININE IN URINE

(Folin's method)

Principle : Creatinine develops an orange red colour when treated with picric acid in the presence of strong alkali. This colour is due to the formation of creatinine picrate. The colour so produced depends on the amount of creatinine present.

Reagents :

1. **Creatinine standard solution :** Dissolve 0.1 gram creatine (0.1611 gram of creatinine Zinc Chloride) in 100 ml. of 0.1 N. HCl. Keep in a cool place in a well-stoppered bottle. 1 ml. of the solution contains 1 mg. of creatinine.
2. Saturated solution of Picric acid (about 1%).
3. Sodium hydroxide solution (10%).
4. Hydrochloric acid solution (0.1N).

Procedure :

In three 50 ml. volumetric flask, the following contents are taken :

	Blank (B)	Unknown (U)	Standard (S)
Urine	—	0.5 ml.	—
Distilled water	0.5 ml.	—	—
Standard solution of creatinine	—	—	0.5 ml.
10% NaOH	1 ml.	1 ml.	1 ml.
Saturated Picric acid solution	10 ml.	10 ml.	10 ml.

Make up the volume by distilled water.

Mix and allow them to stand for 10 minutes for the development of colour. Bring up the volume in each flask to the mark. Mix well and make photoelectric colorimetric reading using green filter at 520 mμ.

Calculation

Let U, S, and B the optical densities of the unkown, standard and Blank respectively.

$$\therefore \frac{\text{Creatinine in unknown}}{\text{Creatinine in standard (0.5 mg)}} = \frac{U-B}{S-B}$$

or, creatinine in 0.5 ml. Urine

$$= \frac{U-B}{S-B} \times 0.5 \text{ mg.}$$

0.5 ml. urine was taken for estimation and 0.5 ml. Standard creatinine taken which contains 0.5 mg creatinine.

$\therefore$ Creatinine in 100 ml. Urine

$$= \frac{U-B}{S-B} \times 0.5 \times \frac{100}{0.5}$$

$$= \frac{U-B}{S-B} \times 100 \text{ mg.}$$

Discussion :

1. Creatinine in urine appears from creatine of the skeletal muscle.

2. Normal adults excrete creatinine from 1 to 1.8 grams daily. Under certain conditions it may be higher. But the value is constant from day to day for a given normal individual, being influenced by the diet which contains significant amounts of creatinine e.g., heavy meat diet.

3. Nitrogen metabolism in the body does not influence creatinine excretion.

4. Obese individuals excrete more creatinine than persons with less body fat.

5. Creatinine coefficient is defined as the number of milligram of creatinine excreted daily per kg. body weight. The normal range is 19 to 30.

6. Exercise can cause the increase in creatinine excretion due to the release from muscle stores.

7. Increased creatinine excretion occurs in wasting diseases, prolonged starvation, fever, etc. due to the break-down of muscle.

8. Creatinine excretion is also increased in the diseases of the muscle such as progressive muscular dystrophy (myopathy), amyotonia congenita etc.

ESTIMATION OF INORGANIC PHOSPHATE IN URINE

(Fiske and Subbarow)

Principle : The inorganic phsophate of urine reacts with molybdic acid to form phosphomolybdic acid, which is reduced to a blue coloured compound (molybdenum blue, which is a mixture of lower oxides of molybdenum) by 1, 2, 4-amino naphthol sulphuric acid. The intensity of blue colour depends on the amount of inorganic phosphate in urine.

Reagents :

1. Sulphuric acid (10N).
2. **Molybdic acid solution :** Dissove 2.5 grams of ammonium molybdate in about 20 ml. of water in a 100 ml. volumetric flask. Add 30 ml. of 10 N. Sulphuric acid to it. Make up the volume upto 100 ml. with distilled water.
3. **1, 2, 4—aminonaphthol sulphonic acid (ANSA) :** Take 195 ml. of 15 per cent sodium bisulphite solution in a glass-stoppered cylinder. Add 0.5 gram of 1, 2, 4- aminonaphthol sulphonic acid and 5 ml. of 20 per cent sodium sulphite. Stopper and shake until the powder is dissolved. If solution is not complete, add more sodium sulphite (1 ml. at a time with shaking.) Transfer the solution to a brown glass bottle and store in the cold. This solution can be used for four weeks.
4. **Standard inorganic phosphate solution (Stock) :** Dissolve 439 grams of Potassium orthophosphate in 100 ml. water (100 mg%).

Working standard : Prepare as if 5 ml. contains 0.4 mg. of phosphate.

Procedure :

In three 100 ml. volumetric flask, the following contents are taken :

	Blank (B)	Unknown (U)	Standard (S)
Urine	—	1 ml. + 70 ml. distilled water	—
Blank	70 ml. water	—	—
Standard solution	—	—	5 ml. + 65 ml. water
Molybdic acid solution	10 ml.	10 ml.	10 ml.
1, 2, 4-Amino-naphthol sulphonic acid	4 ml.	4 ml.	4 ml.

Make up the volume by distilled water. Mix well by inversion. Allow to stand for 5 minutes. Take readings in the photoelectric colorimeter using red filter at 660 mμ.

Calculation :

Let U be the O.D. of the unknown, S the O.D. of the standard and B the O.D. of the blank.

$$\therefore \frac{\text{Phosphate in unknown}}{\text{Phosphate in standard (0.4 mg.)}} = \frac{U-B}{S-B}$$

$\therefore$ Inorganic phosphate in 100 ml. Urine

$$= \frac{U-B}{S-B} \times 0.4 \text{ mg.}$$

Discussion :

1. The inorganic phosphate (expressed as P) of urine is mainly of acid sodium phosphate (NaH_2PO_4) and basic sodium phosphate (Na_2HPO_4) besides other salts of phosphoric acid.
2. The phosphate content of normal urine varies with diet and lies between 1 to 3 grams in 24 hours. The greater part of the phospate excretion arises from ingested food.
3. The phosphate excretion is increased in fevers.
4. Its excretion is decreased in low phosphorus intake, in vitamin D deficiencies resulting in rickets and osteomalacia and in hypoparathyroidism.
5. In hyperparathyroidism, its excretion is increased.
6. Phosphaturias always represent decreased acidity and not increased phosphate content of the urine.

ESTIMATION OF UREA IN URINE

(Doremus)

Principle : Urea in the urine is decomposed when treated with an alkaline solution of hypobromite. The nitrogen thus formed is a measure of the quantity of urea present.

$CO(NH_2)_2 + 3NaBrO + 2NaOH = N_2 + Na_2CO_3 + 3NaBr + 3H_2O$. One molecule of urea (60 g) produces one molecule of nitrogen (28 g).

Reagent :

Sodium hypobromite solution (freshly prepared) : 2.5 ml. of bromine is added to 25 ml. of 40% of NaOH. Use fume cupboard for handling bromine.

Procedure :

1. Close the stop cock of the Daremus Ureometer. Fill the closed limp with a freshly prepared alkaline hypobromite solution bearing no air bubbles at the top. Remove the air bubbles by tilting the Ureometer.

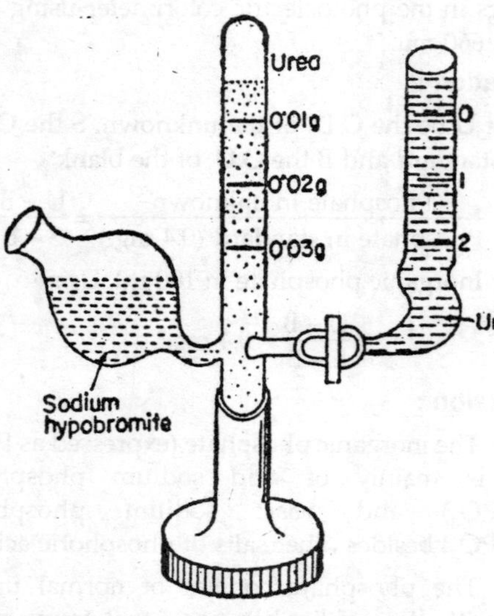

Fig. 10.1 : Doremus Ureometer

2. Add urine into the open side tube upto the zero mark by the help of a glass tubing.

3. Open the stop cock a little at a time and allow 1 ml. of urine to run slowly into the hypobromite solution taking about 10 to 15 minutes time. In case, the urine is allowed to flow too rapidly, the CO_2 produced is not completely absorbed by NaOH present in the hypobromite solution which will lead the result too high.

4. When the evolution of gases has stopped and the hypobromite solution becomes clear, the reading on the closed limb should be taken.

5. Add another 1 ml. of the urine slowly into the open side tube and take the reading again as before. The mean of these two readings gives the result.

Calculation :

The Doremus apparatus is graduated to give direct readings in grams of urea contained in the volume of urine reacted. The gas collected in the apparatus is nitrogen which is liberated by the amount of urea marked on the apparatus.

Discussion :

1. Urea is the end product of protein metabolism and hence its excretion depends on the amount of protein ingested. The average daily excretion of urea by normal adults is usually about 30 grams. On a low protein diet, it may be as low as 3 grams per day.

2. In increased tissue catabolism, as in fevers, the excretion of urea is increased.

3. Urea excretion may be diminished in pronounced kidney and liver disorders due to decreased formation and decreased power of elimination.

4. In marked acidosis, its excretion may be decreased due to the utilization of ammonia to trap hydrogen ion.

ESTIMATION OF PROTEIN (ALBUMIN) IN URINE

(Aufrecht's method)

Principles : The protein of urine is completely precipitated with the help of an alkaloid reagent. The precipitate is centrifuged down in a tube (Aufrecht's tube) which is graduated to indicate the amount of protein present in 100 ml. of the urine.

Reagents :

Esbach's reagent : Prepare 1% solution of citric acid and saturate it with picric acid. It is a saturated solution of picric acid containing 1% citric acid.

Procedure :

1. Filter the urine using a dry filter paper until the filtrate is clear.

2. Fill up the Aufrecht's tube to the mark 'U' with the clear filtered urine.

3. Add Esbach's reagent to the mark 'R'.

4. Stopper and invert the tube a few times to mix the contents thoroughly, but do not shake. A turbidity appears due to the precipitation of protein.

5. Stand for 5 minutes and then centrifuge for 10 minutes. At the end of it take the readings of the horizontal level of the precipitated protein.

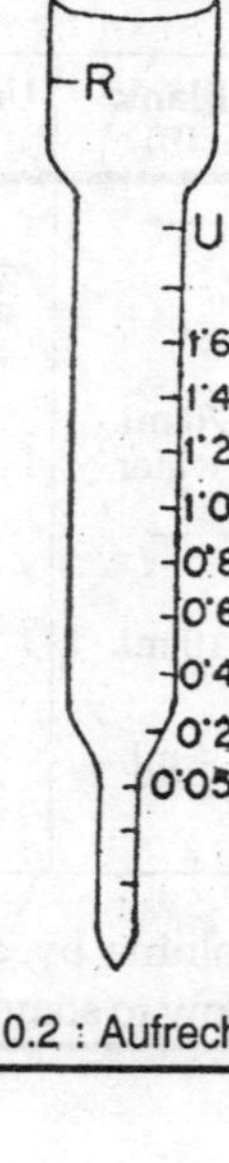

Fig. 10.2 : Aufrecht's tube

6. Centrifuge again for 5 minutes and read.

7. If the two readings are not the same, recentrifuge till the last two readings of the protein is constant.

Discussion :

1. Detectable amount of protein is not found in normal urine.

2. The presence of protein in urine (proteinuria) is abnormal. The types of protein in pathological urine are plasma albumin, globulin and fibrinogen, hemoglobin and methemoglobin, and sometimes Bence Jones proteins and proteoses, or proteins from pus and mucus discharged from urinary tract infection.

3. In kidney diseases (Nephritis and nephrosis), large amount of protein is excreted. In many other conditions where kidney is secondarily involved, there is moderate to heavy loss of protein in urine. In such conditions, the loss of protein is essential to assess by measuring the quantity of it in urine.

ESTIMATION OF REDUCING SUGAR IN URINE

• (Benedict's method)

Preciples : $CuSO_4$ of the quantitative Benedict's reagent is reduced to Cu_2O by the reducing sugar present in urine when boiled. The cuprous oxide then reacts wth Potassium thiocyanate (KCNS) to form white precipitate of cuprous thiocyanate. Potassium ferrocyanide of quantitative Benedict's reagent helps to prevent the precipitation of cuprous oxide for which the red coloured cuprous oxide does not appear as precipitate. The alkali Na_2CO_3 has the advantage over NaOH that it does not cause destruction of small amounts of sugar.

The same principle does apply on the estimation of fructose, lactose and maltose too.

Reagents :

1. Sodium Carbonate (anhydrous crystalline salt).

2. **Benedict's quantitative reagent** : Take weight of 200 g. of sodium citrate, 75 g. of anhydrous sodium carbonate (or 200 g. of crystalline salt), and 125 g. of Potassium Thiocyanate. Dissolve them one by one in sufficient hot distilled water upto a total volume of about 800 ml. Filter and cool to room temperature. Now, accurately weigh 18 g. of crystalline copper sulphate ($CuSO_4 . 5H_2O$) and dissolve it separately in about 100 ml. of distilled water. Pour the copper sulphate solution into the other solution slowly and with constant stirring. Add 5 ml. of 5% Potassium ferrocyanide solution and make up the volume to a litre with distilled water. If the solution is not clear, filter.

Procedure :

Qualitative Benedict's test should be performed for the sugar of the urine to assess the concentration of sugar in urine. The following dilution of urine should be made for the different coloured precipitate by qualitative Benedict's test :

Green precipitate	3 times dilution
Yellow precipitate	6 times dilution
Brick red precipitate	10 times dilution

Measure accurately 10 ml. of quantitative Benedict's reagent into a porcelain basin. Add 2 grams of anhydrous sodium carbonate (or 5 grams of crystalline salt) and a pinch of pumice stone.

Fill a 10 ml. microburette with the urine (or its dilution if necessary). Now, heat the porcelain basin on a low flame. When the solution of the procelain basin begins to boil, add urine from the burette rapidly first but continuously until white precipitates begin to appear. At this stage, add urine drop by drop at an interval of about 10 seconds until the last trace of blue colour is discharged.

Allow the titration mixture to cool. If the mixture still shows bluish tinge, boil the contents again and add a few drops of urine until the trace of blue colour disappears. Take the readings in the burette. Repeat it again twice. All the readings should be very nearer.

Precaution :

1. The boiling should never be interrupted.

2. If the mixture in the porcelain basin becomes too concentrated during boiling, some quantity of hot distilled water should be added to replace the loss by evaporation.

3. Reading in the burette must be taken carefully.

Calculation :

10 ml. quantitative Benedict's reagent is reduced by 0.02 g glucose, 0.0212 g. fructose, 0.027 g. lactose, 0.03 g. maltose.

10 ml. quantitative Benedict's reagent ≡ 'Y' ml. of urine.

∴ 'Y' ml. of urine contain 0.02 gm. of glucose

$$\therefore \text{ 100 ml. of urine contain } \frac{0.02 \times 100}{Y} \text{ gm.} = \frac{2}{Y} \text{ gm.}$$

Discussion :

1. The normal urine contains a very small amount of glucose which cannot be detected by ordinary routine test.

2. The presence of sufficient amount of reducing sugar in urine under pathological condition is said to be glycosuria (Glucosuria). Besides glucose, urine may contain lactose (in pregnant and lactating mothers), Galactose, Pentose.

3. Persistent glucosuria is the common indication of diabetes mellitus. The volume of urine excreted per day in this disease is usually large.

4. The estimation of sugar in urine aids us to assess the severity of the disorder as well as to follow its course under treatment.

ESTIMATION OF CHLORIDE IN URINE
(Whitehorn method)

Principle : A measured volume of urine is treated with a known volume of a standard silver nitrate solution containing concentrated nitric acid. As a result, the chloride of the urine is precipitated as silver chloride. The unreacted silver nitrate is estimated by titrating it against a standard potassium thiocyanate solution using ferric alum as indicator. Ferric alum yields red colour with potassium thiocyanate which disappears in presence of silver nitrate.

Silver nitrate combines with potassium thiocyanate forming silver thiocyanate. Silver chloride cannot react with KCNS due to the presence of excess of nitric acid. The amount of silver nitrate used up for the precipitation of chloride as silver chloride is a measure of the amount of chloride in urine.

Reagents :

1. **Acid silver nitrate reagent :** 29.061 grams of pure silver nitrate is dissolved in water and the volume made upto 1 litre with distilled water. The solution is further mixed 1 litre of pure nitric acid. 1 ml. of this silver nitrate solution is equivalent to 0.005 grams of NaCl.

2. **Standard Potassium thiocyanate solution :** A 2% solution of KCNS is titrated against the acid silver nitrate solution with ferric alum as indicator and its concentration is adjusted so that 1 ml. of the thiocyanate solution corresponds to 2 ml. of the acid silver nitrate solution.

3. Ferric alum (finely powdered).

Procedure :

1. Take exactly 5 ml. urine in a porcelain basin and add 10 ml. of acid silver nitrate reagent to it. Mix well slowly with the help of a glass stirrer and allow it to stand for 5 minutes for the precipitate of silver chloride to aggregate.

2. Add about 0.3 gm. of finely powdered ferric alum (or 6 ml. of 5% solution of ferric alum). Stir slowly again and let the precipitate settle down.

3. Titrate the supernatant liquid against KCNS solution from a microburette.

4. Stir carefully during titration without disturbing the precipitate and continue titration until a red colour is obtained which remains for at least 30 seconds after careful stirring.

5. Take exact reading in the microburette and repeat the experiment thrice avoiding error. All the three readings must be very nearer for a better result to be obtained.

Precaution :

1. During titration the mixture in the porcelain basin should not be stirred vigorously. Vigorous stirring tends to break up the silver chloride aggregate which also take up some KCNS and make the colour disappear leading to unusual high titration value.

$$AgCl + KCNS = AgCNS + KCl.$$

2. If the permanent red colour of the whole mixture appears by the addition of one or two drops of KCNS solution from the microburette, it indicates that 10 ml. of silver nitrate solution taken is just barely sufficient for the precipitation of all the urinary chlorides.

Therefore, a further 10 ml. of acid silver nitrate reagent and 0.3 gram of ferric alum should be added to the mixutre in the porcelain basin. Mixed, allowed to stand for 5 minutes and the titration carried out further. In calculating the result, particular attention must be given for the increased volume of silver nitrate used.

Calculation :

Volume of acid silver nitrate reagent originally used = 10 ml.

Volume of KCNS solution used for titration of remaining $AgNO_3$ = A ml.

$\therefore$ Volume of $AgNO_3$ reagent remaining unused = 2A ml. = B ml.

[Because 1 ml. of KCNS = 2 ml. of acid $AgNO_3$ reagent].

$\therefore$ Volume of $AgNO_3$ used up by chlorides present in 5 ml. of urine = (10 – B) ml.

$\therefore$ Chloride in 100 ml. urine

$$= \frac{(10 - B) \times 0.005 \times 100}{5}$$

$$= (10 - B) \times 0.1 \text{ gms. NaCl.}$$

[1 ml. of acid $AgNO_3$ reagent ≡ 0.005 gm. of NaCl]

$\therefore$ Daily excretion of chloride

$$= (10 - B) \times 0.1 \times 15 \text{ gms. NaCl.}$$

The average daily excretion is assumed to be 1,500 ml.

Discussion :

1. Chloride is the chief anion of urine. The amount of chloride excreted per day by normal adult on a normal diet varies from 10 to 15 gms. (expressed as NaCl).

2. In fasting, the chloride excretion falls rapidly and on high water ingestion, it is increased.

3. In pneumonia and certain other acute infectious diseases, the excretion of chloride is significantly diminished.

4. The decreased excretion of chloride has been found in nephritis associated with edema.

5. Chloride excretion increases in adrenal cortical hormone deficiency as in Addison's disease.

ESTIMATION OF TOTAL TITRATABLE ACIDITY IN URINE
(Folin method)

Principle : The titratable acidity of urine is expressed in terms of the ml. of standard alkali necessary to bring the urine from its original P^H to phenolphthalein end point P^H 8.5 or 9 by the use of the phenolphthalein indicator.

In normal urine, the "acid" titrated consists entirely of $H_2PO_4^-$. The reaction is as follows :

$$H_2PO_4^- + OH^- = HPO_4^{--} + H_2O$$

The urine is titrated with standard 0.1 N. NaOH using phenolphthalein as an indicator. Potassium oxalate is added to precipitate calcium as calcium phosphate on neutralization.

Reagents :

1. 0.1 N. NaOH solution.
2. 1% Phenolphthalein solution.
3. Potassium oxalate.

Procedure :

1. Take 25 ml. of urine in a conical flask.
2. Add to it 5 gms. of finely Powdered Potassium Oxalate and 0.5 ml. of phenophthalein solution.
3. Shake vigorously and titrate immediately with 0.1 N. NaOH solution from a burette till a permanent pink colour appears.
4. Take burette reading and take such readings thrice. Take the mean of the three readings for calculation.

Calculation

$$\frac{25}{a} = \frac{b}{x}$$

$$\therefore \ x = \frac{a \times b}{25}$$

where, a = ml. of 0.1 N. NaOH solution used.
b = Volume of urine excreted in 24 hours.
x = Total acidity of the 24–hour urine specimen.

The result is expressed as ml. of 0.1 N. NaOH.

Discussion :

1. The total acid excreted per day under normal condition by normal adults varies from 200 to 500 ml. 0.1 N NaOH.

2. The urine is acidic on meat diet and on vegetable diet, it is less acidic or alkaline.

3. During active gastric digestion the urine may temporarily be acidic.

4. In renal diseases, the urine becomes acidic due to the incapability of neutralizing acids.

5. In fasting, the titratable acidity may rise to 800 in a few days. Values upto 1,500 may be found in severe diabetic acidosis.

6. The acidity of the urine may increase by the administration of mineral acids; while it is diminished on the administration of sodium bicarbonate.

7. The acidity of the urine is decreased, if the urine is allowed to stand for sometimes, due to the formation of ammonia from urea by bacterial decomposition.

ESTIMATION OF TOTAL NITROGEN IN URINE
(Kjeldahl method)

Principle : The various nitrogenous constituents of urine are converted into ammonium sulphate by digestion (boiling) with concentrated sulphuric acid. Traces of copper sulphate is added to catalyse the oxidation and potassium sulphate is used to raise the boiling point of sulphuric acid. The ammonium sulphate thus formed is heated with a strong solution of sodium hydroxide and the liberated ammonia is distilled into a measured volume of standard acid. At last, this partly

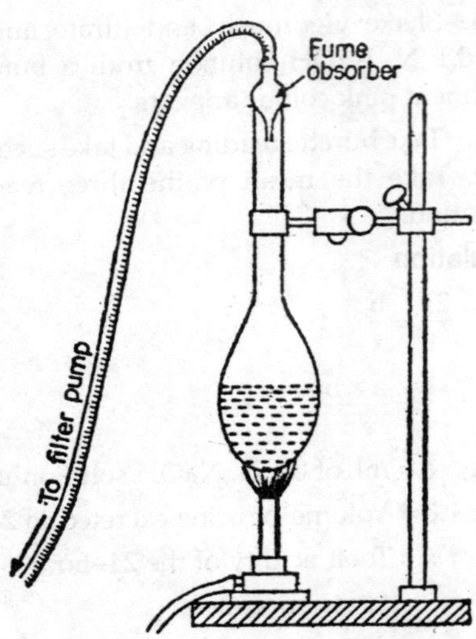

Fig. 10.3 : The arrangement for the digestion of nitrogenous compound

neutralized acid solution is titrated with standard alkali solution. The amount of acid neutralized by ammonia is a measure of the nitrogen content of urine.

Reagents :

1. Sodium hydroxide (40%).
2. Methyl red (0.2%).
3. Saturated solution of copper sulphate.
4. Potassium sulphate.
5. NaOH (0.1N).
6. H_2SO_4 (0.1N).

Procedure :

1. Digestion :

(a) Put 1 ml. urine in a 300 ml. Kjeldahl flask and also 2 ml. conc. H_2SO_4, 2 drops of saturated solution of copper sulphate, 1 gm. of potassium sulphate.

(b) Heat the flask in a fume cupboard over a small flame of microburner. The fume should just touch the bottom of the flask. Use of fume absorbed (Fig. 10.3). Continue the gentle boiling until the solution turns bluish green.

(c) Disconnect the flask and swirl the contents gently to dislodge any black particles which may be sticking to the walls of the flask. Connect again and heat as before for another 5 minutes.

2. Distillation :

(a) Allow the flask to cool till it is just warm. Add 2 ml. water and few pieces of broken procelain and shake immediately to prevent separation of potassium hydrogen sulphate.

(b) Take 20 ml. of standard 0.1 N. H_2SO_4, 10 ml. water and 3 drops of methyl red in a 250 ml. conical flask. Assemble the apparatus according to the diagram (Fig. 10.4). Take care that the delivery tube of the condenser just dips below the surface of the acid. Start running water through the condenser.

(c) Take down the Kjeldahl flask and gently run 10 ml. of 40% sodium hydroxide down its side so that it sinks to the bottom of the flask.

(d) Clamp the flask into position and connect it to the condenser through the still-head.

(c) Fix the burner in such a position that the top of the flame just touches the bottom of the flask. Start heating the flask till boiling commences. Continue boiling for another 10 minutes. At the end of the distillation, lower the conical flask till the tip of the condenser is out of solution. Then remove the burner.

(f) Disconnect the condenser from the flask. Wash down the inside of the condenser and outside of the delivery tube with a jet of distilled water receiving the washing into the conical flasks.

(g) Titrate the contents of the conical flask against the standard 0.1 N. NaOH solution provided. Also titrate 20 ml. 0.1 N. H_2SO_4 with 0.1N. NaOH and the two solutions should closely match.

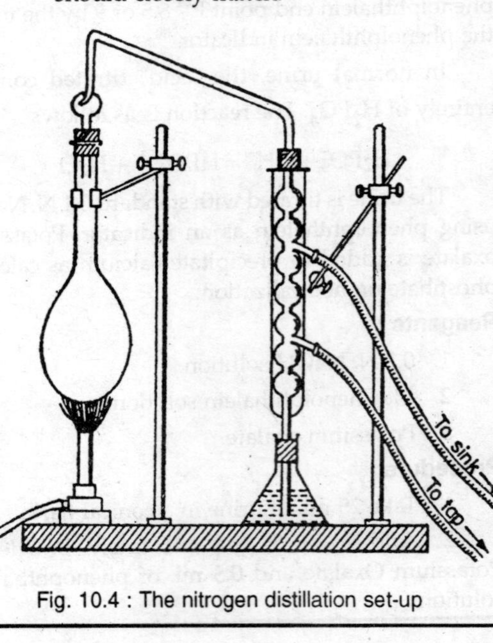

Fig. 10.4 : The nitrogen distillation set-up

Calculation :

Since 1 litre of N/1 acid	=	1 litre of N/1 Nitrogen.
∴ 1 ml. of N/10 acid	=	0.0014 g. nitrogen.
The acid used for neutralization of the ammonia in the above experiment.	=	The total acid originally taken in the conical flask—unneutralized acid that remains in the conical flask at the end of distillation.
	=	(20 ml. acid—the volume of NaOH used for the back titration).
	=	γ ml. N/10 acid.
Hence the total nitrogen in 1 ml. urine	=	($\gamma \times 0.0014$ g.).
The daily excretion of total nitrogen in urine	=	($\gamma \times 0.0014 \times 1500$) g.

Discussion :

1. The daily excretion of total nitrogen in urine by a normal adult on a mixed diet varies from 12 to 18 gms. On a low protein diet, the urinary nitrogen may fall to 5 gms. per day and on a high protein diet, it may rise to 22 to 25 gms.

2. The chief nitrogenous constituents of normal urine are urea, uric acid, creatinine, ammonium salts, amino acids etc.

3. In a normal adult, the total nitrogen of the urine and of the feces are often almost exactly equal to the total nitrogen of the diet. Such a condition is said to be "Nitrogen equilibrium". The feces usually contain very little nitrogen.

4. Persons when excrete more nitrogen than what he ingests is said to be "negative nitrogen balance". But if intake of nitrogen exceeds nitrogen excretion, it is said to be "Positive nitrogen balance".

ESTIMATION OF AMYLASE (DIASTASE) IN URINE

(Wohlgemuth's method)

Principle : Different amounts of urine are incubated with a fixed amount of 0.1% starch solution for 30 minutes. The minimum amount of urine necessary for coverting 2 ml. of starch solution into erythrodextrin at 37°C in 30 minutes is an index of the quantity of enzyme in the urine.

Reagents :

1. **Stock Starch solution (2% in 10% NaCl) :** Make 2 gms. of starch into a paste with a few ml. of distilled water and pour into it about 60 ml. of boiling water. Add 10 gms of sodium chloride and dissolve. Cool, transfer 100 ml. in Volumetric flask and make up to the mark. This keeps for a few weeks.

2. **0.1 per cent starch solution for use :** Dilute the stock solution 20 times with distilled water. Prepare each day.

3. **N/50 iodine solution approximately :** Prepare by diluting a small amount of N/10 iodine 1 in 5.

Procedure :

1. All pipettes for this experiment should be lightly plugged on the top with a little cotton wool to prevent contamination with saliva.

2. Dilute the urine 1 in 10.

3. Take 10 clear and dry test tubes and number them from 1 to 10. Arrange the tubes in serial order in a rack.

4. Measure different amounts of undiluted or diluted urine according to the table given below into ten test tubes by 1 ml. graduated pipette.

5. Add required amounts of distilled water as indicated in the table to make the total volume in each test tube to 1 ml. See that the first five test tubes contain diluted urine and the other five undiluted urine.

6. Place all test tubes in a water bath at 37°C and after 2 minutes add 2 ml. of 0.1% soluble starch solution to each and shake gently to mix. Note the time and leave the test tubes in the bath for exactly 30 minutes.

7. Remove them from the bath and put them in a tumbler of ice cold water for 2 minutes to stop the reaction. Arrange the test tube again in the rack.

8. Add 2 drops of dilute iodine solution to each of the test tubes, mix and note the colour. If the colour fades away on shaking, add a drop or two of the iodine solution to the test tubes. Too much iodine should be avoided.

9. The test tubes which show blue or purple colour still contain starch or amylodextrin. But the tube which does not show blue or purple colour but show reddish colour contains erythrodextrin. The amount of urine in this tube is to be recorded.

Test tube numbers	1	2	3	4	5	6	7	8	9	10
Ml. of diluted urine (1 : 10)	0.5	0.6	0.7	0.8	0.9	..	..	..	..	..
Ml. of undiluted urine	..	..	..	..	..	0.1	0.2	0.3	0.4	0.5
Ml. of distilled water	0.5	0.4	0.3	0.2	0.1	0.9	0.8	0.7	0.6	0.5
Calculated diastatics index	40	33.3	28.6	25	22.2	20	10	6.7	5	4

Calculation :

Noting down the minimum amount of urine necessary for the digestion of 2 ml. of 0.1 per cent starch solution, calculate what volume of the starch will be digested to erythrodextrin by 1 ml. of urine under the specified condition of temperature and time.

Thus, $d\,\frac{37°}{30'}$

$$= \frac{2}{\text{Smallest volume of urine noted to digest the starch to erythrodextrin}} \text{ units.}$$

As for example, if tube No. 7 shows red colour, the smallest volume of urine required is 0.2 ml. undiluted urine. Therefore, diastatic index is

$$d\,\frac{37°}{30'} = \frac{2}{0.2} = 10 \text{ units.}$$

Discussion :

1. The enzymes pepsin, trypsin, amylase and lipase have been identified in urine.

2. Normally, very small amylase activity is present in urine. But its concentration increases highly in acute pancreatitis and parotitis.

3. Amylase is normally secreted by salivary glands and pancreas. It digests starch into maltose through amylodextrin, erythrodextrin and achrodextrin.

4. The acitivity of amylase in urine is usually expressed as "diastatic index" which is defined as the number of millitre of 0.1% starch digested into erythrodextrin by 1 ml. of urine in 30 minutes at 37°C.

5. The diastatic index of a normal urine ranges between 5 to 30 units. Concentrated urine may show higher diastatic index.

ESTIMATION OF VITAMIN 'C' (ASCORBIC ACID) IN URINE

(Harris and Ray, 1935)

Principle : L-ascorbic acid is a strong reducing agent and it reduces the dye 2 : 6 dichlorophenolindophenol into the colourless leucobase. This dye is red in acid solution and becomes colourless when fully titrated with vitamin C. Ascorbic acid in urine is determined by titrating it directly into a known amount dye to the colourless end point.

Reagents :

1. Glacial acetic acid.

2. **2 : 6 dichlorophenolindophenol solution :** Weigh out accurately 40 mg. of the dye and dissolve it in water. Make the volume to 100 ml. with water 0.5 ml. of this solution is equivalent to 0.1 mg. of ascorbic acid. It should be stored at 4°C. Its keeping qualities are poor. Hence, it is to be prepared at frequent intervals (at least weekly).

Procedure :

Take 0.5 ml. of 2 : 6 dichlorophenolindophenol solution in a test tube and add 1 ml. of glacial acetic acid to it.

Collect your urine and immediately fill a 10 ml. microburette with it. Titrate the dye with the urine shaking well the test tube during the titration till the red colour is just discharged. Take the burette reading. Repeat it thrice. Take the mean of the three readings for calculation.

Precaution :

Ascrobic acid is a labile substance and is quickly oxidized to dehydroascrobic acid by atmospheric oxygen. Therefore, it is wise to estimate ascorbic acid of urine sample within a few minutes after passing it.

Sometimes acetic acid is added immediately in the proportion of 1 part to 9 parts of urine. Such acidified urine keeps well for several hours.

If a 24-hour sample of urine is collected with acetic acid as preservative, hydrogen sulphide (H_2S) must be passed through the urine sample prior estimation with a view to reduce any dehydroascrobic acid present for its conversion to ascorbic acid.

Calculation :

Suppose, burette reading is 'X' ml.

∴ 'X' ml. of urine contains 0.1 mg. ascorbic acid.

∴ The daily excretion of ascorbic acid in urine.

$$= \frac{0.1 \times 1{,}500}{X} \text{ mg.}$$

[The average daily excretion of urine assumed to be 1,500 ml.]

Note : Besides ascorbic acid, the urine contains other substances which decolorise the dye.

Although the result obtained for ascorbic acid is not very accurate it is still sufficiently accurate for clinical purposes.

Discussion :

1. Food is the source of vitamin C in the body. Vitamin C is distributed in blood, tissue fluids and in all secretions and excretions of the body including urine. The excretion of vitamin C in urine depends upon the level in blood.

2. The daily urinary output of ascrobic acid is about half its intake. Normally, the daily excretion of it in urine ranges from 6 to 18 mg.

3. Urinary ascorbic acid is much reduced and may even disappear in the deficiency of vitamin C. The values may increase in urine when the intake is high.

CHAPTER 11

FOOD AND FUNCTION TEST

HYDROLYSIS OF STARCH BY SALIVA

(Rebert's achromatic point method)

Principle : A given amount of saliva containing amylase is added to a measured amount of starch solution at 40°C and pH 6.8. The digested samples are removed and treated with dilute iodine at intervals until no further colouration is produced. The time taken to reach this achromatic point is determined. NaCl acts as an activator for amylase.

Reagents :

1. Starch (1%).
2. NaCl (0.5%).
3. **Buffer solution (pH 6.8) :** Treat 50 ml. of 0.2 N. acid potassium phosphate with 21 ml. of 0.2 N. NaOH and dilute to 200 ml. with distilled water.
4. Dilute iodine (0.01N).

Procedure :

1. Measure 5 ml. of 1% starch into a clean test tube.
2. Add 2 ml. of buffer solution (pH 6.8) and 2 ml. of 0.5% NaCl to it.
3. The tube is placed in a water bath at 40°C to warm it and in the same water bath, diluted saliva (1 in 20) is also kept in a test tube.
4. A series of test tubes are prepared, each containing 2 ml. of distilled water and 2 drops of diluted iodine (0.01N).
5. Add 2 ml. of saliva to the warm starch solution. Mix well and note time. Transfer at intervals one to two drops of digested mixture to one of the iodine tubes by a glass tube taking care that no iodine is transferred back to the digested mixture. Shake the sample and note the colour which is blue at first.
6. Note the time at which no colour is produced when a drop of the digested mixture is added to an iodine tube. This is said to be "Chromic period" which is an inverse measure of the activity of the enzyme.

Note : Accurate results can be obtained if the saliva is so diluted as to give chromic period of about 10 to 15 minutes and the digest is taken every 30 seconds towards the end of the period and compared with the blank iodine to serve as control.

Discussion :

Salivary amylase acts upon boiled starch yielding maltose at the optimum temperature and pH. The activity of this enzyme is increased in presence of an activator (Chloride). Its activity is decreased in absence of the suitable conditions.

ESTIMATION OF REDUCING SUGAR IN MILK

(Benedict's method)

Principle : Lactose, the reducing sugar of milk, can be obtained by precipitating the protein of the milk by copper sulphate and sodium hydroxide solution. It may be estimated by using Benedict's quantitative solution just like reducing sugar in urine.

Reagents :

1. Zinc sulphate solution (10%).
2. Sodium hydroxide (1N).
3. Benedict's quantitative reagent.
4. Sodium carbonate (anhydrous).
5. Purmice stone.

Procedure :

Pipette 10 ml. milk into a 100 ml. Volumetric flask and add to it 2 ml. of 10% $ZnSO_4$ solution and 1 ml. of 1N. NaOH solution. Mix well. Make up the volume upto 100 ml. by adding distilled water. Allow to stand for 1 to 2 minutes and then put it in boiling water bath for a few minutes till the casein is precipitated. Then filter. Pour the filtrate into a burette.

Titrate just like reducing sugar in urine using Benedict's quantitative reagent. As such, take three burettee readings and take the mean for calculation.

In the case of cow's milk, it is wise to take 20 ml. of milk with 4.8 ml. of $ZnSO_4$ and 2.4 ml. of NaOH.

Calculation :

10 ml. of Benedict's quantitative solution is reduced by 0.027 gms. of reducing sugar of milk.

Suppose, the mean burette reading is 'X' ml.

Hence, 10 ml. of Benedict's quantitative solution = 'X' ml. filtrate.

∴ 'X' ml. of milk contain (0.027 × 10) gms. of reducing sugar (10 being the dilution factor).

∴ 100 ml. of milk contain $\frac{0.027 \times 10 \times 100}{x}$ gms. of reducing sugar.

∴ 100 ml. of milk contain $\frac{27}{x}$ gms. of reducing sugar.

ESTIMATION OF FAT CONTENT OF MILK

(Babcock's Centrifugal method)

Principle : Concentrated sulphuric acid when added to the milk charrs organic matter and breaks emulsion of fat. The centrifugation of the acid-treated milk in a special tube (Babcock tube, Fig. 11.1) brings the fat in the narrow calibrated part of the tube from which fat content is directly read.

Reagents :

1. Concentrated sulphuric acid (Sp. gr. 1.83-1.84).

2. Acid-alcohol mixture : Mix equal volumes of conc. HCl and amyl alcohol.

Procedure :

1. Introduce milk into the Babcock tube upto 5 ml. mark by means of a narrow pipette.

2. Add conc. H_2SO_4 to fill the body of the tube, 1 ml. at a time. Mix well by rotation after each addition, cool, and then add further acid to fill the body once again.

3. Fill the neck of the tube to the zero mark with an acid-alcohol mixture.

4. Centrifuge the tube for 2 to 3 minutes and read the the percentage of fat from the calibrated part of the neck. If the top of fat column is not at zero, it may be brought there by addition of hot water and then recentrifuge.

5. In case fat content is over 5 per cent, the milk should be diluted with an equal volume of water and then proceed as above, multiplying the observed reading by 2.

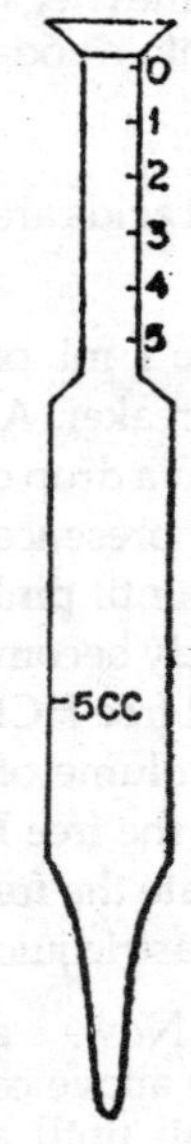

Fig. 11.1 : Babcock tube

Discussion :

Cow's milk and human milk contain 4 per cent fat. Presence of less than 3.5 per cent fat in cow's milk indicates either dilution of milk or removal of fat from it.

GASTRIC ANALYSIS

Collection of Gastric juice for fractional test meal :

1. The subject is advised not to eat or drink anything during the 12 hours preceding the test.

2. Ryle's tube lubricated with a little paraffin is passed to the stomach through oral or nasal passage upto the double mark on the tube. Saliva should not be swallowed afterwards.

3. The resting juice is removed with a 50 ml. syringe. The volume is noted and the contents are kept for further examination. This is called 'Zero sample'.

4. The person is given a pint of oat meal to drink. The time is noted. About 10 ml. of the stomach contents are removed every 15 minutes for $2\frac{1}{2}$ hours into different test tubes marked from 1 to 10. At the end of this period, the stomach is emptied completely and the volume of the residual contents are measured. The tube is then withdrawn gently.

Examination of the specimen :

1. **Colour** : Greenish colour indicates bile pigments.

Reddish or brownish colour indicates blood. Confirm blood by benzidine test.

2. Add a drop or two of dilute iodine solution to a drop of the contents from each tube on a white porcelain tile. Blue colour indicates the presence of

starch. Starch test becomes negative when all the oat meal has passed out into duodenum.

Acid titration :

Both free and total acids are determined in each specimen.

Free acid : Pipette 1 ml. of the filtered gastric content into a small breaker. Add about 2-3 ml. of distilled water and then a drop of Topfer's indicator. It will turn pink in the presence of free HCl. Titrate it with N/100 NaOH until pink colour disappears and the colour ultimately becomes yellowish orange (pH 4.0). At this pH all free HCl is titrated. Take the burette reading. The volume of the alkali required for titration represents the free HCl present in 1 ml. gastric content. Calculate the free HCl as N/10 acid present in 100 ml. of gastric juice.

Total acid : Now, add a drop of phenolphthalein to the above content and continue titration with the alkali until a definite red tinge reappears (pH 8.5). Take the second reading of the burette. The difference between this reading and the initial reading represent the total acid present in 1 ml. of the gastric contents.

The difference between the total acid and free HCl is due to the combined acid (HCl neutalised by weak bases) and organic acid.

Calculate the total acid present in 100 ml. of gastric juice.

In this way find out the free HCl and total acid for each specimen. Draw the results of the gastric analysis on a graph paper.

Abnormal acidity :

1. **Hyperchlorhydria :** If the free HCl is above the normal range it is termed hyperchlorhydria. This is commonly seen in cases of duodenal ulcer.

2. **Hypochlorhydria :** If the free HCl does not exceed 20 ml. N/10 HCl in any specimen is termed hypochlorhydria.

3. **Achlorhydria :** It is a condition in which free HCl is absent from all specimens.

Reagents :

1. N/100 NaOH.

2. **Topfer's indicator :** Dissolve 0.5 gms. of Topfer's indicator (dimethylaminoazobenzene) in 100 ml. of 95% alcohol.

3. Phenolphthalein indicator (1% solution in ethanol).

4. Dilute iodine solution (N/50).

A graphical representation of normal acidity :

In normal individuals there is found a small amount of free HCl in the resting contents. This is usually between 0 and 30 ml. of N/10 acid per 100 ml.

Concentration above 50 ml. N/10 acid probably indicates hyperacidity. The acid is also expressed in terms of unit (1 unit per litre = 1 ml. N/10 acid per 100 ml. juice).

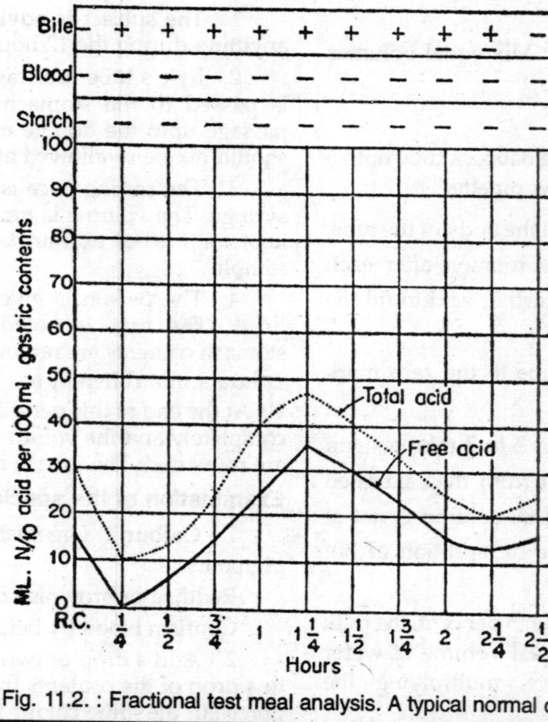

Fig. 11.2. : Fractional test meal analysis. A typical normal curve

After taking oat-meal, acidity falls immediately and starts rising in a quarter of an hour. In about $1\frac{1}{4}$ hour, acidity reaches its maximum. At this time, the free HCl in normal individuals is usually below 50 ml. N/10 acid per 100 ml. juice. The total acidity of the juice is about 10 ml. higher than the free HCl at all times.

In normal cases, the total acidity of the gastric contents is mainly due to the presence of free HCl and partly due to the presence of combined HCl and some organic acids such as lactic acid and butyric acid. Organic acids and combined acid are usually about 10 ml. of N/10 acid per 100 ml. Higher amount of these organic acids are found in cases of pyloric obstruction and gastric carcinoma.

Discussion

1. Gastric juice is collected after stimulating gastric secretion by giving oatgruel or by administration of histamine etc.

2. The chemical examination of gastric contents gives us a valuable information regarding the secretory and motor functions of stomach. The important secretory products of stomach are hydrochloric acid and the enzyme pepsin.

3. Hyperchlorhydria is associated with duodenal or gastric ulcer.

4. Hypochlorhydria is associated with pernicious anemia or gastric carcinoma.

5. Sometimes no hydrochloric acid is secreted in cases of gastric carcinoma and pernicious anemia even after histamine injection (achylia gastrica).

LIVER FUNCTION TEST :

Van den Bergh Reaction :

This consists of two parts—the direct (Qualitative) and the Indirect (Quantitative) reaction.

Reagents :

The diazo-reagent is prepared freshly before use by adding 0.8 ml. of solution B to 10 ml. of solution A.

Solution A: Dissolve 1 gm. of sulphanilic acid in 15 ml. of conc. HCl and make upto 1 litre with water.

Solution B : Dissolve 0.5 gm. of sodium nitrite in water and make upto 100 ml. Prepare freshly at frequent intervals.

The Direct Reaction :

Procedure : Pipette 0.3 ml. of serum in a test tube and add 0.3 ml. of diazo reagent. Mix and observe the colour change.

Results :

Direct Immediate Positive	:	Pink or Permanganate colour develops at the junction of two fluids, within *one minute*.
Direct Delayed Positive	:	Pink or Permanganate colour develops at the junction of two fluids, between 1 *and* 10 minutes.
Indirect Positive	:	The above colour does not develop in 10 minutes but the colour develops after addition of 1 ml. alcohol.
Negative	:	No colour develops even after the addition of alcohol.

The indirect reaction is done for the estimation of serum bilirubin.

Thymol Turbidity Test :

Principle : When serum mixes with thymol buffer, the degree of turbidity is measured colorimetrically.

Reagent :

Thymol buffer : Add 1.38 gms. of barbitone, 1.03 gms. of sodium barbitone and 3 gms. of thymol to 500 ml. of distilled water. Heat the mixture, shake well and cool thoroughly. It will be turbid. Add a small amount of powdered thymol crystals, shake, and allow to stand overnight at a temperature of 20° to 25°C. Shake well and filter. Store at about 20°C.

Procedure : Pipette 0.1 ml. serum into a dry test tube and add 6 ml. of thymol buffer. Allow to stand for $\frac{1}{2}$ hour. Compare with the set of Protein standards containing 10, 20, 30, 40, ... 100 mg. per 100 ml.

Result :

A turbidity equivalent to 10 mg. per 100 ml. protein = 1 unit.

A turbidity equivalent to 20 mg. per 100 ml. protein = 2 units.

A turbidity equivalent to 100 mg. per 100 ml. protein = 10 units.

The result is better obtained from a graph prepared by the protein standards. For photoelectric colorimetric reading, use red filter at 680 mμ.

Discussion :

1. The normal range by this test is 0-4 units. In, infective hepatitis, the thymol-turbidity is usually highest soon after the onset of the jaundice.

2. The developemnt of turbidity is due to liver disease. γ-globulin is responsible for the turbidity rise. β-globulin also takes part in the degree of turbidity. Normal serum albumin inhibits the effect of the γ-globulins.

Flocculation test :

This may be either qualitative or quantitative and involves one or more of the globulin fractions.

Procedure : Place 1 ml. of 0.9% NaCl solution into eight small tubes. Add 1 ml. of serum to the first, mix, transfer 1 ml. to the second tube, mix and proceed similarly along the row of eight tubes so that a series of dilution of serum from $\frac{1}{2}, \frac{1}{4}, \frac{1}{8}, \ldots \frac{1}{256}$ is obtained.

To each tube add 0.25 ml. of 10% Na_2CO_3 solution and 0.3 ml. of a solution made by mixing equal volumes of 0.5 per cent mercuric chloride and 0.025 per cent fuchsin in water.

Results : Normal serum only precipitates from tube 4 onwards. A precipitate in tube 3 or before that constitutes a positive reaction.

Discussion : The test is positive in cirrhosis of liver and is not specific for liver dysfunction.

PROTHROMBIN TIME

(Quick, 1939)

Principle : Prothrombin time is the time required for clotting to take place in citrated plasma to which optimum amounts of thromboplastin and calcium have been added.

Reagents :

1. Sodium citrate solution (3.8 per cent).
2. Thromboplastin (commercial preparation available in the market). Add 0.5 gms. of the dried powder in 5 ml. of 0.85% saline containing 0.5%. phenol and incubate for 15 mts. at 37°C. Centrifuge & remove the supernatant. Prepare serial dilutions 1 in 2, 1 in 4, 1 in 8 with the saline.
3. Calcium chloride (0.025 M) solution : Dissolve about 3 gms. of the anhydrous salt in water and make upto a litre. Standardize.

Procedure : Collect venous blood in a siliconized syringe and add 9 volumes of blood to one of citrate in a plastic container. Mix by inverting several times and centrifuge for 10 minutes.

Add 0.1 ml. of plasma to 0.1 ml. of thromboplastin prepared by the method mentioned above. warm the mixture for 15-30 seconds in a water bath to 37°C. Add 0.1 ml. of the 0.025M. $CaCl_2$ solution, also warm to 37°C using a graduated pipette and start stop watch.

Note the time required for coagulation.

Precaution : Glass containers should be avoided since contact activation of Factor VII shortens the prothrombin time.

Calculation : Prothrombin time is expressed in seconds. A dilution 1 in 2 has an activity of 50% and of 1 in 4 of 25%.

Discussion :

1. The normal prothrombin time is 10 seconds.
2. Prothrombin in formed by the liver cells and vitamin K being required. When bile salts are not present in the intestine, the absorption of vitamin K is impaired. So, in jaundice and liver cell disease, the prothrombin time is prolonged.
3. In obstructive jaundice, the prothrombin time is more. If adequate vitamin K is given parenterally, the prothrombin time returns to normal rapidly in uncomplicated obstructive jaundice, but in liver disease, the response is less.
4. Determination of prothrombin time is also used to decide whether there is a danger of bleeding at operation in biliary tract disease. In normal persons, the prothrombin acitivity is within the range of 70 to 100 per cent. Below this level (at 60%), bleeding may occur.

CHAPTER 12

BLOOD

ESTIMATION OF SERUM ALKALINE PHOSPHATASE AND ESTIMATION OF SERUM INORGANIC PHOSPHATES

(Method of King, Abul-Fadl, and Walker, 1951)

Principle : The enzyme phsophatases catalyse the splitting of phosphate group from certain monophosphoric esters. The enzyme activity is measured by the inorganic phsophate liberated in a given time at a particular temperature.

King and Armstrong have defined 1 unit (K.A. Unit) as the acitivity of the enzyme which will liberate 1 mg. of phenol or equivalent amount of phosphate (1/3 mg.) from phenyl phosphate substrate in 15 minutes at 37°C.

The inorganic phosphate liberated is estimated by reacting it with molybdic acid to form phosphomolybdate which on reduction by reducing agent such as 1 : 2 : 4 aminonaphthol sulphonic acid forms a blue coloured soluble compound (molybdenum blue). The blue colour thus produced is proportional to the amount of inorganic phosphate present.

Reagents :

1. **Sodium Carbonate bicarbonate buffer (0.1 N., pH_{10}) :** Dissolve 0.636 gm. of anhydrous sodium carbonate and 0.336 gm. of sodium bicarbonate in distilled water and make upto 100 ml. with distilled water. Store in a refrigerator.

2. **Substrate disodium phenyl phosphate (0.01 M.) :** Dissolve 0.218 gm. of disodium phenyl phosphate in distilled water and make upto 100 ml. Bring it quickly to boil, cool and add about 2 ml. chloroform and store in a refrigerator.

3. Trichloracetic acid, 20% (W/V).

4. Molybdic acid : Dissolve 5 gms. of ammonium molybdate in 100 ml. of 5N. H_2SO_2 (Add 14 ml. Conc. H_2SO_2 slowly to 86 ml. H_2O to prepare 5N. H_2SO_4).

5. **Aminonaphthol sulphonic acid solution (0.25 per cent) :** Add 0.25 gm. of 1 : 2 : 4 amino-naphthol sulphonic acid to 97.5 ml. of 15% sodium bisulphite and 2.5 ml. of the 20% sodium sulphite. Stopper and shake until dissolved. If the solid remains undissolved, add sodium sulphite solution, 0.5 ml. at a time, shaking well after each addition unitl complete solution is made. Store in the cold in a brown bottle. Renew every month.

6. **Stock phosphate standard (1 mg. per ml.) :** Dissolve 0.439 gm. of pure potassium dihydrogen phosphate (KH_2PO_4) in water and make to 100 ml.

7. Phosphate standard for use (0.004 mg. per ml.) : Dilute 0.4 ml. of stock to 100 ml. 5 ml. of standard contains 0.02 mg. phosphate.

Procedure :

Buffer + Substrate :

Test (T)	Control (C)	Blank (B)
(3 ml. + 3 ml.)	(3 ml. + 3 ml.)	(3 ml. + 3 ml.)

Put all the test tubes in a water bath at 37°C for about 5 minutes to attain the temp. Then add 0.3 ml. serum to 'T', mix for 30 seconds and place it into the bath. After exactly 15 minutes stop the reaction of the enzyme by adding 1.2 ml. trichloracetic acid which will cause the precipitation of all proteins.

Now, add 0.3 ml. serum to 'C' and immediately add to it. 1.2 ml. trichloracetic acid. Mix well.

After a few minutes, filter both 'T' and 'C' to have a clear filtrate.

Add 0.3 ml. water to 'B' and add 1.2 ml. trichloracetic acid to it and mix.

	Test (T)	Control (C)	Blank (B)	Standard (S)
Filtrate	5 ml.	5 ml.	5 ml.	..
Standard solution	..	..	..	5 ml.
Molybdic acid	0.8 ml.	0.8 ml.	0.8 ml.	0.8 ml.
Aminonaphthol Sulphonic acid	0.2 ml.	0.2 ml.	0.2 ml.	0.2 ml.

Mix well and keep these four tubes in the dark for 10 minutes to develop colour. Read the optical densities in a photoelectric colorimeter using red filter at 680 mμ.

Calculation :

Serum alkaline phosphatase per 100 ml. Serum

$$= \left(\frac{T-B}{S} \times \text{amount of inorganic phosphate in standard} \times \frac{100}{\text{ml. of serum used}} \times 3\right) \text{K.A. Units.}$$

[3 is the factor. Because 1 mg. of phosphate is liberted by 3 K.A. Units of the enzyme]

$$= \left(\frac{T-B}{S} \times 0.02 \times \frac{100}{0.2} \times 3\right) \text{K.A. Units.}$$

$$= \left(\frac{T-B}{S} \times 30\right) \text{K.A. Units.}$$

Serum inorganic phosphate per 100 ml. Serum.

$$= \left(\frac{C-B}{S} \times \text{amount of inorganic phosphate in standard} \times \frac{100}{\text{ml. of serum used}}\right) \text{mg.}$$

$$= \left(\frac{C-B}{S} \times 0.02 \times \frac{100}{0.2}\right) \text{mg.}$$

$$= \left(\frac{C-B}{S} \times 10\right) \text{mg.}$$

Note : 5 ml. filtrate contains 0.2 ml. serum and 5 ml. standard contains 0.02 mg. of inorganic phosphate.

Discussion :

1. Serum contains acid and alkaline phosphatases depending upon whether they are active in acid (pH_5) or alkaline (pH_{10}) medium, respectively. Alkaline phosphatase is found in a number of organs such as small intestine, bones, liver and kidneys.

2. The normal range for serum alkaline phosphatase by the King Armstrong method is from 3 to 13 units per 100 ml. (23-92 I.U./litre). With higher values for children, during periods of most active growth, usually upto 20 units, occasionally between 20 and 30 in infants.

3. In hypophosphatasia, serum alkaline phosphatase value is low, even 3 units per 100 ml.

4. There is an increase in alkaline phosphatase in rickets, osteomalacia and obstructive jaundice but a marked increase in rickets in children.

5. A moderate increase in alkaline phosphatase has been found in the impaired absorption of Vitamin D and of Calcium.

6. The normal serum inorganic phosphate ranges from 2.5 to 5 mg./100 ml. in adults and 4 to 6 mg./100 ml. in children.

7. Serum inorganic phosphate level is low in rickets, Osteomalacia and in hyperparathyroidism.

8. An increase in serum inorganic phosphate is found in hypoparathyroidism and in renal failure.

ESTIMATION OF BLOOD SUGAR

Besides blood glucose other reducing substances such as glutathione, ergathione, and glucuronic acid and its compounds are present in red cells. These non-glucose reducing substances are equivalent to 30 mg. glucose per 100 ml.

Several workers found that if the blood is put into an isotonic solution of sodium sulphate in which the cells are not hemolysed, diffusible substances pass into solution and the cell envelops which contain much of the non-glucose reducing substances are taken down with the precipitate.

Asatoor and King used isotonic sodium sulphate-copper sulphate with the addition of sodium tungstate to have the product copper tungstate which gives results near to the true glucose.

The high degree of accuracy in results very close to the true glucose has been obtained by the use of glucose oxidase which oxidizes glucose to gluconic acid without affecting any other sugar.

Various methods have been adopted for the estimation of blood sugar. The titrimetric method gives also better results. The original Folin-WU method still appears to be widely used. This technique gives results nearer to the true glucose content. When the blood sugar is abnormally low, it is better to use a more specific technique. Accordingly, a glucose oxidase method is preferred. Some of the titrimetric methods are well suited but they are not so sensitive as the colorimetric.

Preservation of blood : Blood is to be collected in a tube containing 6 mg. of Potassium oxalate as anticoagulant and 2 mg. of NaF. as antiglycolytic agent per ml. of blood. Mix thoroughly.

Glucose Oxidase method
(Method of Marks, 1959)

Principle : Glucose is oxidized by glucose oxidase to gluconic acid.

Glucose + H_2O + O_2 → Gluconic acid + H_2O_2.

The hydrogen peroxide formed is broken down to water and oxygen by a peroxidase in presence of an oxygen acceptor which is converted to a coloured compound which is read colorimetrically.

Reagents :

1. NaCl Solution (0.9 per cent).
2. $ZnSO_4$ Solution (5 per cent).
3. NaOH Solution (0.3 N).
4. O-Tolidine solution (1 per cent in absolute ethanol).
5. "Ferm Cozyme", a stable liquid preparation of glucose oxidase containing 750 units per ml. from Hughes and Hughes Ltd., London.
6. **Acetate buffer (0.15 M. pH5) :** Add approximately 3 volumes of 0.15M. acetic acid to 7 volumes of 0.15M. Sodium acetate (17.7 gms sodium acetate per litre). Adjust to pH5.
7. **Peroxidase :** 20 mg. per 100 ml. of acetate buffer. This is refrigerated for months.
8. **Glucose oxidase reagent :** Add 0.5 ml. of Fermocozyme to about 80 ml. of acetate buffer. Then add 5 ml. of the peroxidase solution. Mix and add 1 ml. of the O-Tolidine. Make to 100 ml. with buffer. Keep in the refrigerator in a dark bottle.
9. **Standard glucose solution :** Prepare a solution containing 100 mg. glucose per 100 ml. saturated benzoic acid and dilute with the acid to obtain solution containing 2.5, 5, 7.5 and 10 mg. per 100 ml. These are equivalent to 50, 100, 150, 200 mg. glucose per 100 ml.

Procedure : Add 0.4 ml. of 5% Zinc sulphate solution, 0.4 ml. of 0.3N. NaOH, 1.1 ml. of 0.9% NaCl and 0.1 ml. blood to a test tube. For low values use 1 ml. of saline and 0.2 ml. blood and for high values use 0.05 ml. blood and 1.5 ml. saline.

Mix well, centrifuge and transfer 1 ml. of the supernatant to a test tube. Measure 1 ml. water as blanks and 1 ml. of standard glucose solution equivalent to 200 mg. glucose per 100 ml.

Add 3 ml. glucose oxidase reagent to each at $\frac{1}{2}$ minutes intervals, mix gently for not more than ten seconds, and read the colour developed exactly 10 minutes later at 625 mμ or using an orange filter.

Notes :

1. Colour development reaches a maximum in about 10 minutes and then declines slowly. Therefore, reading should be taken immediately after 10 minutes.
2. This technique is the same for cerebrospinal fluid glucose estimation.

Calculation :

Blood glucose per 100 ml.

$$= \left(\frac{\text{Reading of unknown}}{\text{Reading of standard}} \times 200 \text{ mg.}\right)$$

Method of Folin and WU :

Principle : The protein-free blood filtrate is heated with alkaline copper sulphate solution, using a special tube to prevent reoxidation. The cuprous oxide thus formed is treated with a phosphomolybdic acid solution, a blue colour formed is read in a photoelectric colorimeter and compared with the standard.

Reagents :

1. Sodium tungstate solution (10%).
2. Sulphuric acid (2/3N).
3. **Alkaline Copper sulphate :** Dissolve 40 grams of anhydrous Na_2CO_3 in 400 ml. of water in a litre flask. Add 7.5 grams of tartaric acid and dissolve. Add 4.5 grams of crystalline copper sulphate, dissolve and dilute the solution to 1,000 ml.
4. **Phosphomolybdic acid :** Place 35 grams of molybdic acid and 5 grams of sodium tungstate in a 500 ml. beaker. Add 200 ml. of 10% NaOH and about 200 ml. of water. Boil for about 30 minutes (volume should be reduced to about 350 ml.). Cool and add 125 ml. of 85% phosphoric acid. Make upto 500 ml. with distilled water.
5. NaOH solution (10 per cent).
6. Phosphoric acid (85 per cent).
7. **Stock standard sugar solution (1 per cent) :** Dissolve 1 gram of pure dextrose in 100 ml. saturated benzoic acid (0.25%). Benzoic acid acts as a preservative.

Working Standards :

Std. 1 :	Dilute 1 ml. of stock standard to 100 ml. by saturated benzoic acid	(1 ml. = 0.1 mg.)
Std. 2 :	Dilute 2 ml. of stock standard to 100 ml. by saturated benzoic acid.	(1 ml. = 0.2 mg.)

Procedure : Take 0.5 ml. of oxalated blood in a hard glass test tube. Add to it 3.5 ml. of water, 0.5 ml. of soidum tungstate (10%), and 0.5 ml. of 2/3N. H_2SO_4. Mix by shaking. Wait for 5 minutes and then filter by Whatmann filter paper No. 1.

	Unknown (U)	Standard I (S_1)	Standard II (S_2)	Blank (B)
	2 ml. filtrate	2 ml.	2 ml.	2 ml. of water
Alkaline Copper Sulphate Sol. :	2 ml.	2 ml.	2 ml.	2 ml.

Keep the tubes in a boiling water bath for 8 minutes. Cool in a running water without shaking.

Phosphomolybdic acid	2 ml.	2 ml.	2 ml.	2 ml.

After about 1 minute dilute to the mark with distilled water and mix. Mixing should be done properly as the blue colour is formed in the bulb of the tube. Transfer the solution to suitable containers and take reading using red filter at 680 mμ.

Calculation :

Mg. of glucose per 100 ml. blood.

$$= \frac{\text{Reading of unknown}}{\text{Reading of standard}} \times \text{mg. of glucose in standard} \times \frac{100}{0.2}$$

$$= \frac{\text{Reading of unknown}}{\text{Reading of standard}} \times 0.2 \times \frac{100}{0.2}$$

[Conc. Standard_2 in 2 ml. = 0.2 ml.]

Discussion :

1. The normal fasting blood sugar level ranges from 60 to 100 mg. per 100 ml. After a carbohydrate meal or 50 grams glucose drinks, blood sugar level reaches a maximum in about 1 hour time and then it returns to normal fasting level in about $1\frac{1}{2}$ to 2 hours time. If the blood sugar does not return to near normal in 2 hours and if the difference shows 50 mg. above the fasting level, it indicates low glucose tolerance, as seen in diabetes mellitus.

2. In mild diabetes, values of 140 to 300 mg. are obtained, and in severe diabetes, values upto 1200 mg. per 100 ml. have been noted. Hyperglycemia is also observed in severe nephritis, pancreatic disease, hyperthyroidism, and in certain hepatic disorders. Low blood sugar values are found in insulin adminstration, Addison's disease, hypopituitarism, cretinism, myxedema.

ESTIMATION OF BLOOD UREA

(Urease Nesslerization method)

Principle : The enzyme urease present in soyabean splits urea into ammonia and carbondioxide which combine to form ammonium carbonate at an optimum temperature. This produces a compound after reacting with Nessler's reagent developing yellow colour.

Reagents :

1. Soyabean powder.

2. **Isotonic sodium sulphate solution :** Dissolve 3 gms. of $Na_2SO_4.10H_2O$ in distilled water and make upto 100 ml.

3. Sodium tungstate solution (10 per cent).

4. Sulphuric acid (2/3 N).

5. Gum Ghatti solution; Suspend 1 gm. of Gum ghatti, loosely tied in muslin, for twenty four hours in a litre of distilled water. Preserve the solution with a few drops of chloroform. Discard the residue in the muslin.

6. **Nessler's reagent :**

(a) Koch (1924).

A. Dissolve 22.5 gms. of iodine in a solution of 30 grams of potassium iodide in 20 ml. of water. When solution is complete, add 30 grams of mercury. Shake well, keeping cool by immersing in cold water from time to time, until the supernatant has lost all yellow colour due to iodine. Decant the supernatant and test a few drops with starch for the presence of iodine. If starch gives no colour, add an iodine solution in small quantities. Dilute to 200 ml. and mix.

B. Sodium hydroxide solution (10 per cent).

Mix 200 ml. of A with 975 ml. of B. Allow any precipitate to settle and use the clear supernatant.

(b) **Harrison** (1947) : Mix 150 ml. of A, 700 ml. of B and 150 ml. of water.

A. Dissolve 150 grams of potassium iodide in about 100 ml. of distilled water. Add 200 grams of mercuric iodide, when dissolved, dilute to about a litre. Filter and make to 2 litres.

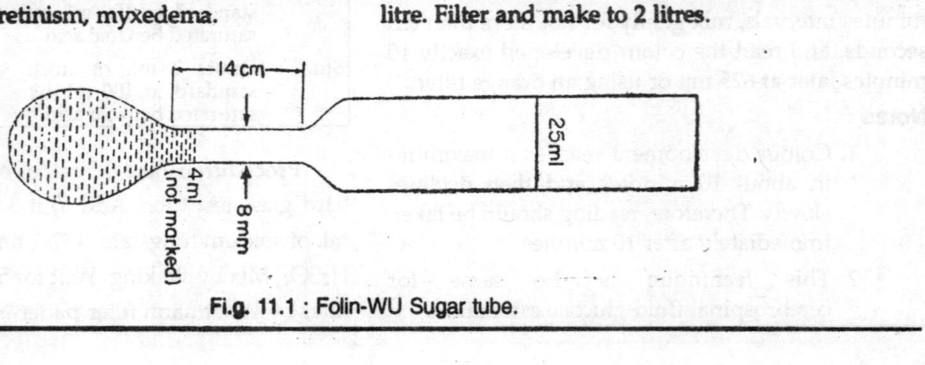

Fig. 11.1 : Folin-WU Sugar tube.

B. Sodium hydroxide solution (10 per cent).

7. **Stock ammonium sulphate solution standard :** Dissolve 2.2 grams of the pure dry salt in distilled water and make upto 1 litre.

Standard ammonium sulphate for use : Dlilute 5 ml. of the stock standard to 100 ml. with water. 1 ml. of this is equivalent to 0.05 mg. urea.

Procedure :

Add 0.2 ml. of blood to 3.2 ml. of isotonic sodium sulphate solution. Then add about 20 mg. of powdered soya bean, stopper the tube and incubate for at least 15 minutes at 40° to 50°C. Add 0.3 ml. of 10 per cent sodium tungstate and 0.3 ml. of $\frac{2}{3}$ N. H_2SO_4. Mix well, stand for a few minutes and centrifuge.

	Test (T)	Standard (S)	Blank (B)
	2 ml. Supernanatant	2 ml. Standard	7 ml. water
Water (ammonia free) or Gum ghatti	5 ml.	5 ml.	....
Nessler's reagent	1 ml.	1 ml.	1 ml.

Take Reading using blue filter at 480 mμ.

Calculation : 2 ml. supernatant is equivalent to 0.1 ml. blood.

Mg. of urea per 100 ml. of blood

$$= \frac{\text{Reading of unknown}}{\text{Reading of standard}} \times 0.1 \times \frac{100}{0.1}$$

$$= \frac{\text{Reading of unknown}}{\text{Reading of standard}} \times 100.$$

For accurate results, a curve should be drawn with different concentration of standard solution. From this, unknown can be determined.

Discussion :

1. The normal range of urea on a full ordinary diet is from 20 to 40 mg. per 100 ml. of blood. The urea content is influenced by the amount of protein in the diet and tends to be lower in people on low protein diets.

2. Blood urea is lower in preganancy, it is common between 15 and 20 mg. per 100 ml. Blood urea decrease is rare. In some cases of severe liver disease, the value is decreased.

3. Blood urea level rises in conditions in which excretory functions of kidneys are impaired such as in kidney diseases, in pre-renal conditions such as dehydration, shock etc., and in post-renal conditions such as obstruction in urine flow, enlarged prostrate gland, stricture of urethra due to stone in ureter etc.

ESTIMATION OF SERUM URIC ACID

(Method of Caraway)

Principle : Uric acid in blood reduces phosphotungstic acid in presence of Sodium Carbonate to blue coloured reduced phosphotungstate. This colour has been found to be proportional to the amount of uric acid present.

Reagents :

1. **Tungstic acid :** Add 25 ml. of 10% sodium tungstate, 25 ml. of 2/3 N. H_2SO_4 and drop of phosphoric acid with mixing to 400 ml. of distilled water. Keep in a brown bottle and store in refrigerator. Discard when cloudy.

2. **Phosphotungstic acid (Stock) solution :** Dissolve 50 grams of sodium tungstate in about 400 ml. of water. Add 40 ml. of 85% phosphoric acid and reflux gently for 2 hours. Cool, transfer to 500 ml. flask and make to the mark with water. Keep this in a brown bottle.

Dilute solution for use : Dilute 10 ml. of the stock solution to 100 ml. with water and store in a brown bottle. This keeps well for months.

3. Sodium Carbonate solution (10 per cent). Keep in a well stoppered polythene bottle.

4. **Uric acid standard solution (Stock) :** 10 mg. per 100 ml. Dissolve 60 mg. of lithium carbonate in 15 to 20 ml. of water in a test tube. Heat the solution to about 60°C and pour on to 100 mg. of uric acid taken in a small breaker. Stir until dissolved, heat further if necessary. When dissolved, tansfer with washing to a 100 ml. flask. Add 2 ml. of 40% formalin and then slowly with shaking, add 1 ml. of 50% (W/V) acetic acid. Make to the mark with water and mix. Keep in a well stoppered brown bottle away from light. This keeps at least for a year.

Uric acid standard for use : Dilute 1 ml. of the stock to 20 ml. with water. Store in a brown bottle. Renew every week. This contains 0.005 mg. uric acid per ml.

Procedure : Add 9 ml. of dilute tungstic acid to 1 ml. of serum in a centrifuge tube and mix well by inversion. Centrifuge for 5 to 10 minutes.

	Test (T)	Standard (S)	Blank (B)
	5 ml. of Supernatant	5 ml. standard	5 ml. distilled water
Sodium Carbonate solution	1 ml.	1 ml.	1 ml.
Dilute Phosphotungstic acid	1 ml.	1 ml.	1 ml.

Mix and place them in a water bath at 25°C for 3 minutes. Take readings in a photoelectric colorimeter using a red filter at 700 mμ.

Calculation : 5 ml. of the supernatant is equivalent to 0.5 ml. of serum.

Mg. of uric acid per 100 ml. serum

$$= \frac{\text{Reading of unknown}}{\text{Reading of standard}} \times 0.025 \times \frac{100}{0.5}$$

$$= \frac{\text{Reading of unknown}}{\text{Reading of standard}} \times 5.$$

Discussion :

1. Uric acid is formed endogenously from nucleoprotein metabolism and exogenously from the metabolism of purines taken in the food. The normal range of uric acid in serum is **3 to 9 mg.** per 100 ml.

2. The most important condition in which uric acid level is raised in blood is gout, a disease of the joints. Urates are deposited as crystals in the solid form in and around the joints causing swelling and pain. The increase in this disease is from 9 to 15 mg. per 100 ml. serum. In impaired renal function in which blood urea level is raised, the uric acid level ranges from 4 to 20 mg. per 100 ml. of serum. In gout only, no blood urea level will rise.

3. Increases in serum uric acid are also found in leukemia, toxemia of pregnancy, lobar pneumonia, and large abscesses. In these conditions, there is excessive breakdown of cells resulting in increased metabolism of nucleoproteins which are oxidized to uric acid in the body.

4. In rheumatic conditions, in which uric acid is normal, Cortisone tends to produce an increase in serum uric acid.

5. Low values are found in Wilson's disease and other cases of the Fanconi Syndrome.

6. Uric acid estimations give little information to the clinician except in cases of gout.

ESTIMATION OF SERUM (PLASMA) PROTEINS

(Biuret method)

Principle : This method is based on the sensitive biuret reaction given by all proteins including polypeptides. The purple colour produced by reacting protein with dilute copper sulphate in alkaline medium is proportional to the amount of protein present.

Reagents :

1. **Sodium sulphate solution (23.68 per cent) :** Dissolve 23.68 grams of anhydrous Na_2SO_4 in 80 ml. water and make upto 100 ml.

2. **Stock biuret reagent :** Dissolve 22.5 grams of Rochelle salt (Potassium sodium tartarate) in about 200 ml. of 0.2N NaOH and add 7.5 grams of Copper sulphate stirring continously unitl solution is complete. Add 2.5 grams of Potassium iodide and make upto 500 ml. with 0.2N. Sodium hydroxide.

3. **Biuret solution for use :** Dilute 50 ml. of stock biuret reagent to 250 ml. with 0.2 N. NaOH containing 5 grams of Potassium iodide per litre.

4. **Standard solution of Protein :** Prepare it from bovine albumin which will contain 0.5 gram protein per 100 ml. Measure accurately 2 ml. portions into tubes and store at —20°C until required. Take out one tube when required. Then add 3 ml. of biuret reagent directly.

Procedure :

1. For albumin estimation, precipitate the globulin with sodium sulphate solution and remove it by filtration.

To 6 ml. of 23.68 per cent solution of sodium sulphate add 0.4 ml. of serum (plasma), cork and gently mix by inversion several times. Final sodium sulphate concentration becomes 22.2%. At this stage, white turbidity is seen due to precipitation of globulin. Keep the test tube at 37°C in a water bath for at least 15 minutes (longer time is preferable) to allow formation of larger aggregate of globulin precipitate. Filter through Whatman filter paper No. 1 to get a clear filtrate which contains albumin.

2. For total protein estimation, dilute the serum (plasma) with sodium chloride solution.

Take 0.2 ml. serum into 3 ml. of 0.9 per cent NaCl solution, cork and mix gently by inversion several times.

3. Dilute 0.2 ml. of the standard with 3ml. of Saline solution.

	Total Protein (T)	Albumin (A)	Standard (S)	Blank (B)
	2 ml. dilute total protein	2ml. Albumin filtrate	2 ml. diluted standard	2 ml. water
Working biuret reagent	2 ml.	2 ml.	2 ml.	2 ml.

Mix and put all the tubes in a water bath at 37°C for 10 minutes. Take readings in the photoelectric colorimeter using a green filter at 540 mμ.

Calculation :

Grams of total protein in 100 ml. of serum (plasma)

$$= \frac{\text{Reading of total protein} - \text{Reading of Blank}}{\text{Reading of Standard} - \text{Reading of Blank}} \times$$

Grams of total protein in 100 ml. of Standard protein

Grams of albumin in 100 ml. serum

$$= \frac{\text{Reading of Albumin} - \text{Reading of Blank}}{\text{Reading of Standard} - \text{Reading of Blank}} \times$$

Grams of total protein in 100 ml. of Standard protein

Grams of Globulin in 100 ml. serum = [Grams of total protein per 100 ml. – grams of albumin per 100 ml.]

$$\therefore \text{ A : G ratio} = \frac{\text{Grams of Albumin Per 100 ml.}}{\text{Grams of Globulin Per 100 ml.}}$$

Discussion :

1. The chief proteins of serum (plasma) are albumin and globulin. Fibrinogen is also an additional fraction of the protein of plasma (about 300 mg/100 ml). The total protein in normal serum (plasma) varies from 6 to 8 grams per 100 ml.

Albumin .... 3.5 to 5.1 grams per 100 ml.

Globulin .... 1.8 to 3.1 grams per 100 ml.

Normally, the ratio of albumin to globulin lies between 2 : 1 to 1.2 : 1.

2. Liver cells synthesize albumin. A low serum albumin may be due to heavy loss of albumin in the urine, malabsorption of protein from the alimentary canal or decreased formation in the liver owing to liver diseases.

3. A decrease in total proteins is always due to a low albumin level accompanied by no increase in globulin. Therefore, there is a fall in albumin-globulin ratios.

4. An increase in total proteins may occur in dehydration. Both albumin and globulin are increased due to hemoconcentration and the albumin-globulin ratio remains unaltered.

5. A reduction in total protein occurs in edema in which more albumin is excreted in urine. The total protein is reduced with low albumin values in severe hemorrhage, shock, extensive burns, negative nitrogen balance due to increased protein breakdown, acute infectious diseases, untreated diabetes mellitus, hyperthyroidism.

6. In severe liver disease, although albumin is found to be low with a low total protein, it is more common for the globulin to be increased so that the total protein is normal or increased.

7. An increase in globulins may be observed in many chronic infections, rheumatoid arthritis, tuberculosis, Kala-azar, etc.

ESTIMATION OF SERUM CALCIUM

(Method of Clark and Collip, 1925)

Principle : Serum Calcium is precipitated as calcium oxalate on the addition of ammonium oxalate. Calcium oxalate, on treatment with Sulphuric acid, liberates equivalent amount of free oxalic acid which is then titrated against a standard Potassium permanganate solution in presence of H_2SO_4. The oxalic acid is oxidized to CO_2 and H_2O.

$$2KMnO_4 + 3H_2SO_4 + 5(COOH)_2 \rightarrow K_2SO_4 + 2MnSO_4 + 10CO_2 + 8H_2O$$

Reagents :

1. Ammonium oxalate solution (4 per cent).
2. Ammonia solution (2 per cent, V/V) : Dilute 2 ml. of ammonia to 100 ml. with water.
3. **Stock 0.1 N. Potassium permanganate solution :** Dissolve 0.3162 grams of $KMnO_4$ in 100 ml. distilled water. Store in a brown bottle at 4°C.

N/100 $KMnO_4$ solution : Dilute the stock before use.

4. **Stock N/10 oxalic acid :** Dissolve 0.6304 grams of dry oxalic acid in 100 ml. of distilled water.

N/100 oxalic acid : Dilute the stock ten times before use.

5. Sulphuric acid (1N).

Procedure : Take 2 ml. of serum, 2 ml. of distilled water, and 1 ml. of ammonium oxalate solution in a centrifuge tube. Mix with the help of a thin glass rod. Allow the tube to stand for at least 30 minutes. Centrifuge the tube for 10 minutes. A small amount of calcium oxalate can be found deposited at the bottom of the tube. Pour away the clear supernatant by inverting the tube and stand it on a pad of filter paper in the inverted position for 5 minutes for the liquid to drain completely.

Then wipe the mouth of the tube and run in 5 ml. of ammonia solution by the side of the tube. Stir the precipitate using the same rod. Centrifuge again. Pour off the supernatant and repeat the draining on the filter paper.

Add 2 ml. of 1N. H_2SO_4. Mix by the same glass rod. Put the tube in water to 70° - 75°C in a beaker to have free oxalic acid completely and titrate the contents with N/100 $KMnO_4$ keeping the tube in the water bath. A faint pink develops at the end point which persists for about a minute. Use a microburette for the titration.

For standard, take 1 ml. of N/100 oxalic acid and 2 ml. of 1N. H_2SO_4 and titrate with $KMnO_4$ solution keeping the tube in the water bath at 70°C - 75°C.

Calculation :

1 ml. of N/100 oxalic acid

= 1 ml. of N/100 Calcium.

$$= 20 \times \frac{1}{100} = 0.2 \text{ mg. Calcium.}$$

If 'X' ml. of permanganate is required to titrate the oxalic acid liberated from calcium, and 'Y' ml. for 1 ml. of N/100 oxalic acid,

∴ Calcium present in 2 ml. Serum

$$= \left(\frac{X}{Y} \times 0.2\right) \text{mg}$$

∴ Calcium present in 100 ml. Serum

$$= \left(\frac{X}{Y} \times 0.2 \times \frac{100}{2}\right) \text{mg}$$

$$= \left(\frac{X}{Y} \times 10\right) \text{mg}$$

Discussion :

1. Blood calcium exists in three forms—ionized (diffusible), protein bound, and complexed with citrate (non-diffusible). Total calcium is required for bone and teeth formation but ionic calcium is important for neuro-muscular irritability, blood clotting and tissue permeability.

2. Normal serum calcium ranges from 9 to 11 mg. per 100 ml. (4.5 to 5.5 m Eq/L). In children, the normal value tends to be towards the upper limit of the range.

3. Serum calcium level is influence by the level of its absorption from the intestine, by alterations in the amount of Parathyroid hormone, and changes in serum inorganic phosphate and serum proteins.

4. Lower values are found in hypoparathyroidism, rickets, osteomalacia, steatorrheas and renal failure.

5. Higher values are observed in hyperparathyroidism, hypervitaminosis D, and in myeloma (raised plasma protein level).

ESTIMATION OF SERUM CHOLESTEROL

(Sackett's method)

Principle : Cholesterol of serum is extracted by ether-alcohol mixture. The extract is evaporated and cholesterol is dissolved in chloroform. Cholesterol then reacts with acetic anhydride and sulphuric acid developing a green colour (Libermann-Burchard reaction) which is proportional to the amount of cholesterol present.

Reagents :

1. Alcohol (Rectified spirit).
2. Ether.
3. Chloroform.
4. **Stock Standard solution of Cholesterol :** Dissolve 50 mg. of pure cholesterol in chloroform and make upto 25 ml. with chloroform. Standard cholesterol for use (5 ml = 0.4 mg. of cholesterol) : Dilute 1 ml. of the stock standard to 25 ml. with chloroform.
5. Ethanol—ether mixture : Mix 3 volumes of ethanol and 1 volume of ether.
6. Acetic anhydride-sulphuric acid mixture : Mix 20 ml. of acetic anhydride with 1 ml. of concentrated sulphuric acid. Make freshly just before use.

Procedure : Take 10 ml. of ethanol-ether mixture in a centrifuge tube and add 0.2 ml. serum to it. Cork the tube tightly and shake vigorously. Allow the tube to lie horizontally so that the precipitate is fairly distributed evenly along the tube. Stand for 30 minutes and then centrifuge for a few minutes to get a firm deposit. Decant the supernatant completely into a suitable test tube. Evaporate to dryness in a water bath.

Dissolve the residue in 5 ml. of chloroform. Into another tube take 5 ml. of the standard cholesterol. Add 2 ml. of acetic anhydride-sulphuric acid mixture to each of the tubes. Mix and stand in the dark at 25°C for 15 minutes. In the blank, take 7 ml. chloroform.

Take reading using a red filter at 680 Mμ.

Calculation :

Mg. of cholesterol per 100 ml. serum

$$= \frac{\text{Reading of unknown}}{\text{Reading of standard}} \times 0.4 \times \frac{100}{0.2}$$

$$= \frac{\text{Reading of unknown}}{\text{Reading of standard}} \times 200.$$

Discussion :

1. Cholesterol, a derived lipid, is an important constituent of blood. It exists in free and esterified form in blood.

2. The normal range of cholesterol in adults is from 150 to 220 mg. per 100 ml. The value rises with age more in men than in women. In middle age, the value is raised. The value does not alter after meals. In pregnancy, the value rises by 20 to 25 per cent of the normal value.

3. Hypercholesterolemia is found most characteristically in nephrosis, diabetes mellitus, obstructive jaundice, myxoedema, and in Xanthomatosis. Values upto 400 and 500 mg. Per 100 ml. are commonly found in diabetes mellitus when treatment is inadequate. In myxoedema, values upto 500 to 700 mg per 100 ml. are seen which gives a valuable help in diagnosis. Thyroxine administration causes a marked fall in serum cholesterol.

4. The effect of hyperthyroidism is to reduce the serum cholesterol as low as 80 to 10 mg. per 100 ml. Sometimes in hemolytic jaundice, the value falls below 100 mg. per 100 ml. in some cases. Similar findings may be seen in malabsorption syndrome, in severe wasting, in acute infections.

ESTIMATION OF SERUM BILIRUBIN

(Lathe and Ruthven, 1958)

(*Method of Malloy and Evolyn modified)

Principle : Serum is diluted with water and methanol added in an amount insufficient to precipitate the proteins, yet sufficient to permit all the bilirubin to react with the diazo-reagent developing colour. This colour is proportional to the amount of bilirubin present.

Reagents :

1. Absolute methanol.

2. Hydrochloric acid (1.5 per cent, V/V with water).

3. **Diazo-reagent** : Prepare freshly before use by adding 0.3 ml. of solution B to 10 ml. of solution A.

Solution A : Dissolve 1 gram of sulphanilic acid in 15 ml. of 0.2 N. HCl and make upto 100 ml. with water.

Solution B : Dissolve 0.5 gm. of sodium nitrite in water and make upto 100 ml. Prepare freshly at frequent intervals.

4. Standard solution of Biliburin : Prepare a solution containing 10 mg. per 100 ml. chloroform. It may be necessary to reflux the mixture gently to dissolve the bilirubin.

Procedure : Add 0.2 ml. serum into 5.4 ml. of water in a test tube and mix. Pipette 2.8 ml. of this into a second test tube for use as the blank.

To the test add 0.7 ml. of diazo-reagent and to the blank 0.7 ml. of Sulphanilic acid solution. Mix, stand for 5 minutes and read using green filter at 540 mμ. This gives the *conjugated bilirubin*.

To obtain the *total bilirubin* add 3.5 ml. of methanol to each tube and read again after standing 5 minutes.

For *standard* add 0.2 ml. of bilirubin standard to 3.5 ml. methanol. Then add 0.7 ml. of diazo-reagent and after mixing 2.6 ml. of water and read against a water blank after 5 minutes at 540 mμ.

Note : For values above 15 mg. per 100 ml., 0.1 ml. of serum and 5.5 ml. of water can be used.

Calculation :

Mg. conjugated bilirubin per 100 ml. serum = $\frac{\text{Reading of unknown}}{\text{Reading of standard}} \times 10 \times 1.05$ conjugated bilirubin has a lower extinction in water so the factor 10.5 has to be inserted.

$$\text{Total bilirubin} = \left(\frac{\text{Reading of unknown}}{\text{Reading of standard}} \times 20\right) \text{mg.}$$

The difference of the two gives the indirect bilirubin.

Discussion :

1. The serum bilirubin of most normal persons appears to be in the range of 0.2 - 0.6 mg. per 100 ml. serum.

2. Serum bilirubin determination gives a measure of the intensity of the jaundice. Higher values are found in obstructive jaundice than in hemolytic jaundice. In obstructive jaundice when the obstruction is complete, the serum may reach about 20 mg. per 100 ml. Values between 20 and 30 mg. Per 100 ml. may sometimes be reached in acute infective and toxic hepatitis.

3. In pernicious anemia, the serum bilirubin rarely exceeds 3 mg. per cent. In the chronic hemolytic anemias, increases are also small and reaches upto 10 mg. Per 100 ml.

4. In sub-clinical jaundice, serum bilirubin increases between 1 and 3 mg. per 100 ml. which is of diagnostic value.

5. In obstructive jaundice, if the obstruction is due to some malignant growth, the serum bilirubin rises to a high level and remains almost constant at that level.

6. Determination of serum bilirubin is useful in jaundice of the newly born. High values upto 30 to 40 mg. Per 100 ml. may be reached especially in babies who have developed Rh antiboides.

ESTIMATION OF CREATININE IN BLOOD

Principle : Same as for creatinine in urine.

Reagents :

Working standard : 1 ml. of the stock solution (mentioned in urine creatinine) to 500 ml. (1 ml. = 0.002 mg. creatinine), other reagents are the same as in creatinine in urine.

Procedure :

	Unknown (U)	Standard (S)	Blank (B)
	5 ml. Folin-WU filtrate	5 ml. Standard	5 ml. water
Picric acid	2 ml.	2 ml.	2 ml.
NaOH	0.5 ml.	0.5 ml.	0.5 ml.

Allow to stand for 15 minutes and take readings in a photoelectric colorimeter using green filter at 540 mμ.

Folin-WU filtrate is prepared by taking 1 ml. of oxalated blood, 7 ml. water, 1 ml. sodium tungstate (10%), and 1 ml. of 2/3 N. H_2SO_4.

Calculation :

Mg. of Creatinine present in 100 ml. serum

$$= \frac{\text{Reading of unknown}}{\text{Reading of standard}} \times 0.01 \times \frac{100}{0.5}$$

$$= \frac{\text{Reading of unknown}}{\text{Reading of standard}} \times 2.$$

Discussion :

1. Creatinine is the least variable nitrogenous constituent of the blood. The normal range varies from 1 to 2 mg. per 100 ml. serum (90 to 175 m mol/L).

2. Values from 2 to 4 mg per 100 ml. are noted in early nephritis, and in chronic hemorrhagic nephritis with uremia, values from 4 to 35 mg per 100 ml. serum are seen.

3. Creatinine is more readily excreted by the kidneys than urea or uric acid and an increase of creatinine to 4 to 5 mg. or over per 100 ml. is the evidence of marked impairment of kidney function. Such high creatinine values in chronic hemorrhagic nephritis indicate an unfavourable prognosis.

IN VIVO TESTS OF IODINE UPTAKE

These determine the ability of the thyroid to take up iodine. After an oral dose of radioactive iodine in the form of inorganic iodide, the rate of accumulation of the isotope in the thyroid is measured using an external gamma-counter placed over the gland. I^{131} with a half-life of about 8 days has been mostly used but the use of I^{132} is increasing. This has a half-life of 2.6 hours so that repeat determinations can be carried out after a short period of time. The test is most useful for investigating hyperthyroid states in which uptake is high but is of less value in hypothyroid conditions.

DETERMINATION OF PROTEIN-BOUND IODINE

Protein-bound iodine includes T_3 and T_4 which are almost entirely protein bound.

Reagents :

1. Zinc sulphate solution, 100 gms $ZnSO_4$, $7H_2O$/I.

2. Sodium hydroxide, 0.5 mol/I. Ten ml. of the $ZnSO_4$ diluted to about 60 ml. with water.

3. Potassium hydroxide, 2 mol/I.

4. Sodium arsenite solution, 6.50 gms/I.

5. Sulphuric acid-hydrochloric acid mixture. Pour 98 ml. concentrated H_2SO_4 into about 350 ml. distilled water, cool, add 27 ml. Conc. HCl and make to 500 ml. with water.

6. Ceric ammonium sulphate solution. 12.65 gms. of ceric ammonium sulphate/I in 1.6 mol/I sulphuric acid.

7. Iodide standards. Dissolve 120 mg. sodium iodide in water and make to a litre.

Make all the solutions with double distilled water.

Procedure :

1. Measure 1 ml. serum in a centrifuge tube and dilute it with 7 ml. double distilled water.

2. Add 1 ml. zinc sulphate. Mix and add 1 ml. sodium hydroxide and mix well.

3. Allow to stand for 15 minutes and centrifuge for 10 minutes.

4. Decant the supernatant fluid, add 10 ml. distilled water. Stir properly. Centrifuge again and discard the supernatant. Carry out two further washings in the same way.

5. After pouring off the supernatant fluid, add 1 ml. KOH and stir and wash the rod with 0.5 to 1.0 ml. water.

6. Place the tubes overnight in an oven at 100° to 105°C to drive off water.

7. After thorough drying, ash in a muffle furnace. Place in the cold furnace, close the oven door to eliminate draught during heating.

8. Heat at 600°C for one hour. Open the door at 5, 25, 45 min. after reaching 600°C.

9. Remove and allow to cool to room temperature.

10. Centrifuge for 10 minutes after adding 10 ml. water.

11. Pipette duplicate 0.4 ml. serum of the clear supernatant fluid into pyrex test tubes.

12. To each add 0.5 ml. arsenite, then slowly add 1 ml. sulphuric hydrochloric acid.

13. Mix and place the tubes in a constant temperature bath at 37° ± 1°C for 10 minutes.

14. Warm the ceric ammonium sulphate in the same way, then at 1 min intervals add 1 ml of this to each of the tubes mixing quickly.

15. Read the extinction at 420 nm in a spectrophotometer using a 10 mm path length.

16. Run a reagent blank in duplicate with each set of tests using 1 ml. water instead of 1 ml. serum.

17. It is also useful to run a reagent blank with an internal standard, for example 1 ml. of the standard containing 160 nmol I_2/I (4.06 µg of I/100 ml.).

18. For a calibration curve put up a set of standards as follows :

Serum PBI (nmol I_2/I)	0	200	400	600
Serum PBI (µgI/100 ml.)	0	5.08	10.15	15.23
Standard (ml.)	0	0.5	1.0	1.5
Water (ml)	3.6	3.1	2.6	2.1
Add 0.4 ml KOH and proceed with the addition of arsenite etc as for the test				

Calculation :

Read the iodine concentration for the test from the calibration curve and subtract from this the blank value found. This is usually of 20 nmol I_2/I (0.5 µg I/100 ml). It should not exceed 40 nmol I_2/I. If it happens so, look for a source of possible contamination.

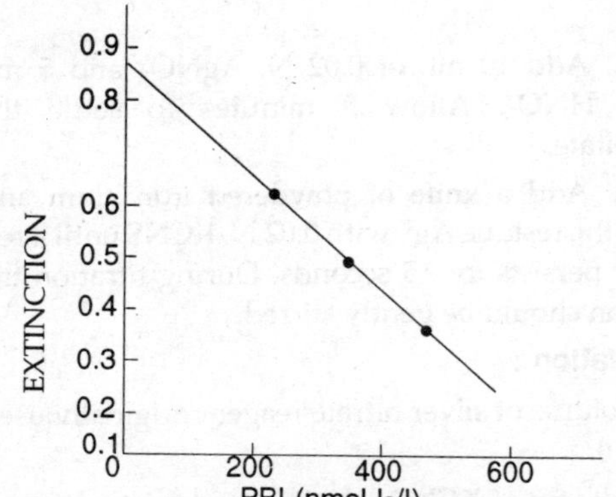

Fig. 12.2 : Calibration curve for protein-bound iodine

Interpretation of PBI

1. **Drugs causing low T_4 values.**

(a) **By inhibiting T_4 formation in the thyroid.**

antidiabetic drugs	— tolbutamide
antithyroid drugs	— Methimazole, thiouracil
anti TB drugs	— Isoniazid, PAS.
anticonvulsants	— aminoglutethimide
analgesic	— Tanderil
others	— Sulphonamides, Lithium salts.

(b) **By reducing the amount of TBG.**

anabolic steroids	— norethandrolone, oxymethalone
androgens	— testosterone and methyl testosterone.
corticosteroids	— cortisone

(c) **By displacing T_4 from binding sites on TBG.**

analgesics	— aspirin, butazolidin
anticonvulsants	— Epanutin, mephenytoin

2. **Drugs causing high T_4 values**

By increasing T_4 output from the thyroid	— Iodides, Lysergic acid derivatives.
By increasing the amount of TBG	— Diethylstilboestrol and ethanyloestradiol, the contraceptive pill

Hyperthyroidism :

1. The more marked degrees of thyrotoxicosis are reflected in increased levels of PBI or T_4.

2. The high T_4 concentrations are associated with high saturation of TBG and a reduction in binding sites for T_3 in the T_4 uptake methods.

3. When thyrotoxicosis is strong but the T_4 concentration and T_3 uptake are normal a likely diagnosis is T_3 thyrotoxicosis in which T_3 is produced preferentially. This is sometimes the early stage of the typical thyrotoxic state in which there is normally over-production of both T_3 and T_4.

Hypothyroidism :

1. When the clinical symptoms of myxoedema are apparent, the circulating hormone levels fall considerably. This decrease is associated with increased serum CK and cholesterol making the diagnosis almost certain.

2. Hypothyroidism occurring in the infant leads to critinism with impaired skeletal and mental development.

3. A group of genetic disorder involving an enzyme required for thyroid hormone synthesis found in childhood goitre with hypothyroidism. The disorders are defect in the iodide carrier, defect in coupling of iodotyrosine, defective release of T_3 and T_4.

4. In adult hypothyroidism the differentiation is between primary thyroid failure and malfunction secondary to pituitary insufficiency. Primary thyroid failure is associated with high incidence of thyroid antibodies, secondary hypothyroidism is associated with insufficiency of adrenocortical and gonadal function. In primay thyroid failure there is decreased excretion of 17-hydroxycorticosteroids and 17-oxosteroids.

5. The milder degrees of hypothyroidism ,are less easily detected by the simple thyroid function tests and may be difficult to diagnose clinically. The TRH–TSH test is currently the most satisfactory investigation. It is positive very early and can antedate the need for therapy.

6. The commonest form of non-toxic goitre is endemic goitre. Causes include iodine deficiency, natural goitrogens, constitutional factors, excessive iodine intake. The considerable increase in gland size is the consequence of the attempt to synthesise sufficient T_3 and T_4. The extra needs at puberty in the milder cases lead to enlargement at that age. Ir. most cases, the output to T_3 and T_4 are normal. In many of the cases the uptake of radio-iodine by the gland is increased. Thyroid antibodies are absent.

7. Thyroid nodules comprise adenomas and carcinomas. The majority metabolise iodine less than normal tissue. Some medullary carcinomas of the thyroid occur in families in association with bilateral phaeochromocytomas.

8. Increased TBG concentration occurs in pregnancy. Decreased concentration occurs in starvation and in nephrotic syndrome. The nephrotic syndrome causes the increased urinary T_3 and T_4 excretion.

DETERMINATION OF CHLORIDE IN BLOOD OR SERUM

Principle : Blood or serum is deproteinised by the Folin and WU method. Inorganic chloride is determined in the filtrate by the whitehorn procedure.

Qualitative tests for small molecules and ions in protein-free blood filtrate :

Preparation of Folin & WU : Dilute 6 ml. of blood with 42 ml. water in a flask. Allow the RBC to hemolyse. Add 6 ml. of 10% sodium tungstate solution (chloride-free) followed by 6 ml. of 0.66 N. H_2SO_4. Cork the flask and shake vigorously for one minute. Filter and divide the filtrate into several portions for the following tests :

1. **Biuret test :** If the filtrate is acid, neutralise before testing. If deproteinisation is properly done, biuret test will be negative.

2. Acidify a second portion with 1 ml. Conc. HNO_3. Add 1 ml. of $AgNO_3$ solution. Shake and allow to stand. A precipitate of AgCl should form. Decant the supernatant and confirm by dissolving the precipitate in NH_4OH solution.

3. The pH of the third portion of the filtrate is adjusted between 5 and 8. Test for amino acids by the ninhydrin reaction.

4. The fourth portion of the filtrate is to be made alkaline and apply a test for reducing sugar.

5. The fifth portion is tested for urea by urease test.

6. Test the sixth portion for inorganic phosphate.

Procedure :

1. Measure 2 ml. of blood (or serum) into a 100 ml. conical flask and add 14 ml. of distilled water. Allow to stand until the blood is hemolysed.

2. Add 2 ml. each of 10% sodium tungstate solution and 0.66 N. H_2SO_4.

3. Cork the flask and shake for one minute. Filter.

4. The filtrate contains 1:10 dilution of the original blood or serum.

5. Take 10 ml. of the filtrate into a porcelain dish.

6. Add 10 ml. of 0.02 N. $AgNO_3$ and 5 ml. Conc. HNO_3. Allow 5 minutes to settle the precipitate.

7. Add a knife of powdered iron alum and titrate the residue Ag^+ with 0.02 N. KCNS until a red colour persists for 15 seconds. During titration the solution should be gently stirred.

Calculation :

Volume of silver nitrate reagent originally used = 10 ml.

Volume of KCNS solution used for titration of remaining $AgNO_3$ = A ml.

∴ Volume of $AgNO_3$ reagent remaining unused = 2A ml. = B ml.

[Because 1 ml. KCNS = 2 ml. of acid $AgNO_3$ reagent].

∴ Volume of $AgNO_3$ used up by chloride present in 2 ml. of blood = (10 – B) ml.

∴ Chloride in 100 ml. of blood (or serum)

$$= \frac{(10 - B) \times 0.005 \times 100}{2}$$

[Since 1 ml. of acid $AgNO_3$ solution ≡ 0.005 gm. of NaCl]

$$= (10 - B) \times 0.25 \text{ gms. NaCl.}$$

Interpretation :

1. The thiocyanate solution should always be standardised against the standard $AgNO_3$ solution.

2. Hypochloraemia appears in the loss of electrolytes due to continued vomiting and gross watery diarrhoea.

3. Chloride analysis is very important in intestianl diseases, severe prolonged vomiting or severe burns as a guide to treatment with NaCl.

4. The higher concentration of the whole blood chlorides in severe anaemia is due to the high proportion of plasma analysed.

5. The normal range of whole blood chloride is 270–320 mg./100 ml.

6. The high values of whole blood chloride are found in some cases of nephritis, eclampsia, prostatic obstruction, cardiac disease, severe anaemia.

7. The low values of whole blood chloride are observed in intestinal obstruction, severe prolonged vomiting, severe burns, pneumonia, some cases of diabetes and fevers, addison's disease.

DETERMINATION OF CHLORIDE IN SERUM OR PLASMA BY SCHALES & SCHALES METHOD

Reagents :

1. **Mercuric nitrate solution (8.5 mmol/l) :** Dissolve 3.0 gms of $Hg(NO_3)_2$, $2H_2O$ in a few hundred ml. water, add 20 ml. 2 mol/l HNO_3 and make to one litre with water.

2. **Diphenylcarbazone indicator :** Dissolve 100 mg in 100 ml. ethanol. Store in the dark at 4°C.

3. **Standard chloride solution (100 mmol/l) :** Dissolve 585 mg. NaCl in water and make to one litre.

Procedure :

1. Add 0.2 ml. serum or plasma to 1.8 ml. water followed by 60 μl of the indicator.

2. Titrate with a mercuric nitrate solution using microburette.

3. Repeat the titration with 2 ml. of standard chloride solution; the expected titrant volume is about 2.3 ml.

4. The end point is better with protein-free solutions which give an intense violet-blue colour on adding the first drop of excess mercuric nitrate.

5. There is an immediate colour change to deep violet in case of protein-containing solutions. On adding more titrant this becomes pale-yellow or colourless until a sharp change to pale-violet denotes the end-point.

Calculation :

Plasma (or Serum) chloride (mmol/l)

$$= \frac{\text{ml. titrant needed for unknown}}{\text{ml. titrant needed for standard}} \times 100$$

The normal range of serum chloride is 96 to 106 mmol/l.

DETERMINATION OF SERUM TOTAL PROTEINS, ALBUMIN AND GLOBULIN FRACTIONS.

(By The Biuret Method)

Reagents :

1. **Sulphate-sulphite solution (278 gms/l) :** Dissolve 208 gms anhydrous sodium sulphate and 70 gms anhydrous sodium sulphite in about 900 ml. water to which 2 ml Conc. H_2SO_4 is added. Make it to a litre in a volumetric flask of one litre. Check that the pH is above 7.0. Keep it in a stoppered bottle at above 25°C – 37°C in an incubator.

2. **Stock biuret reagent :** Dissolve 45 gms Rochelle salt in about 400 ml. of 200 mmol/l sodium hydroxide and add 15 mgs copper sulphate stirring continuously until the solution is complete. Add 5 gms potassium iodide and make to a litre with 200 mmol/l sodium hydroxide.

3. **Biuret reagent for use :** Dilute 200 ml. of stock reagent to a litre with 200 mmol/l sodium hydroxide containing 5 mgs of potassium iodide/l.

4. **Tartrate-iodide solution :** Dissolve 9 gms Rochelle salt in 200 mmol/l sodium hydroxide containing 5 gms potassium iodide/l.

5. Ether (Analar).

Procedure :

Carry out all the steps up to the final colour development at a temperature above 25°C, otherwise sodium sulphate may crystallise out.

A. **Total protein :**

1. Pipette 6.0 ml. sulphate-sulphite solution into a centrifuge tube and add 0.4 ml. serum to it. Invert to mix.

2. At once remove 2 ml. of the mixture and add to it 5 ml. biuret reagent in a test tube.

B. **Albumin :**

1. Add about 3 ml ether to the rest of the serum sulphate-sulphite mixture, stopper and shake 40 times. Do not shake vigorously; otherwise the albumin may be denatured, centrifuge for 5 minutes.

2. Tilt the tube and insert a pipette into the clear solution below the globulin layer. Take care not to disturb the precipitate.

3. Pipette 2 ml. and add to it 5 ml. biuret reagent in a test tube.

In addition set up the following tubes :

C. **Serum blank :**

1. Ad 2 ml. serum - sulphate - sulphite mixture to 5 ml. tartrate-iodide solution and mix.

2. This should be put up whenever the serum is opalescent or abnormally pigmented.

D. **Biuret blank :**

Add 2 ml. sulphate-sulphite solution to 5 ml. biuret reagent.

E. **Standard :**

Pipette 0.4 ml. standard serum into 6.0 ml sulphate-sulphite solution as above and transfer 2 ml of the mixture into 5 ml. biuret reagent in a test tube.

F. **Standard serum blank :**

Prepare this as test serum (c) and this can be omitted with clear sera.

Shake, then place the tubes in a water bath at 37°C for 10 minutes. Allow to cool for 5 minutes at room temperature, then read at 555 ηm. Read the serum blank against the tartrate-iodide solution and the tests and standard against the biuret blank.

Calculation :

Serum proteins (gms/I)

$$= \frac{\text{Reading of unknown} - \text{Reading of unknown serum blank}}{\text{Reading of standard} - \text{Reading of standard serum blank}} \times \text{Concentration of Standard}$$

Total protein and albumin are obtained. Then difference between these gives the total globulins.

DETERMINATION OF PLASMA FIBRINOGEN

Collection of the fibrin clot :

1. Mix 0.5 ml. plasma, 14 ml. water and 0.5 ml. calcium chloride solution in a small beaker, place a fine glass rod in the liquid and allow to stand in an incubator at 37°C until a clot is formed.

2. Usually 30 minutes are sufficient for clot formation but it may take longer time if the fibrinogen levels are low.

3. Rotate the rod to collect the clot and press against the side of the beaker to squeeze liquid out of the clot.

4. Clotting may not occur in case of low prothrombin content even when plasma fibrinogen level is normal. In such cases clotting is initiated by adding 0.1 ml. thrombin (500 units/ml) to 0.5 ml. plasma and 10 ml. sodium chloride solution (9 gms/I). A fibrin clot forms within a few minutes.

Tyrosine Method

1. Wash the clot in water and then dissolve it in 5 ml. 250 mmol/I sodium hydroxide.

2. Neutralise it with 1 mol/I sulphuric acid and add 0.5 Folin-ciocalteu reagent followed by 3 ml. sodium carbonate solution (200 gms/I).

3. Make up to 25 ml. and place in an incubator at 37°C for an hour.

4. For standard use 0.5 ml. of a tyrosine solution (200 mg/I of 100 mmol/I hydrochloric acid) treated as the test.

5. Read at 680 nm with water as blank.

6. One mg tyrosine is equivalent to 16.4 mg fibrinogen.

Plasma fibrinogen (gms/I)

$$= \frac{\text{Reading of unknown}}{\text{Reading of standard}} \times \frac{200}{1,000} \times 16.4$$

$$= \frac{\text{Reading of unknown}}{\text{Reading of standard}} \times 3.28$$

COLORIMETRIC METHOD FOR TOTAL SERUM GLOBULINS USING HOPKINS–COLE REACTION

Method of Goldenberg and Drewes

Reagents :

1. **Glyoxylic acid reagent :** Add 400 ml. glacial acetic acid and 1 gm glyoxylic acid to a solution containing 1 gm copper sulphate dissolved in 90 ml. water. Mix at once, then slowly and carefully—while stirring—add 60 ml. concentrated sulphuric acid. After cooling to a room temperature make to 1 litre with glacial acetic acid and again mix. Keep it in a refrigerator which is stable at least a year.

2. **Standards :** Two are used.
   (a) A protein mixture containing 75 gms/I total protein made up of 45 gms/I of albumin and 30 gms/I of human gamma globulin.
   (b) A standard of N-acetyl-DL-tryptophan : Dissolve 1.75 gms in 10 ml. 750 mmol/I sodium hydroxide and dilute to a litre with water. This is equivalent to a serum globulin of 30 gms/I.

Procedure :

1. Add 4 ml. of the glyoxylic acid reagent to 20 µl serum in a test tube. Mix well, cap, and place in bath of boiling water for 4 minutes, then cool in a running water.
2. Treat the standard in the same way.
3. Read both against glyoxylic acid reagent at 550 nm.

Calculation :

Serum total globulins (gms/I)

$$= \frac{\text{Reading of unknown}}{\text{Reading of standard}} \times 30$$

INTERPRETATION

1. The concentration of total protein may be altered by changes in plasma volumes without altering the albumin-globulin ratio. An increase may occur due to dehydration and a decrease due to overloading with water.

2. Vigorous exercise also increases the total protein concentration by 5 to 10 per cent without changing the albumin-globulin ratio.

3. The increse in total globulins—most often of the γ-globulins—is observed in disease while the concentration of albumin remaining normal.

4. The concentration of total protein is decreased as a result of fall in albumin concentration while globulin concentrations do not change or increase by a smaller amount. Although the changes in albumin and globulin concentration may cancel eath other, the albumin-globulin ratio may range from below 0.5 to well over 2.5

5. Plasma albumin formation is reduced in case of vomiting and diarrhoea which may impair intestinal absorption of amino acids.

6. The plasma albumin level is decreased in excessive losses :
   (a) in the urine in nephrotic syndrome.
   (b) from wide-spread burns.
   (c) in defective anabolism due to reduced synthesis in liver disease, impaired intake of protein in malnutrition, defective digestion or malabsorption, increased catabolism of protein in fever and in severe hemorrhage.

7. The plasma albumin concentration is also reduced in insufficient protein intake, in carcinoma of stomach or pancreas, and in B group vitamin deficiencies.

8. Increased amounts of globulin are found in maternal plasma and amniotic fluid in pregnancies. The plasma globulin concentration is also increased in ulcerative colitis and Crohn's disease and in viral hepatitis.

9. Fibrinogen concentration is increased in pneumonia, in active rheumatic disease, and in tuberculosis. Increase in fibrinogen is also observed in the nephrotic syndrome and a slight increase in pregnancy.

QUANTITATIVE ANALYSIS OF BLOOD

Blood, a fluid containing red colour, circulates in blood vessels. It consists of erythrocytes, leucocytes and thrombocytes. The red colour is due to the presence of hemoglobin.

Blood of 0.1 to 0.2 ml. may easily be obtained from finger or ear. Larger amount (1–10 ml.) are taken from a suitable vein in the arm by means of a sterilised syringe. The suitable site of the skin is also made sterilised by swabbing with alcohol or ether. Blood should always be collected in an oxalated vial for prevention from clotting. Blood clots within 3 to 5 minutes after withdrawal from blood vessels. Plasma fibrinogen is converted to fibrin (clot) by the effect of thrombin and calcium ions (Ca^{++}). The fibrin afterwards contracts and the clot becomes small producing yellowish fluid known as *serum*. Fibrinogen and Ca^{++} are mainly involved in clotting. Therefore, clotting is prevented by removing Ca^{++} by the addition of sodium oxalate which forms precipitates of calcium oxalate or by the addition of sodium citrate which makes calcium non-ionisable. Defibrinated blood is that from which fibrin is removed as soon as it is formed by whipping the blood vigorously with a bunch of wires.

The substancces of blood may be classified into three main groups :

1. **Large organic molecules :**
   (a) A mixture of hemoglobin and oxyhemoglobin of RBC.
   (b) Serum proteins (Albumin and globulin)
   (c) Small amount of blood-clotting protein, fibrinogen.
   (d) Small amounts of enzymes together with leucocytes, blood platelets etc.

2. **Small organic molecules :**

(a) Glucose, various amino acids, fatty acids, α-keto acids, organic phosphates, steroids and vitamins.

(b) Urea, carbon dioxide.

3. **Inorganic constituents :**

(a) Ions of Sodium, Potassium, Calcium, Phosphate, Sulphate.

(b) Traces of iron, copper, and other metals.

DETERMINATION OF HAEMOGLOBIN

Principle : Heme is detached from the globin when haemoglobin is treated with alkali; iron is oxidised to the ferric state. The final solution of haematin is compared colorimetrically with a standard haematin solution.

Reagents :

1. **Haematin Standard :** Crystalline haematin (75 mg of pure haematin of 8.57% iron) is dissolved in 100 ml. sodium borate buffer and kept for 2 days near 0°C and then diluted to one litre with borate buffer. This solution gives the same colour intensity as blood of 14.8 gms Hb/100 ml., when diluted 1 in 100 with 0.1 N. NaOH and compared in a visual colorimeter using a green filter. The standard keeps for 6 months.

2. **Buffer :** 900 ml. of 0.05 M Borax (19.07 gms $Na_2B_4O_7, 10H_2O$ per litre) mixed with 100 ml. of 0.1 M NaOH.

Procedure :

1. Measure 9.9 ml. of 0.1 N. NaOH into a clean test tube.

2. Add 0.1 ml blood with pipette by washing it twice with the solution.

3. Heat in the boiling water bath for 5 minutes.

4. Cool and compare against the standard using green filter in the colorimeter.

Results :

The blood has been diluted 100 times but the standards are equivalent to similarly diluted blood of 14.8% Hb.

Express as grams of Hb/100 ml. of blood.

GLUCOSE TOLERANCE CURVE

1. The person undergoing the test should not take any lunch in the previous night and breakfast in the morning of the day on which the test for glucose tolerance is to be done.

2. 0.1 ml. blood should be taken from his vein.

3. The person is then to be given 50 grams of glucose, after dissolving it in water, to drink. The blood sugar level, therefore, rises rapidly. In time it falls to normal level again.

4. At intervals of 15, 30, 60, 90, 120 minutes blood samples should be taken and at the same time his urine sugar should be tested at the intervals of the above periods.

5. The results obtained from blood sugar determinations should be recorded by a graph.

6. The slope of the rising and falling portions of the graph indicate the capability of the person to utilise sugar.

7. In case of a diabetic person the additional sugar given during the test may cause the blood sugar level shoot up the renal threshold value resulting glycosuria. The renal threshold value varies from person to person.

(Add this at the end of the determination of all blood constituents).

PRECIPITATION OF SERUM PROTEINS

1. To about 3 ml. serum add an equal volume of saturated solution of ammonium sulphate. Shake well. A turbidity will be noticed due to the precipitation of *globulin*. Filter and test the filtrate again to have precipitation by the similar way. Wash the precipitate on the filter paper with water.

2. The fltrate from above experiment is saturated fully with ammonium sulphate crystals. Turbidity will again be seen due to the precipitation of *albumin*.

NON-PROTEIN CONSTITUENTS OF BLOOD OR SERUM

Mix 3 ml. serum with double the quantity of water. Boil and add 2% acetic acid drop by drop to the boiling solution until the coagulated protein appears as a coarse suspension. Filter and adjust the pH of the filtrate to 5.4 (tested by chlorphenol red indicator), boil again and filter if necessary. The protein-free filtrate can be used for the detection of the following substances :

1. **Test for chloride :**

Add 2 drops of conc. nitric acid and about 1 ml. silver nitrate to a part of the above filtrate. A white precipitate of silver chloride indicates the presence of chloride.

2. **Test for sulphate :**

Add a drop of conc. hydrochloric acid and 2 ml. barium chloride solution to another portion of the above filtrate. The formation of white precipitate of barium sulphate shows the presence of sulphate.

3. **Test for calcium :**

To another portion of the above filtrate add 1 ml. potassium oxalate. The appearance of white precipitate of calcium oxalate indicates the presence of calcium.

4. **Test for phosphate :**

Add 1 or 2 drops of conc. nitric acid and 1 ml. ammonium molybdate to the other portion of the above filtrate. Formation of lemon coloured precipitate after heating indicates the presence of phosphates.

5. **Test for glucose :**

Add 5 drops of glycerol, a large pinch of anhydrous sodium carbonate and 2 drops of 2.5% copper sulphate to another portion of the filtrate. Boil for one minute. A yellow precipitate of cuprous oxide indicates the presence of reducing sugar in blood.

BLOOD PIGMENTS

The haemoglobin of RBC and its derivatives such as haematin, haemochromogen etc. are called blood pigments. These can be detected by benzidine test. Oxygen, formed from hydrogen peroxide by the catalytic activity of blood pigments, oxidises benzidine to the blue-coloured compound.

Benzidine test :

Dilute a drop of blood with 10 ml. water. Boil and cool. Use this solution for this test.

Dissolve a knife point of benzidine powder in about 2 ml. glacial acetic acid in a clean test tube. Divide the solution into two parts. Keep one part as control. To the other part add a few drops of blood solution. Immediately deep or blue or greenish-blue colour will be seen. If the blue colour develops in the control, the test should be discarded and repeated.

The blue colour exists for a few minutes only and then changes to brown.

Questions to answer :

1. Discuss the principle for photoelectric colorimeter.
2. Why is photoelectric colorimeter termed as absorptionmeter and how will you select filters?
3. State the operation of photoelectric colorimeter, the importance of blank, and name the types of photoelectric colorimeter.
4. What precautions will you take to avoid hemolysis of blood?
5. Name the method and state the principle and procedure of estimation of creatinine in urine.
6. State the conditions in which creatinine excretion is increased.
7. Discuss the principle for the estimation of phosphate in urine and the conditions in which phosphate excretion is increased.
8. Name the method and state the principle of estimation of urea in urine.
9. State the composition of Benedict's quantitative solution and why is pumice stone added in the procedure of the estimation of reducing sugar in urine? Discuss the principle of Benedict's method for estimation of reducing sugar in urine.
10. What precautions should be taken in the estimation of chloride in urine? Name the method and state the principle for the chloride estimation in urine.
11. Discuss the precautions to be taken in the estimation of Vitamin C in urine.
12. State the procedure for the gastric analysis. What do you mean by hyperchlorhydria, hypochlorhydria and achlorhydria?
13. Discuss the liver function tests.
14. Name the method for the estimation of true glucose in blood. What is the normal level of fasting true glucose in blood? Under what condition blood glucose is increased and decreased?

CHAPTER 13

KIDNEY AND LIVER FUNCTION TESTS

Kidney functions in the elimination of waste products and toxic substances from the body. It regulates water balance and maintains normal crystalloid and colloid osmotic equilibrium between plasma and tissues. It also maintains normal acid-base equilibrium of the body.

The following tests are required to be done to assess the kidney function under diseased conditions :

1. **Concentration test :** This is based on the determination of the concentrating power of the kidney.

The patient is given high protein content meal with 200 ml. of fluids at 6 p.m. in the previous day. No food or drink is allowed till the test is completed. Empty the bladder at bed time. Discard this urine and any other sample passed in the night. Collect the urine sample at 8 a.m. and 10 a.m. separately. Determine the specific gravity of each sample.

In case the kidneys are normal, at least one specimen shows a sp. gr. above 1.022, often reaching 1.032. In hypofunction, the sp. gr. is nearer 1.010.

2. **Dilution test (Water test) :** In presence of edema this test should not be done.

The bladder is emptied early in the morning and urine discarded and breakfast not allowed. The patient is advised to drink 1,200 ml. of water in 20 to 30 minutes (at 8 a.m.). The hourly urine samples are collected from 9 a.m. to 12 noon. Measure the sp. gr. of each sample.

In normal persons, the water elimination is 1,200 ml. in 4 hours. The sp. gr. of at least one sample is 1.002 to 1.003. Water elimination is significantly decreased and the sp. gr. falls below 1010 in impaired state of kidney. Renal or extra renal (edema or cardiac failure) may be the cause of the decreased output. If the concentrating power is normal, the cause must be extra renal.

3. **Urea clearance test :** This test is a sensitive index of renal function in children as well as in adults. Under normal conditions urea excretion is constant and the blood cleared of urea per minute serves as a measure of the kidney function.

In the normal adult, the urea excretion is directly proportional to the level of urea in blood when 2 ml. or more is excreted per minute. This is called maximum clearance (Cm). If the urine output is less than 2 ml. per minute, it is clear that reabsorption of urea by the tubules is increased. When the urea excretion is diminished and varies with the square root of output of urine per minute, it is called standard clearance (Cs.).

Procedure : A light breakfast without tea or coffee is given on the day of test. Drug therapy is stopped a day before. The patient is advised to empty his bladder. Note the time. Allow him to drink 500 ml. of water to have free flow of urine. Collect blood for urea estimation at about 50 minutes. At 60 minutes collect all the urine. Measure the volume. Ask him to drink about 500 ml. of water more to drink. At 120 minutes collect separately all the urine and measure the volume. Estimate the blood urea nitrogen (urea) and urine urea nitrogen (urea) of the two urine samples.

Calculation :

$$Cm = \frac{U \times V}{B} \text{ ml.} \qquad Cs = \frac{U \times \sqrt{V}}{B} \text{ ml.}$$

whereas, U = mg. of urine urea nitrogen per 100 ml.

V = urine volume per minute in ml.

B = Mg. of blood urea nitrogen per 100 ml.

The clearance may be expressed in ml. but generally it is expressed in percentage of the normal clearance. The values for average normal Cm is 75 ml. (100%) and for Cs is 54 ml. (100%).

$$\therefore \text{ Per cent of Cm} = \frac{U \times V}{B} \times \frac{100}{75} = \frac{1.33UV}{B}$$

$$\therefore \text{ Per cent of Cs} = \frac{U \times \sqrt{V}}{B} \times \frac{100}{54} = \frac{1.85U\sqrt{V}}{B}$$

If the surface area of the patient varies widely from the normal (1.73 sq. m.), then the true 'V' is obtained by multiplying the observed 'V' by 1.73/A. Where A is the surface area in sq. m. This correction is necessary particularly in case of children.

Discussion : The clearance test is of great help in following the progress of advanced cases. The elimination or deposition of edema does not affect the test, so it is done in such cases.

Normal clearance in unimpaired kidney function is 75% or above (in Indians 40-70%). If all extra renal causes are excluded, a clearance of 10% or less indicates severe kidney damage.

The urea clearance test serves as an indicator of glomerular function. If the clearnace value falls, it is the indication that the glomeruli are destroyed. The concentration tests are very sensitive to slight degree of renal damage but they fail to show difference between moderate and severe renal impairment. If the concentration test is normal, further studies are not necessary. Urea clearnace test assures whether the decreased excreting power is serious or not. Excretory power and clinical condition follow the clearance test but not the specific gravity.

LIVER FUNCTION TESTS

A. Abnormalities in Bile pigments.

1. **Determination of serum bilirubin.**

See page 200.

2. **Icteric Index :**

(a) The icteric index assesses jaundice by measuring the intensity of the yellow colour of the serum.

(b) Serum is diluted with sodium chloride solution (9 g/I) and its colour is compared with a 1 in 10,000 solutions of potassium dichromate which has an icteric index of one. The normal range is from 4 to 6.

(c) Clinically, jaundice should be present when it is above 15.

(d) This test should not now be used as serum bilirubin can be determined simply and quickly to give a more useful result.

3. **Test for Bile pigments in urine :**

Fouchet's test :

(a) Add a few ml. $BaCl_2$ solution (100 gms./I) to about 10 ml. urine in a test tube.

(b) Filter, allowing to drain well. Spread the filter paper on another dry paper and add a drop or two of Fouchet's reagent.

(c) A greenish-blue colour due to the oxidation product of bilirubin is obtained if it is present.

[Fouchet's reagent is prepared by dissolving 25 gms. of trichloroacetic acid in 50 ml. water. Add 10 ml. ferric chloride solution (100 gms/I) and make to 100 ml. with water.]

4. **Qualitative tests for urobilinogen in urine :**

(a) To 10 ml. of fresh urine add 1 ml. Ehrlich's aldehyde reagent.

(b) Allow to stand for 3 to 5 minutes and note any colour produced.

(c) Normal urine gives only a pink colour but a distinctly red colour suggests the presence of increased amounts of urobilinogen.

[Ehrlich's reagent is prepared by dissolving 2 gms. of P-dimethylaminobenzaldehyde in 48 ml. Conc. HCl and 50 ml. water].

5. **Qualitative test for urobilin in urine and feces.**

Urine :

(a) To 5 to 10 ml. of urine in a test tube add two to three drops tincture of iodine to convert urobilinogen to urobilin.

(b) In another test tube shake about a gram of zinc acetate with a volume of absolute ethanol equal to the volume of urine taken.

(c) Mix both well by several pourings and filter.

(d) Urobilin gives a greenish-yellow fluorescence due to the formation of zinc urobilin.

(e) Examine spectroscopically. Zinc urobilin shows a band in the green at the junction of green and blue.

(f) Before applying this test it is advisable to remove other bile pigments by adding one-fifth volume of $CaCl_2$ solution (100 gms/I). Filter and carry out the test on the filtrate.

Feces :

(a) Extract a portion of faeces with a mixture of 20 volumes ethanol and 1 volume HCl.

(b) Mix thoroughly and allow to stand overnight or several hours.

(c) Examine the extract spectroscopically.

(d) To do this neutralize with NaOH and mix with an equal volume of ethanolic zinc acetate prepared as above and filter.

Interpretation

1. Urobilinogen is formed in the intestine by the reduction of bilirubin. Therefore, *fecal urobilinogen* depends on the amount of bilirubin entering the intestine. Fecal urobilinogen is increased in conditions in which there is increased breakdown of haemoglobin to bilirubin as in

haemolytic jaundice. The dark coloured stool found in these conditions are due to the large amount of urobilins present.

2. Faecal urobilinogen is less if there is obstruction to the entry of bile into the intestine. Low figures occur if the intestinal flora is disturbed by the administration of antibiotics.

3. In infective hepatitis even in high degree of jaundice, values for faecal urobilinogen are not nearly so low as in obstruction due to tumour. In cirrhosis the low values are due to impaired excretion of bilirubin by the liver.

B Abnormalities of bile salt metabolism :

Simple urine test :

Hay's test :

(a) Sprinkle a little finely divided flowers of sulphur on to the surface of 5 to 10 ml. urine cooled to below 15°C.

(b) Bile salts reduce the surface tension so that the particles of sulphur sink to the bottom of the liquid. Such urine also more easily form a froth on shaking.

Interpretation

1. In clinical practice bile salts are concerned in gallstone formation. Bile salt metabolism is altered in diseases of the liver, biliary tract or small intestine.

2. Cholelithiasis is associated with reduced formation of bile salt.

3. Cholestasis leads to an increased concentration of bile salts in serum.

4. In disease of the terminal ileum, there is failure of the enterohepatic circulation of bile salts and steatorrhoea develops with increase in faecal bile acids.

C. Tests based on carbohydrate metabolism.

Determination of blood galactose

Reagents :

1. Phosphate buffer (200 mmol/I, pH 7.5)

2. NAD (13 mmol/I; the equivalent of 9 mg. pure monohydrate per litre).

3. Galactose dehydrogenase (5 mg/ml).

4. Perchloric acid : Dilute 2.85 ml. perchloric acid with water to 100 ml.

Procedure :

1. Add 200 µl fresh blood to 1 ml. perchloric acid in a centrifuge tube sucking up and down several times and centrifuge.

2. Pipette into a spectrophotometer cuvette 3 ml. buffer, 0.1 ml. NAD^+, and 0.2 ml. supernatant.

3. Mix and read the extinction (E_1) at 340 nm.

4. Add 20 µl galactose dehydrogenase. Mix and when the reaction stops (30 to 40 minutes) again read the extinction (E_2).

5. To obtain the extinction due to the enzyme (E_e) mix 3.3 ml buffer and 20 µl enzyme and read against the buffer.

Calculation :

Blood galactose (mmol/I) = $(E_2 - E_1 - E_e) \times 15.6$ or × 280.5 as mg/100 ml.

Interpretation :

1. The normal blood value for galactose index is 0 to 6.1 mmol/I (0 to 110 mg/100 ml.)

2. In infective and toxic hepatitis galactose indices upto 28 to 33 mmol/I (500 to 600 mg./100 ml) are found.

3. In cirrhosis increased values are also obtained upto 28 mmol/I (500 mg/100 ml.) and over depending upon the severity of the disease: Increased values are also found in hyperthyroidism.

D. Tests based on the abnormalities of plasma proteins.

1. Thymol turbidity test :

Reagent :

Buffered thymol solution : Add 500 ml water to 1.38 grams of barbitone, 1.03 gms sodium barbitone, and about 8 gms of thymol. Heat the mixture just to boiling, shake well and cool thoroughly. It will be turbid. Add a small amount of powdered thymol crystals. Shake, allow to stand overnight at 20 to 25°C. Shake well and filter. Store in a slightly warm place at about 20°C.

Procedure :

1. Pipette 100 µl fresh serum into a dry test tube and add 6 ml. thymol buffer. Allow to stand for 30 minutes.

2. A turbidity similar to that given by a protein concentration of 0.1 g/I is said to be 1 unit, that of 0.2 g/I. 2 units and so on to 10 units for 1 gm./I. If the turbidity exceeds this, dilute with thymol buffer to bring the reading under 10 units and make the necessary adjustment in the calculation. Subsequently standards may be prepared from serum of known total protein content. The serum, diluted to contain 1g/I is used to prepare a set of standards as follows :

Thymol turbidity (Units)	0	1	2	3	4	up to	10
Diluted serum (ml)	0	0.2	0.4	0.6	0.8		2.0
Water (ml)	2.0	1.8	1.6	1.4	1.2		0
Add 6 ml. sulphosalicylic acid (30 g/I) and mix well.							

3. These and the test can be read in the absorptiometer at 680 nm.

Normal range : This is 0 to 4 units. The majority of persons give results between 1 and 2 units.

2. **Thymol flocculation test :**

(a) After measuring the turbidity leave the tube to stand in the dark overnight and read the flocculation. This may be graded : Negative (no flocculation); 1+, 2+, 3+, 4+.

(b) The flocculation test proves to be positive in some cases in which turbidity is normal and may thus increase the sensitivity of the test.

Interpretation

1. The γ-globulin has a precipitating effect as have some β-globulins. On the other hand, albumin has a protective action. So, increase in the γ-globulins and decrease in the albumin increase flocculation.

2. The tests are generally negative in post-hepatic jaundice, mainly in the early stages, and positive in those types of hepatic jaundice in which there is hepatocellular damage so they have been most used in differentiating infective and toxic jaundice from post-hepatic jaundice.

3. In infective hepatitis values between 10 and 20 units are commonly reached for the thymol turbidity and can be up to 40 units.

4. Positive tests are also obtained in a number of diseases other than those involving the liver in which there is an increase in γ-globulins. These include rheumatoid arthritis, chronic infections, myelomatosis, sarcoidosis, glandular fever.

E. TESTS BASED ON CONJUGATION AND EXCRETION.

Intravenous test : Because of impaired absorption of the benzoate and nausea formation in case of oral test, an intravenous form of the test has been used. 1.77 gms of sodium benzoate dissolved in 20 ml. water is given intravenously. Shortly before injecting this the patient empties the bladder and drinks a cupful of water. The bladder is then emptied one and two hours after the injection.

Determination of hippuric acid in urine.

The hippuric acid is precipitated from acidified urine by saturation with sodium chloride or ammonium sulphate. It is then filtered, washed, dissolved in hot water and titrated with sodium hydroxide.

Reagents :

1. Sulphuric acid (250 ml/I i.e. about 5 mol/I)
2. Solid sodium chloride.
3. Saturated sodium chloride solution.
4. Sodium hydroxide (500 mmol/I).
5. Barium chloride solution.
6. Phenolphthalein as indicator.

Procedure :

1. Add 3.0 gms. of chloride to 100 ml. urine in a flask and shake well to dissolve. Add 3.0 ml. sulphuric acid. The mixture should be acid to congo red.

2. Stir with a glass rod to encourage the precipitation of hippuric acid.

3. Stand at least half an hour. Then filter.

4. Wash several times with saturated NaCl solution until the filtrate gives no precipitate with $BaCl_2$ solution.

5. Dissolve the precipitate in hot water.

6. Titrate with NaOH using phenolphthalein as indicator.

Calculation :

1 ml. NaOH ≡ 0.072 gms. sodium benzoate or 0.061 gms benzoic acid. It is essential to do something for some hippuric acid which remains in solution when the urine is saturated with NaCl. For this add 0.1 gm/100 ml. urine. So,

$$\text{Equivalent sodium benzoate excretion (g)} = (0.072 \times \text{ml NaOH} + 0.1) \times \frac{\text{Urine Volume (ml.)}}{100}$$

Interpretation

1. With the oral test at least the equivalent of 3.5 gms. sodium benzoate (3.0 gms. benzoic acid) should be excreted as hippuric acid. Smaller amounts are found when there is liver failure. In infective hepatitis less then 1 gm. may be excreted.

2. With the intravenous test at least 0.85 gm. sodium benzoate (0.7 gm. benzoic acid) should be excreted as hippuric acid in the first hour or up to 1.36 gms. (1.15 gms. benzoic acid) in the two hours, with smaller amounts in liver failure.

Note : Dilute any specimen with a volume less than 100 ml. with water to 100 ml. and proceed as above.

F. PROTHROMBIN TIME

1. Prothrombin time is the time required for clotting to take place in fresh citrated plasma to which optimal amounts of thromboplastin and calcium have been added.

2. Prothrombin time is related inversely to the concentration not only of prothrombin but also of factors V, VII and X required for clotting and it can

be more sensitive to a lack of VII and X than of prothrombin.

Measurement of Prothrombin time

Reagents :

1. Sodium chloride solution (9 gms/I containing phenol 5 gms./I).

2. Sodium citrate solution (38 gms./I).

3. **Thromboplastin** : Prepare from human brain. Remove superficial vessels and meninges, mancerate with three to four volumes of acetone in a mortar. Decant off the acetone and repeat the manceration with further acetone four times. Dry the powder obtained on a suction filter. Keep the powder in screw-top bottles in a vacuum desiccator at 5°C when it will keep for at least a year.

4. Calcium chloride solution (25 mmol/I) : Dissolve about 3 gms of the anhydrous salt in water and make upto a litre. Standardise.

Procedure :

1. Collect venous blood and add nine volumes of blood to one of citrate solution in a plastic container. Glass containers should be avoided since contact activation of factor VII shortens the prothrombin time.

2. Mix by inverting several times and centrifuge for 10 minutes at 3,000 rpm.

3. Add 0.1 ml. thromboplastin to 0.1 ml. plasma.

4. Warm the mixture for 15 to 30 seconds in a water bath at 37°C.

5. Add 0.1 ml. $CaCl_2$ also warm to 37°C, quickly using a graduated Pasteur pipette, and start a stop watch.

6. Note the time required for coagulation.

7. Carry out the test in duplicate taking the mean of the two readings.

8. Determine the prothrombin time of normal plasma in duplicate in the same way.

9. Normal plasma clots in 10.5 to 12.5 seconds if the thromboplastin concentration has been correctly determined.

Calculation :

1. Results are expressed as patient's prothrombin time in seconds with the normal value given, as the prothrombin index in per cent which is the ratio of prothrombin time of the normal control to the patient's prothrombin time multiplied by 100.

2. Results may also be expressed as the prothrombin acitivty (i.e. the percentage dilution of normal serum with the sodium chloride solution which gives a prothrombin time equal to that of the patient.

3. For each batch of prothrombin a curve relating the prothrombin time to dilution is prepared using the mean of six normal plasmas.

4. A dilution of 1 in 2 will have a prothrombin activity 50 per cent, a 1 in 4 dilution of 25 per cent and so on.

Interpretation

1. Prothrombin is used in controlling anticoagulant therapy as well as used in liver disease and jaundice.

2. Vitamin K is required for the formation of prothrombin in the liver. So, prothrombin time is increased when there is a deficiency of vitamin K or when there is impaired liver synthesis of it.

3. In the malabsorption of fat or in the liver cell damage there is less formation of prothrombin.

4. Determination of prothrombin time is used to decide the danger of bleeding at operation in biliary tract disease.

5. In normal persons the prothrombin activity should be within the range of 70 to 100 per cent. Bleeding may occur below this level.

G. DETERMINATION OF BLOOD AMMONIA

Principle : Blood collected under anerobic conditions is transferred immediately to a trichloroacetic-hydrochloric acid mixture, centrifuged and the ammonia concentration determined in the supernatant using the Berthelot reaction.

Reagents :

1. **Ammonia-free water** : Stand over mixed bed ion exchange resin in a stoppered glass bottle.

2. Trichloroacetic acid-hydrochloric acid blank reagent : Dilute reagent$_3$ with an equal volume of water.

3. **Trichloroacetic acid-hydrochloric acid precipitating reagent** : Dissolve 100 mgs of trichloroacetic acid in 500 ml. hydrochloric acid, 1 mol/I and dilute to 1 litre.

4. Methyl orange indicator (Stock solution 1 g/I) : Dissolve 0.1 methyl orange in water and make up to 100 ml. Dilute 1 : 20 for use.

5. Sodium hydroxide (2.5 mol/I) : Dissolve 10 gms NaOH in water and make up to 100 ml.

6. **Phosphate buffer (pH_{12})** : Dissolve 9.5 gms. Na_3PO_4, $12H_2O$ or 3.55 gms Na_2HPO_4 in water and make up to 100 ml.

7. **Phenol-nitroprusside reagent :** 50 gms phenol and 0.25 gm nitroprusside/I. Dilute 1 to 5 for use.

8. **Alkaline hypochlorite reagent :** 25 gms of NaOH and 2.1 gms hypochlorite per litre. Dilute 1 to 5 for use.

9. **Stock standard ammonia solution (5 mmol/I) :** Dissolve 33.0 mg dry ammonium sulphate in water and dilute to 100 ml.

10. **Working standard ammonia solution (50 µmol/I) :** Dilute stock standard 1 : 100 with reagent 2.

Procedure :

1. Prepare a 1 : 1 extract of blood by adding an equal volume to an unknown volume of precipitating reagent. Stopper and shake. The blood should be transferred immediately after collection. The extract is then and there prepared after receiving. Centrifuge for 5 minutes at 3,000 rpm.

2. The determination of ammonia is best carried out in calibrated plastic tubes. After use they are well washed by ammonia-free water and air-dried.

3. Pipette in duplicate 1 ml. blood extract, 1 ml. working standard and 1 ml. acid blank reagent into separate tubes.

4. Add 1 drop working methyl orange indicator and titrate with sodium hydroxide solution until the colour changes to yellow; approximately 0.2 ml. is required.

5. To each tube add 1 ml. phosphate buffer, 1 ml. phenol-nitroprusside reagent and 1 ml. alkaline-hypochlorite reagent mixing well between each addition.

6. Make up to 5 ml. with water and finally mix by inversion.

7. Incubate at 37°C for 30 minutes and read at 635 nm against water.

Calculation :

Blood ammonia (µmol/I)

$$= \frac{\text{Reading of unknown} - \text{Reading of blank}}{\text{Reading of standard} - \text{Reading of blank}} \times 100$$

or × 140 as µg NH_4^+N/100 ml.

Interpretation

1. The normal range for this method is 19 to 37 µmol/I (26 to 52 µg NH_4^+N /100 ml.) for venous blood and 14 to 39 µmol/I (20 to 54 µg NH_4^+N /100 ml) for arterial blood compared with 28 to 53 µmol/I (40 to 75 µg NH_4^+N /100 ml) for the whole blood ammonia by the Conway method.

2. Ammonia is mostly formed in the body from nitrogenous material by bacterial action in the intestine, but is also formed in the kidney from glutamine, and in muscle from adenine. Metabolism of ammonia takes place mainly in the liver where it is converted to urea.

3. In liver disease, ammonia conversion to urea is impaired. Increases in ammonia can be found in more advanced cases of cirrhosis, particularly coma.

H. DETERMINATION OF GLUTAMINE IN CEREBROSPINAL FLUID

Reagents :

1. Sulphuric acid (100 ml. concentrated acid/I with water).

2. Sodium hydroxide solution (100 g/I).

3. Gum Ghatti solution (2 gms./I).

4. Nessler's solution : Mix 200 ml. A, 700 ml. B, 150 ml. water.

Solution A : Dissolve 22.5 gms iodine in a solution of 30 gms potassium iodide in 20 ml. water. Add 30 gms mercury. Shake well, keeping cool by immersing in cold water from time to time until the iodine colour is lost. Test it for starch until it becomes negative. Dilute to 200 ml. and mix.

Solution B : Sodium hydroxide solution (100 gms/I accurately) : Prepare a saturated solution (550 gms/I), stand for three days, dilute to 100 gms/I and standardise.

5. **Standard solution of glutamine (2 mmol/I) :** Dissolve 29.2 mg in water and make up to 100 ml. with water. Prepare freshly.

Procedure :

1. Add 0.2 ml H_2SO_4 to 1 ml. C.S.F. in a test tube.

2. Place in a boiling water bath exactly for 10 minutes and cool in water.

3. Add 0.3 ml. sodium hydroxide, 5 ml. water, 0.5 ml gum ghatti, and 2 ml. Nessler's solution.

4. Read the colour immediately in a colorimeter at 490 nm. Treat the standard and the blank in the same way, but omit the heating in boiling water with blank.

Calculation :

Cerebrospinal fluid glutamine (nmol/I)

$$\frac{\text{Reading of unknown}}{\text{Reading of standard}} \times 2 \text{ or } \times 29.2 \text{ mg/100 ml.}$$

Interpretation

1. The normal range was found to be 0.4 to 1.0 mmol/I (6 to 14 mg/100 ml.).

2. In nine cases of cirrhosis, values ranged from 1.1 to 2.1 mmol/I (16 to 31 mg/100 ml.) and in seven cases of hepatic coma from 2.15 to 3.7 mmol/I (30 to 54 mg/100 ml.).

3. In other types of coma normal values were obtained except in case of hemophilus meningitis which gave 1.23 mmol/I (18 mg/100 ml).

I. SERUM ENZYMES IN LIVER DISEASE AND JAUNDICE

1. **Estimation of serum alkaline phosphatase.**

See page 192.

2. **Estimation of Aspartate and Alanine Aminotransferases (Method of Wilkinson et. al)**

Reagents :

(a) **Phosphate buffer (1 mol/I) :** Dissolve 136 gms. KH_2PO_4 and 33 gms sodium hydroxide in water and make to a litre.

(b) **Phosphate buffer (100 mmol/I) :** Dilute 100 ml. reagent (a) to a litre and adjust the pH to 7.4 at 25°C.

(c) **Reduced NAD (2.5 mg/ml in reagent (b)) :** Prepare small quantities freshly before use.

(d) **2-Oxoglutarate (100 mmol/I) :** Dissolve 730 mg 2-oxoglutaric acid in 35 ml. water, add 5 ml. phosphate buffer reagent (a) and adjust to pH 7.4 at 25°C by adding 1 mol/I sodium hydroxide, then make to 50 ml. with water.

(e) **L-Aspartate (375 mmol/I) :** Dissolve 5 gms L-aspartic acid with gentle warming in 50 ml. water plus 35 ml 1 mol/I sodium hydroxide, cool to room temperature, add 10 ml phosphate buffer reagent (a), adjust to pH 7.4 at 25°C with 1 mol/I sodium hydroxide and make to 100 ml.

(f) **L-Alanine (750 mmol/I) :** Dissolve 6.67 gms. L-alanine in 75 ml water and 10 ml. phosphate buffer reagent (a), adjust to pH 7.4 at 25°C with 1 mol/I sodium hydroxide and make to 100 ml.

Reagents d, e, f may be kept for one month at 4°C.

(g) **Malate dehydrogenase (5 U/ml.) :** Malate dehydrogenase in glycerol and not in ammonium sulphate should be used to avoid glutamate dehydrogenase interference.

(h) **Lactate dehydrogenase (5 U/ml) :** Dilute the rabbit muscle enzyme in glycerol with water to give sufficient for the day's need.

Procedure :

Incubate the followings for 15 minutes at 25° ± 0.2°C.

(a) **For aspartate aminotransferase :** 1.3 ml. phosphate buffer reagent (b), 0.2 ml. serum, 0.2 ml. NADH solution, 1.0 ml L-aspartate and 0.1 ml. malate dehydrogenase suspension.

(b) **For alanine aminotransferase :** 1.3 ml. phosphate buffer reagent (b), 0.2 ml. serum, 0.2 ml. NADH solution, 1.0 ml L-alanine, and 0.1 ml lactate dehydrogenase.

Transfer to a 1 cm cuvette of an ultraviolet spectrophotometer, set at 340 nm with the cuvette temperature controlled at 25° ± 0.2°C. Begin the reaction by adding 0.2 ml. 2-oxoglutarate, mixing carefully. There is usually a lag phase lasting 2 to 5 minutes. After this read at least at minute intervals for 10 minutes to obtain ΔE_{340}/min. Such readings should be made only during a period while zero order kinetics are being obeyed.

Calculation :

Serum AST or ALT activity (U/I)

$$= \frac{\Delta E_{340}/\text{min} \times 3.0 \times 10^3}{6.3 \times 0.2}$$

$$= \Delta E_{340}/\text{min} \times 2380$$

Normal ranges :

Aspartate aminotransferase (AST) : Males 5-15, Females 5-12 U/I.

Alanine aminotransferase (ALT) : Males 5-19, Females 5-12 U/I.

Notes : Pyridoxal-5-phosphate is the coenzyme for aminotransferase activity. It can affect the values of AST and ALT activity. Serum from a case of severe deficiency of the B group of the vitamins must require addition of coenzyme to produce maximal aminotransferase activity. In such rare cases it is required to run the test with such addition to give final concentrations in the substrate mixture of 0, 1 and 10 mg/I.

3. **Estimation of Red Cell Acetyl Choline esterase**

Reagents :

(a) **Buffer :** Dissolve 4.12 gms sodium barbitone, 540 mg anhydrous potassium dihydrogen phosphate and 44.73 gms potassium chloride in about 900 ml. water, adjust to pH 8.1 at 25°C with 100 mmol/I hydrochloric acid and make to a litre with water. This can be kept at 4°C.

(b) Sodium chloride solution (9 gms/I).

(c) **Substrate acetylcholine chloride (20 gms/I)** : : Keep 5 ml aliquots preserved with a few drops of toluene in sealed ampoules at 4°C. Sigma ampoules containing 150 mg. can be dissolved in 7.5 ml water.

(d) Saponin solution (100 mg/I).

Procedure :

1. For preparing the haemolysate, centrifuge 5 ml. heparinised blood in a calibrated centrifuge tube at 3,000 rpm for 10 minutes. Discard the supernatant plasma and wash the cells with NaCl solution. After the final washing discard the supernatant and add a volume of NaCl solution equal to that of the red cell deposit. Mix thoroughly and haemolyse 0.4 ml suspension in 9.6 ml saponin solution.

2. Measure 3 ml haemolysate and 3 ml. buffer into one tube (the test) and buffer only into a second tube (the blank) and incubate for 10 minutes at 25°C and measure the initial pH of both.

3. Add 0.6 ml. substrate to each tube, mix, note the time, and at the end of an hour again measure the pH. The pH of the blank should not have fallen more than 0.01, if it has, prepare fresh buffer.

Calculation :

Δ pH = Initial pH – final pH.

So units acetylcholinesterase = ΔpH × 100.

Normal range : 0.5 to 1.0 units (mean 0.75).

Interpretation

1. The activity of serum alkaline phosphatase is increased in cholestasis. The highest level of it is found in hepato-cellular jaundice. Increase in serum alkaline phosphatase are a poor index of liver cell damage.

2. Increase in both AST and ALT are commonly found in liver disease with particularly in infective hepatitis, the increase in ALT is greater than that in AST.

3. Trauma to skeletal muscle can produce an increase in serum AST activity. Thus, high values are found when there is myoglobinuria.

4. Red cell acetylcholinesterase enzyme is formed in the liver and its activity is reduced in liver cell damage. Normal values are generally seen in obstructive jaundice. Low values are found in uremia and shock, in anaemia, tuberculosis and cancer and also in pregnancy.

5. Serum cholinesterase activity is also reduced in poisoning by organophosphorus compounds used in insecticides.

CHAPTER 14

QUANTITATIVE ANALYSIS OF CEREBROSPINAL FLUID (CSF)

Appearance :

Normal CSF is clear, colourless and gives no coagulum or sediment on standing.

Colour :

1. Normally no red blood cells should be present. If some is present as a result of trauma, the first ml or two of the fluid is to be collected separately. The subsequent fluids are almost clear.

2. Pathologically, blood may be present in haemorrhage into the ventricles and also after neurosurgical operations.

3. Sometimes haemolysis occurs and haemoglobin liberated is converted into bilirubin, so within a day after such a haemorrhage, the supernatant fluid becomes coloured on centrifugation.

Turbidity :

Turbidity is seen when there is a marked increase in the number of polymorphs present and is found in meningitis, more specially in the coccal forms.

Coagula :

1. On standing, fibrin clots may form in pathological fluids containing fibrinogen which is usually only found when there is a considerable increase in protein.

2. In tuberculous meningitis a fine web-like coagulum sometimes forms if the fluid is allowed to stand overnight. Such a web may take up tubercle bacilli.

Cell Count :

1. The cells may be red cells or various forms of leucocytes which may be classified as the lymphocytes, large mononuclear, plasma cell, neutrophil and eosinophil.

2. Occasionally fungi may be visible during the cell count.

3. The presence of red cells indicates haemorrhage into the CSF. An increased number of white cells indicates meningeal irritation. This is most marked in acute meningitis.

4. In bacterial infections the cells are of polymorphs in numbers up to $2,000 \times 10^6$/I. In viral menigitis, the response is mainly lymphocytic with cell counts in the range of 50 to 250 and occasionally up to $1,500 \times 10^6$/I.

5. Acute encephalitis produces only small increases in cells, usually mononuclears.

6. In tubercular meningitis, the response is lymphocytic or mononuclear in the range of 10 to 350×10^6/I.

Total Protein

A. Determination of Total Protein :

Reagents :

1. **Sulphosalicylic acid—sodium sulphate solution :** Dissolve 30 gms sulphosalicylic acid and 70 gms anhydrous sodium sulphate in water and make to one litre. Filter and store in a dark bottle. Discard if any colour or turbidity develops.

2. Sodium chloride solution (8.8 gms/I).

Procedure :

1. Add 0.5 ml centrifuged CSF to 3.5 ml sulphosalicylic acid reagent. Mix immediately and read after 10 minutes against an acid blank in a colorimeter at 660 nm or using a red filter.

2. To prepare a standard curve, dilute serum of known protein content with saline to obtain a solution containing 2 gms/I. Set up a series of tubes as follows :

CSF total protein (g/I)	0.	0.20		0.60	0.80	1.00	1.20	1.60	2.00
Dilute serum (ml)	0	0.1	0.2	0.3	0.4	0.5	0.6	0.8	1.0
Saline (ml)	1.0	0.9	0.8	0.7	0.6	0.5	0.4	0.2	0
Add 0.5 ml to 3.5 ml acid reagent and read as above									

Inerpretation

1. Normal lumber CSF has a protein content of 0.15 to 0.45 g/I, although most are in the range 0.25 to 0.30 g/I. The average figures for cisternal and ventricular fluids are 0.20 and 0.10 g/I, respectively. More than half the protein is albumin.

2. An increase in total protein results from the breakdown of the blood CSF and brain-CSF barriers usually in inflammatory reaction.

3. In many cases the concentration of protein does not exceed 1g/I though increase of up to 4 gms/I are not infrequent in acute menigitis, in polyneuritis and meningioma.

4. Small increases in cells and protein may occur in multiple sclerosis.

Globulins

Although some changes in globulins other than gamma have been associated with neurological disease they have not been consistent and have not been used routinely for diagnostic work. Most interest has centred on the γ-globulin fraction.

B. Qualitative test for γ-globulin.

Reagent :

Pandy's solution : Dissolve 10 gms phenol in 150 ml. water. The reagent should be clear and colourless.

Procedure :

Add 2 drops of CSF to 2 ml reagent. Most normal fluids remain clear and none show more than the faintest degree of opalescence. The test is reported as showing no opalescence, slight opalescence, opalescence, marked opalescence and turbidity.

Interpretation :

1. The test shows a rapid approximate comparison to be made between the γ-globulin increase and that of total protein.

2. γ-globulin increases are great in multiple sclerosis and neurosyphilis; a positive test may occur when the total protein concentration is still within normal limits.

3. The presence of red cells in CSF gives a slight opalescence and any cell should be removed by centrifugation.

C. Determination of Glucose

See page 193.

The glucose concentration in CSF is lower than in blood, and the most important abnormality is a further reduction in glucose content, larger volume CSF is often used than for blood.

Note : 1. CSF which has become contaminated during laboratory sampling may show a fall in glucose content if kept at room temperature.

2. Streptomycin given intrathecally shortly before CSF sampling may interfere with copper reduction methods.

Interpretation

1. The glucose content of normal lumbar fluid is usually between 2.8 and 4.4 mmol/I (50 to 80 mg/100 ml.) though a range of 2.5 to 5.6 mmol/I (45 to 100 mg/100 ml) is often allowed. It is thus slightly lower than the glucose content of blood. Ventricular fluid approximates to that of blood.

2. The decreased glucose value is found in meningitis. In bacterial meningitis the glucose may disappear completely but in tuberculous meningitis it is usually between 10 to 40 mg/100 ml. In viral menigitis the glucose concentration is often normal but is occasionally as low as 20 mg/100 ml.

3. Small increases are found in the mild hyperglycemia which accompanies some infections but are of little diagnostic value.

D. Determination of Chloride

See page 203.

Interpretation

1. The chloride content of normal lumbar, cisternol or ventricular CSF lies between 120 and 130 mmol/I and is thus appreciably higher than the plasma chloride.

2. In meningitis there is a decrease usually to between 120 and 102 mmol/I. If there is much vomiting, particularly in tuberculous meningitis before specific treatment becomes available, the plasma chloride falls and the CSF value follows it. The lowest figures are seen in untreated tuberculous meningitis.

3. An increased value is sometimes found in the hyperchloremic acidosis of chronic renal failure.

E. Other Constituents.

These are rarely investigated in routine practice.

1. The **bicarbonate** concentration at 18 mmol/I is lower than in plasma to compensate for the increased chloride value and normal sodium and potassium concentrations. Too rapid replacement of a bicarbonate deficit in plasma may adversely affect the CSF pH.

2. The urea concentration is similar to that of plasma unless the latter is changing rapidly when the CSF response is delayed. The CSF urea value may be helpful in assessing renal function when collected at postmortem as it is less affected than the plasma figure by agonal changes.

3. The calcium is much lower than the serum figure because of the lower protein concentration. It falls if the ionic calcium value in serum is reduced as in tetany.

4. The phosphate level is somewhat lower than plasma but is unhelpful in diagnosis.

5. The copper concentration is increased in Wilson's disease as expected.

6. Urate is present in only low concentration, 35 μmol/I (0.20 mg/100 ml) but is raised in progressive cerebral atrophy.

7. Lipids are also found in low concentration and the cephalin fractions increase in demyelinating disorders.

8. Vitamin B_{12} concentrations reflect those of plasma, are low in pernicious anaemia.

9. Folate figures, 200 to 250 μg/I, are about five times the plasma figures but give no extra diagnostic information.

10. AST is increased in metastatic carcinoma. The enzyme CK is increased in Duchenne muscular dystrophy and in many other neurological disorders.

CHAPTER 15

ELECTROPHORESIS, CHROMATOGRAPHY AND pH PAPER ELECTROPHORESIS OF SERUM PROTEIN

Introduction : Electrophoresis is a method by which electrically charged particles of a mixture of substances are separated under an electric field. The positively charged particles migrate towards cathode, and the negatively charged particles towards anode. The rate of migration depends on the number of charges each particle carries. As a result of different rates of migration, a mixture of proteins (such as plasma) can be separated into a number of fractions having similar mobility.

Tiselius (1937) first perfomed electrophoresis in a fluid electrolyte. In this method, the differently charged particles moved inside the buffer. This method was called "moving boundary electrophoresis". Few workers only used this method because of the costly and complicated apparatus.

Consden, Gordon, and Martin (1946) introduced filter paper to hold the electrolyte in the electrophoretic technique. Filter paper was the first anticonvection medium to be widely used. Later on, other media such as cellulose acetate, agar gel, starch gel, starch block and acrylamide gel have been used in place of paper with better results.

Principle : A small amount of serum is applied on a Whatmann filter paper No. 1. strip previously soaked in buffer solution, the two ends of the strip dip in two compartments of an electrophoresis tank containing the buffer solution. A direct current is applied to the ends of the paper for a desired period. During this time different fractions of the protein mixture separate into bands depending upon the amount of charge, the shape and size of the protein particles. After that electric current is disconnected, paper is removed from the tank, dried and stained with a protein stain to demonstrate different protein bands. The fastest band is albumin, behind it are α_1-globulin, α_2-globulin, β-globulin and γ-globulin (shown in Fig. 15.2). The stained fractions are eluted in suitable solvent and the eluted colour is measured in a colorimeter.

Two types of electrophoretic tanks are available :

1. Vertical and 2. Horizontal.

1. In the vertical type, there are five compartments. The electrodes are present in two outermost compartments, which is connected to adjoining compartments by wicks. By this arrangement the pH change occurring in the electrode compartment is prevented from affecting the adjoining compartment in which the ends of the paper dip. In the middle, there is a central compartment. The arrangement is shown in Fig. 15.1. All compartments excepting the central one are filled by buffer.

2. In the horizontal type, paper lies in horizontal position supported by strings. The paper ends dip in two compartments on the two sides of the tank. These compartments are connected by means of wicks to adjoining compartments containing electrodes. Buffer is kept in all compartments to equal levels.

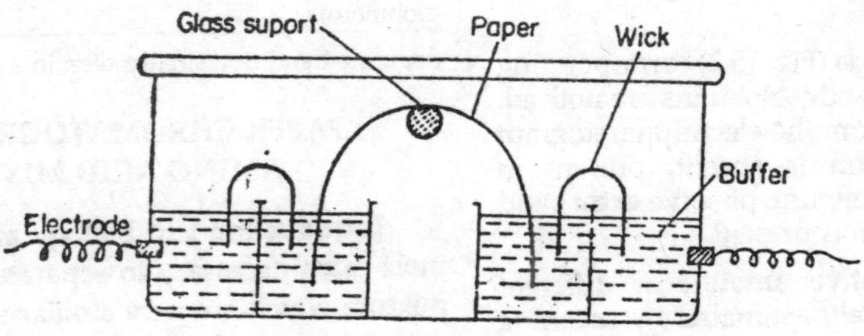

Fig. 15.1. : A vertical electrophoretic tank

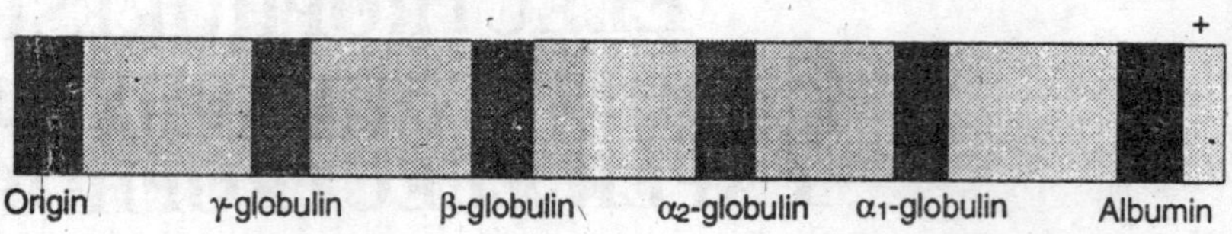

Fig. 15.2. : Serum protein electrophoretogram

Procedure : Pour sufficient buffer into the four compartments upto the same level, leaving the central compartment empty. Cut Whatmann No. 1 paper into strips of 35 cm. long and 5 cm. wide. Mark the centre of the strip with a pencil and mark the point of application sample by another line 2.5 cm. in length parallel to and 5 cm. from the mid line (for the verticle tank) or 7.5 cm. away from the mid line for the horizontal tank. Now, apply the paper in such a way that the two ends of the paper dips in the two buffer compartments adjoining the electrode compartments. Replace the lid of the tank. In about 2 hours time the whole paper is soaked with the buffer. Now, apply 10μl. of the serum (or plasma) by a micropipette on the line marked for sample application. Replace the paper in its place and cover the tank. At a time, several such strips can be set up for different samples. Now, connect one electrode with the cathode and the other with anode of a direct current with constant voltage. The electrode nearer the site of application is connected to the cathode. Switch on the current and allow the electrophoresis to run for 16 to 18 hours. A voltage gradient of 3.2 volts per cm. length of the paper, or a current of 1.5 to 2.0 milli amps for each 5 cm. wide strip is suitable. At the end of the desired length of time, first disconnect the instrument from the mains, remove the strip from the tank, and dry it at 100°C to 110°C for 30 minutes.

Staining : Pour sufficient stain in a large size glass or enamel tray and put the dried paper strips in it for at least 6 hours (overnight staining is always preferable). After removing from staining tray, wash for 6 minutes each time in two changes of wash solution, then in the fixative solution for a further 6 minutes. Dry the paper in air or in an oven at 100° – 110°C.

After staining 5 bands (Fig. 15.2) corresponding to albumin, α1-, α2-, β,- and γ-globulins are noticed. It is possible to guess from the electrophoretogram if some protein fraction is absent, present in decreased or increased amount, or some extra band such as myeloma protein is present.

However, the relative amount of different fractions can be accurately estimated by scanning the paper under photoelectric scanner or different bands are cut and eluted in 6 ml. of 0.01 N. NaOH for 30 minutes. These eluted coloured solutions are read in a photoelectric colorimeter at 540 mμ.

Reagents :

1. **Buffer (pH 8.6, ionic strength 0.083 M) :** Dissolve 3.12 grams of diethylbarbiturate in water and make upto 1 litre.

2. **Dye solution :** Add 25 ml. of 95 per cent ethanol to 0.1 gram of bromphenol blue and 50 grams of $ZnSO_4$. Mix well, add 5 per cent acetic acid (V/V), mix to dissolve and make upto a litre with further 5 per cent acetic acid.

3. **Wash solution :** Acetic acid (5 per cent, V/V).

4. **Fixation solution :** Acetic acid (5 per cent) containing 0.3 per cent of sodium acetate (CH_3COONa, $3H_2O$).

Electrophoretic pattern changes in some conditions.

'—' mark indicate no change.

Disease	Albumin	α1-globulin	α2-globulin	β-globulin	γ-globulin
Normal (gm/100 ml)	3.5 to 5.5	0.1 to 0.4	0.4 to 0.8	0.5 to 1.0	0.7 to 1.5
Infancy	—	high	high		—
Pregnancy	Low	high	high	high	—
Acute infection	—	high	high	—	—
Chronic infection	low	high	high	—	high
Cirrhosis	low	high	—		high
Nephrotic syndrome	low	high	very high	—	low
Myelomatosis	—	—	—	—	high
Hypogamma globinemia	—	—	—	—	very low

Separate band occasinally seen in a α and β positions.

PAPER CHROMATOGRAPHY OF AMINO ACID MIXTURE

Introduction : In biology and chemistry, it is most often necessary to separate components of a mixture which are very similar and are difficult to separate by chemical or physical methods. Chromatography and electrophoresis are two powerful modern methods utilised for such purpose.

Tswett (1906), the Russian biologist, first appreciated the possibilities of chromatography and he put the term 'Chromatography'. Consden, Gorden, and Martin (1944) described paper chromatography in which separations were done mainly by partition.

Principle : The separation of components of a mixture by a chromatographic system depends on multiple partition process. Small differences in partitioning of each component of a mixture are multiplied many folds. The greater such differences, the greater is the ease of separation.

A small drop of solution containing the mixture of compounds is put on a strip of filter paper and allowed to dry. A suitable solvent (mixture of two solvents) is allowed to flow along the filter paper over this spot. The substances in the initial spot are extracted by the flowing solvent and carried forward along the filter paper to a distance which appears related to their partition coefficient between the free and bound solvent phases of the filter paper. After the solvent has run for a suitable distance along the paper, the paper is removed, dried and subjected to suitable tests to locate the various compounds.

Rf value is defined as the ratio of the distance travelled by the component to the distance covered by the solvent. Rf value depends on the nature of the solvent, the temperature, and the presence of other substances.

Procedure : Whatmann filter paper is cut into 35 × 15 cm. sheet. A pencil line is drawn about 3 cm. above the shorter edge of paper and 5 points are marked at equal spacing leaving 2.5 cm from the two edges. On the middle point, a mixture of four amino acids is applied with the help of a fine capillary. It is dried with the help of hot air blown by a hair dryer. Again, another small quantity of the mixture is spotted at the same place and dried. This process is repeated 2 to 3 times more. On the other points, the individual amino acid of the mixture are similarly applied. The positions of these amino acids are marked with a pencil.

The solvent for developing the chromatogram is a mixture of n- butanol : acetic acid : water (12 : 3 : 5), respectively. This mixture is freshly prepared.

If a chromatographic tank is not available, specimen jar (about 40 × 15 cm.) with fitting lid can very well serve the purpose. In such a jar, put about 100 to 150 ml. of the solvent mixture and replace the lid so that the lid is air tight. In case of doubts, apply vaseline to the lid to avoid leakage. In an hour, the

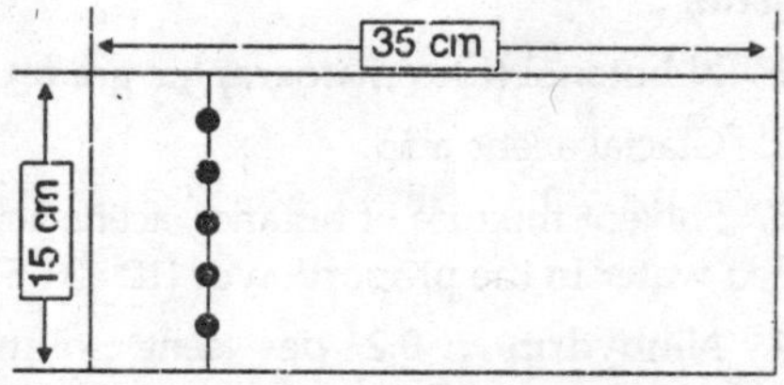

Fig. 15.3 : Serum Protein electrophoretogram

inside atmosphere will be saturated with the solvent vapour. Now, fold the paper in which sample has been applied in the shape of a cylinder and tie the opposing ends of the paper with staples or thread. Open the lid and place this folded paper in upright position in the jar, the pencil line lowermost and about a centimeter above the solvent (Fig. 15.4). Replace the lid. The paper should stand absolutely vertically. Leave the chromatogram to develop for 10 to 15 hours or earlier if the solvent has ascended quite near the upper margin of the paper.

Take out the paper at the desired time, cut the stitches and let it dry completely in the air.

After the paper has dried thoroughly, the location reagent (0.2 per cent. ninhydrin in acetone) is sprayed uniformly on the paper with the help of an all glass sprayer. The paper is then allowed to dry first in the air and then in a hot air oven at 105°C for 3 minutes. Purple coloured amino acid spots are seen on the chromatogram.

Identify the amino acids in the mixture with the help of spots produced by known amino acids. The R_f values of the amino acids can be calculated as :

$$R_f = \frac{\text{Distance travelled by particular amino acid}}{\text{Distance covered by the solvent from the point of origin}}$$

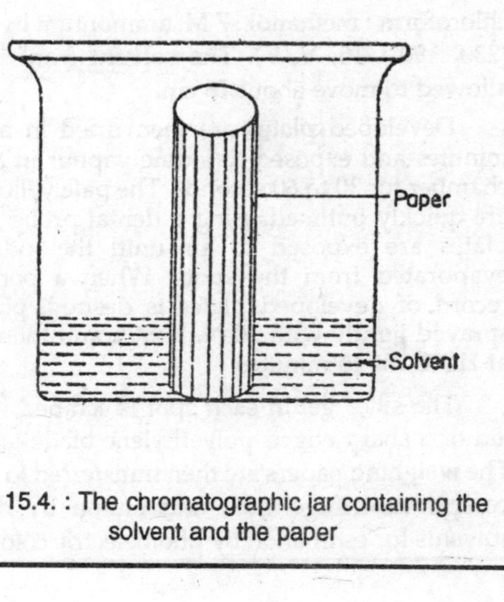

Fig. 15.4. : The chromatographic jar containing the solvent and the paper

Reagents :

1. N-butanol (Chromatographic grade).
2. Glacial acetic acid.
3. Solvent mixture of butanol, acetic acid, and distilled water in the proportion of (12 : 3 : 5).
4. Ninhydrin : 0.2 per cent solution of ninhydrin in acetone. It is prepared just before use.

THIN LAYER CHROMATOGRAPHY (TLC)

Introduction : In recent years, thin layer chromatography has been developed. This technique consists of a thin layer of absorbents (silica gel, alumina, Kiselguhr or cellulose) on a glass plate or plastic sheet. Since absorbent is used, it is also termed as absorption chromatography. This technique also provides superior results than that of paper chromatography. The spots are more compact with better resolution and the run is comparatively of shorter duration. Therefore, quicker run is possible.

Preparation of plates : Chromatographic plates (20 × 20 cm) of 200 μ thickness are prepared by using a suspension of 30 grams of silica gel G in 63 ml. of 0.1 M. Na_2CO_3 solution by shaking vigorously for 90 seconds. Only these plates are used which appear to be uniform in both transmitted and reflected light. These plates are activated at 110°C for 30 minutes immediately prior to use.

Procedure : Samples (5—100μL) are applied as a spot of less than 5 mm. diameter on the lower right corner of the plates under a stream of warm air. Plates are first developed in a standard Brinkmann developing chamber previously saturated with the vapour of the solvent mixture with chloroform : methanol : acetic acid : water (250 : 74 : 19 : 3; V/V). When the solvent front migrates about 15 cm., plates are dried in air for 15 minutes and develop in the second dimension (90° rotation clockwise) with chloroform : methanol : 7 M. ammonium hydroxide (230 : 90 : 15; V/V). The solvent front is again allowed to move about 15 cm.

Developed plates are then dried in air for 5 minutes and exposed to iodine vapour in a sealed chamber for 30 to 60 seconds. The pale yellow areas are quickly outlined using a dental probe and the plates are exposed to air unitl the iodine has evaporated from the spots. When a permanent record of developed plates is desired, plates are sprayed lightly with 10 N. H_2SO_4 and then heated at 110°C for 15 minutes.

The silica gel in each spot is scraped with the aid of a sharp edged polyethylene blade on paper. The weighing papers are then transferred to a 12 ml. conical centrifuge tube and eluted by different solvents for estimation by photoelectric colorimeter.

Discussion : The constituents of the mixture of amino acids, and the constituents of neutral lipids and phospholipids are separated and estimated in a short time.

DETERMINATION OF P^H

The electrometric method is the most accurate method for the determination of pH of a solution. This method measures the electromotive force set up due to the existence of H^+ and OH^- in the solution. the instrument which measures pH in this manner is known as pH meter.

The other method is the colorimetric method which is also fairly accurate and can be used in the absence of a pH meter. The basis of it is that certain organic dyes (indicators) change colour with change in pH. Each indicator in its ionised state gives one colour and in unionised state gives another colour.

Each indicator has its useful range of pH within which it can indicate change in pH by change in colour and beyond which there is no further change in colour.

It is possible to prepare an indicator shose pH range covers almost the entire pH scale. Such an indicator is known as 'Universal indicator'. The universal indicator enables us to determine the approximate pH of a solution and thus helps us to choose an appropriate indicator and buffers for subsequent more exact determination of pH of the solution.

Principle : When equal volumes of the unknown solution and a series of standard buffer solution with closely varying pH values are treated with a suitable indicator, the colour of the unknown solution will be found to match with one of the standard buffer solution. The pH of the unknown is, therefore, the same as the pH of the buffer with which its colour matches.

Permanent standard made from glass discs of the appropriate colour and shade for each indicator may also be used in place of the standard buffer and indicator. Hellige's comparator with its permanent standard discs made of coloured glass is a very convenient instrument for this purpose.

P^H determination of a colourless solution : First determine the approxiamte pH of the solution with the help of a universal indicator and colour chart. Take a small piece of universal indicator paper and just dip its one end in the given solution. After about 30 seconds compare the colour of the paper with the standard colour chart provided and know the approxiamte pH. If universal indicator solution is to be used in place of paper, put a drop of it in about 2 ml. of the solution and compare the colour of the solution with that of the colour chart. After having known the approximate pH of the solution choose from the set of indicators an indicator which more closely covers the pH range of the solution under examination. Then add a drop of this indicator to approximately 2 ml. of the solution, and match the shade of the colour produced against the series of coloured standard buffers. The standard colour solutions are prepared by adding one drop of the same indicator to 2 ml. of each of the buffer solutions (chosen in the pH range of the unknown). The pH of the unknown solution is that pH of the buffer with which the colour matches. In case the colour of the unknown is intermediate between two colours produced by two consecutive buffers, the pH of the unknown lies between the two known standards. The accuracy of this method can be increased by employing set of buffer solutions with closely differing pH values.

pH determination of coloured solution (or Urine) : Place a test tube containing plane urine or coloured solution in front of the coloured buffer standards. View the two test tubes together and then match this against the urine or coloured solution containing the same indicator in front of which a test tube containing water is to be placed. The use of a comparator makes the matching easier, otherwise, a double rowed test tube rack is also convenient.

Reagents :

1. Set of indicators.
2. Standard buffer solutions.

Electrometric method of pH determination :

The switch of the instrument is put on and after a few minutes when the instrument is warm enough, the pointer is adjusted to 0 mv or 7.0 pH position by *set zero* control. A standard buffer solution is taken in a clean glass or polythene beaker. Electrodes are lowered so that they are immersed in the solution to a depth of about one inch. The temperature of the buffer solution is measured by a thermometer and the instrument is again adjusted by *temperature compensate* control to this value. The pointer is again set at 7.0 pH by means of *set control*. The *Selector* switch is now turned to proper range i.e. 0-7 or 7-14. Pointer will now move to show the pH of the buffer and by the help *set buffer* control, the pointer is set to the exact pH value of the buffer. Now, the pH meter is standardized and is ready for determination of pH of any given solution.

To measure the pH of a given solution, the *selector* is brought back to *zero* position and the electrodes are cleaned by distilled water. The given solution is taken in a clean beaker and the electrodes are immersed in it. *Selector* switch is now put on expected pH range and the reading of the pointer on the pH scale is noted. This value is the pH of the given solution.

SPECTROSCOPE

In Biochemistry spectroscopic methods are used in abundance. Simple hand spectroscopes are used frequently. Spectroscopes of different designs are available. These must be treated with care and caution. These must not be put down on the bench in case these are wet or dirty. It is most essential to learn to adjust the instrument properly.

Adjustment of the hand spectroscope :

1. First of all, close the slit.
2. Point the instrument towards a source of day-light and open the slit very slighly.
3. Focus the Fraunhofer lines of the solar spectrum. These are the fine vertical black lines across the spectrum. These are not seen unless a very narrow slit is used.
4. Focus the wavelength scale.
5. Set the scale with reference to the D-line in the figure below. It is far better to use the solar line than the line obtained by the help of a sodium flame near at hand.
6. Adjusting the position of the scale carefully tighten the locking nut and check your setting. A slight final readjustment may be required. Much care should be taken not to strip the threads of the setting screw and locking nut.
7. Parallax should be tested.

Hints for Practical use :

1. It is wise to use always the narrowest possible slit. It is better to dilute a strong solution of a pigment than to use a wide slit; the wider the slit, the greater is the error in the readings.
2. In case of a strong pigment solution to be examined, dilute it cautiously, a little at a time, using the suitable solvent. Broad bands and 'cut-outs' can often be resolved into one or more separate bands in this way.
3. Always record and tabulate the positions of all the bands observed.
4. Always put the instrument down gently, otherwise the scale may be changed or some part of the optical system may even be knocked out of place.

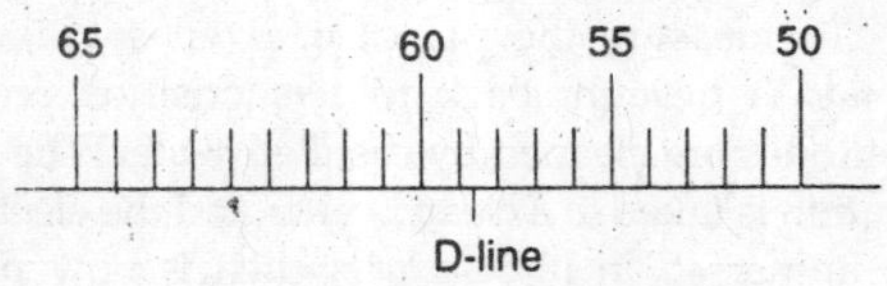

Fig. 15.5 : Hand spectroscope

Units employed in Spectroscopy :

The usual units are the mμ and the Angstrom Unit (A).

$$1\mu = \frac{1}{1000}\text{ mm.};\ 1\text{m}\mu = \frac{1}{1000}\mu;\ 1\text{A} = \frac{1}{10}\text{ m}\mu.$$

Wave length of D-line = 589 mμ and E-line = 527 mμ.

The characteristic light is absorbed by many biological substances. A white light when is allowed to pass through a solution containing a substance which has got the capacity of absorbing a light of particular wave-length, and the emerging rays are viewed through a **spectroscope**. One or more dark bands are observed in the solar spectrum. These dark bands correspond to those rays of light which have been absorbed by the substances of the solution. The substance of the solution can easily be identified by accurately noting the position of the absorption bands of the spectrum. Thus, spectroscope helps us to identify different blood pigments and porphyrins in biological fluids.

Spectroscopes are of different types. A direct vision spectroscope is highly useful for ordinary work (Fig. 15.6). This consists of a prism having an adjustable slit behind and an adjustable eye-piece. A pencil of sunlight, when allowed to pass through the prism, disperses into solar spectrum which can be viewed through the eye-piece. Some of the spectroscope are provided with scale to indicate the wave lengths of dispersed rays in Angstrom units (1A = mμ). The position of different absorption bands can be found out with the help of the scale.

In the solar spectrum a number of black lines are found. These black lines are said to be Fraunhofer's lines which are caused by absorption of certain rays of light from sun rays by the atmosphere. Among them B, C, D, E, b and F are distinctly visible (Fig. 15.7) and helpful in assuring the positions of absorption bands, especially when scale is not along with the instrument.

POSITION OF FRAUNHOFER'S LINES
(Fig. 15.7)

Line	Colour region	Wave length (mμ)
B	Red	687.7
C	Red	656.3
D	Orange-yellow	589.0
E	Green	527.0
b	Green	518.4
F	Blue	486.1
G	Violet	430.8

Operation of the Instrument

1. In good day light the instrument is viewed through its eye-piece holding in hand.

2. The slit is gradually closed with the help of its adjustor until the spectrum is not very bright nor very dark.

3. The eye-piece is then moved forward and backward unitl Fraunhofer's lines are well viewed.

4. If the instrument is provided with a scale, it is adjusted with the help of the scale adjustment screw so that the graduation mark 589 (usually shown as the biggest line) coincides with the D-line present in the orange-yellow region of the spectrum. The scale adjustment screw is then secured by tightening another screw as if the scale is not displaced accidentally.

5. The instrument is then clamped to a stand in horizontal position.

6. The solution to be examined is put in a good glass test tube (1 cm diameter) and is held vertically in front of the instrument with the help of another clamp. In case the spectrum appears dark it is necessary to open the slit slightly.

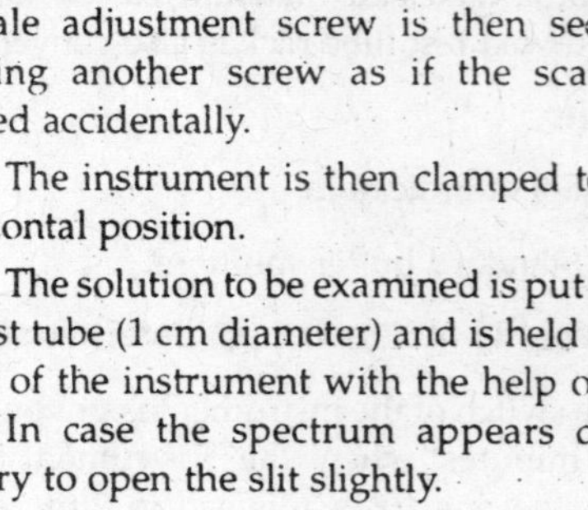

Fig. 15.6. : Direct vision spectroscope

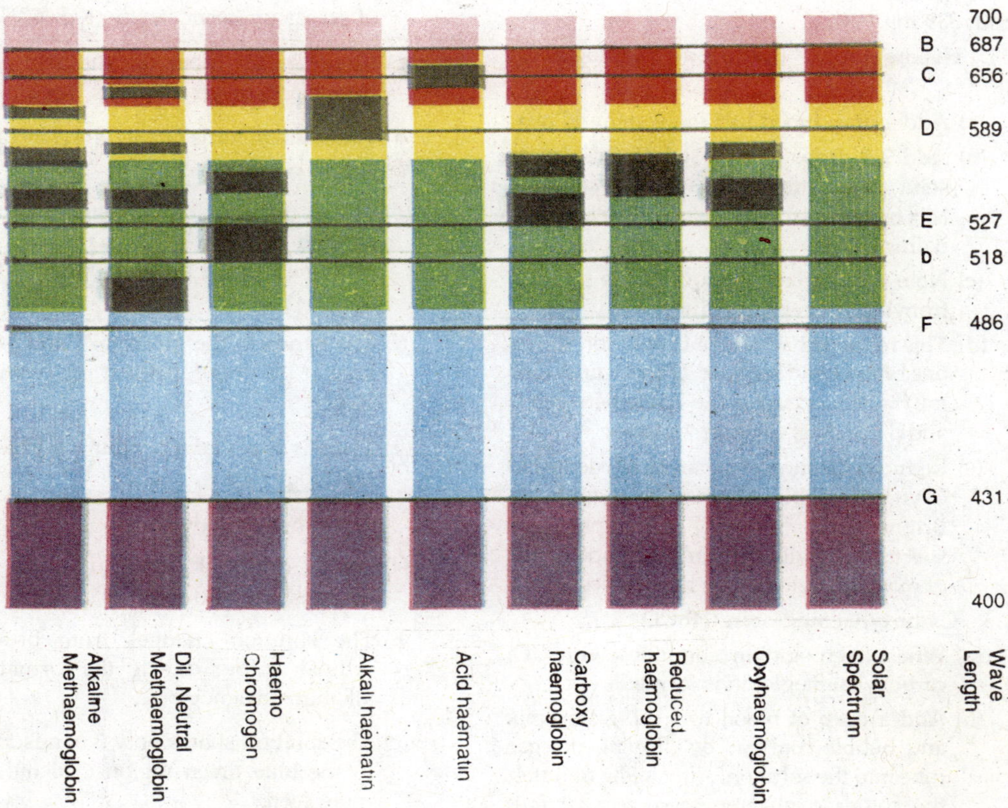

Fig. 15.8. : Blood pigments : Absorption spectra

7. It has been noticed that at one or more positions in the spectrum no colour develops, rather it appears as dark bands. The position of each dark band (centre of band) is read on the scale. If no scale is provided, the position of dark bands are noted in relation to the Fraunhofer's lines.

8. If a pigment shows more than two bands, they are designated α, β, λ and so on (starting from the red end).

The absorption spectra of oxyhemoglobin (HbO_2) :

1. In dilute solutions HbO_2 shows two bands in the green.

2. The alpha-band reaches nearly to the D-line of the solar spectrum and has its centre at about 578 mμ.

3. The β-band nearly reaches the E-line and has a centre at about 540 mμ.

4. As the concentration increases, the two bands get darker and broader and finally merge together to form a single band which stretches from the yellow to beyond the E-line.

5. As the concentration increases there is also more extensive absorption in the blue end of the spectrum. With a very strong solution the whole of the spectrum except the red is cut out.

Spectroscopic examination of oxyhemoglobin :

1. Use a very narrow slit and examine a 1 in 200 dilutions of defibrinated blood by distilled water.

2. Take about 5 ml. of it in a test tube and examine with spectroscope.

3. There are two asborption bands in the green.

4. The one towards the red end of the spectrum is called the alpha-band and is somewhat narrower and darker than the β-band.

5. It is to be noted that the redward edge of the α- band extends upto the D-line and the blueward edge of the β-band up to the E-line.

6. The edges of the bands vary with the concentration and thickness of the layer.

7. The middle of α is about 577 mμ and that of β about 539 mμ.

1. **Haemoglobin (Reduced haemoglobin, Hb)**

(a) Add 1 drop blood to 5 ml. water and mix.

(b) Add 2-3 drops of 15% sodium bisulphite solution or ammonium sulphide solution. Mix by gentle rotation of the tube between palms.

(c) Note the change in colour of the solution from scarlet to purple due to reduction.

(d) This reduced haemoglobin solution shows one broad band between D and E lines (565 mμ) which occupies the space between α and β bands of oxyhaemoglobin.

(e) Reduced haemoglobin can be reoxidised to oxyhaemoglobin by shaking it with air. Bright red colour develops and spectroscopically two characteristic bands of oxyhaemoglobin can now be seen.

3. **Carboxyhaemoglobin (HbCO)**

(a) When haemoglobin combines with CO, carboxyhaemoglobin is formed.

(b) Add a drop of blood to 5 ml. water, mix and bubble coal gas by dipping the gas tube into the solution. After a few minutes, the colour oxyhaemoglobin (scarlet red) changes to cherry red or carmine red or pinkish red (carboxyhaemoglobin colour).

(c) Examine the solution spectroscopically. Two bands in between D and E lines will be seen with their centres at 570 and 535 mμ. These two bands are very similar to those of oxyhaemoglobin (578 and 540 mμ).

(d) It is not so easy to distinguish between the oxy- and carboxyhaemoglobins by the ordinary spectroscope. A reversion spectroscope is suitable in this case.

(e) Using simple tests, it may, however, be possible to distinguish between these two if HbCO and HbO_2 are treated with mild reducing agent such as ammonium sulphide or sodium bisulphite, the two bands of HbO_2 change to one band, whereas those of HbCO remains unchanged

4. **Methaemoglobin (HbO)**

(a) The blood is to be haemolysed by putting one drop of blood to 5 ml. of distilled water.

(b) Add about 3-4 drops of saturated solution of potassium ferricyanide ($K_3Fe(CN)_6$).

(c) Close the test tube with the help of your thumb and invert it several times.

(d) The solution turns brown when methaemoglobin is formed.

(e) The spectrum of methaemoglobin is 4-banded and the most characteristic and prominent one is present in the red (at 633 mμ).

(f) The bands in the green (578 and 540) are similar to those present in oxyhaemoglobin.

(g) There is also a faint band in the blue green zone (500 mμ).

5. **Alkaline haemoglobin**

(a) Add 2 drops of strong ammonia to the above solution to make it alkaline.

(b) The solution changes from brown to reddish colour due to the formation of alkali methaemoglobin.

(c) The solution shows only 3 bands, the one in the blue green region (500 mμ) is no more seen.

(d) The band in the red is now no more most prominent and is seen nearer the D-line in comparison to the neutral methaemoglobin.

Absorption bands of some blood pigments and porphyrins

	Pigments	Number of bands	Position of bands in mμ
1.	Oxyhaemoglobin	2	578, 540
2.	Haemoglobin (Reduced Hb)	1	565
3.	Carboxyhaemoglobin	2	570, 535
4.	Methaemoglobin (Neutral)	4	633, 578, 540, 500
5.	Methaemoglobin (Alkaline)	3	610, 578, 540
6.	Acid haematin (In glacial acetic acid)	4	632, 540, 510, 400
7.	Alkaline haemalin (In 10% aqueous NaOH)	1	580
8.	Uroporphyrin (In 25% HCl)	4	597, 577, 553.5, 410.5